D1195664

STP 1342

Advances in Environmental Measurement Methods for Asbestos

Michael E. Beard and Harry L. Rook, editors

ASTM Stock Number: STP1342

ASTM
100 Barr Harbor Drive
West Conshohocken, PA 19428-2959
Printed in the U.S.A.

Library of Congress Cataloging-in-Publication Data

Advances in environmental measurement methods for asbestos / Michael E. Beard and Harry L. Rook, editors.
p.cm. — (STP; 1342)
"ASTM Stock Number: STP1342."
Papers presented at a symposium held July 13-17, 1997 in Boulder, Colorado.
Includes bibliographical references and index.
ISBN 0-8031-2616-6
1. Asbestos dust—Measurement—Congresses. 2. Asbestos fibers—Measurement—Congresses. 3. Asbestos—Analysis—Congresses. I. Beard, Michael E., 1940- II. Rook, Harry L. III. ASTM Committee D-22 on Sampling and Analysis of Atmospheres. IV. ASTM special technical publication; 1342.

TD887.A8 A34 2000
666'.72—dc21 99-055148

Copyright © 1999 AMERICAN SOCIETY FOR TESTING AND MATERIALS, West Conshohocken, PA. All rights reserved. This material may not be reproduced or copied, in whole or in part, in any printed, mechanical, electronic, film, or other distribution and storage media, without the written consent of the publisher.

Photocopy Rights

Authorization to photocopy items for internal, personal, or educational classroom use, or the internal, personal, or educational classroom use of specific clients, is granted by the American Society for Testing and Materials (ASTM) provided that the appropriate fee is paid to the Copyright Clearance Center, 222 Rosewood Drive, Danvers, MA 01923; Tel: 508-750-8400; online: http://www.copyright.com/.

Peer Review Policy

Each paper published in this volume was evaluated by two peer reviewers and at least one editor. The authors addressed all of the reviewers' comments to the satisfaction of both the technical editor(s) and the ASTM Committee on Publications.

To make technical information available as quickly as possible, the peer-reviewed papers in this publication were prepared "camera-ready" as submitted by the authors.

The quality of the papers in this publication reflects not only the obvious efforts of the authors and the technical editor(s), but also the work of the peer reviewers. In keeping with long-standing publication practices, ASTM maintains the anonymity of the peer reviewers. The ASTM Committee on Publications acknowledges with appreciation their dedication and contribution of time and effort on behalf of ASTM.

Printed in Philadelphia, PA
January 2000

Foreword

This publication, *Advances in Environmental Measurement Methods for Asbestos,* contains papers presented at the symposium of the same name held 13–17 July 1997 in Boulder, Colorado. The symposium was sponsored by ASTM Committee D-22 on Sampling and Analysis of Atmospheres, and by the Environmental Information Association. The conference chairmen and co-editors of the publication were Michael E. Beard, Consultant, Raleigh, North Carolina, and Harry L. Rook, National Institute of Standards and Technology, Gaithersburg, Maryland.

Contents

Overview

ASTM Committee D 22 on Sampling and Analysis of Atmospheres sponsors a variety of conferences and seminars to promote the exchange of information about monitoring various constituents and properties of air. One such conference is held periodically on the campus of the University of Colorado in Boulder and is known as the ASTM Boulder Conference. The 1997 ASTM Boulder Conference on Advances in Environmental Measurement Methods for Asbestos was held 13–17 July 1997 at the University of Colorado. This conference was co-sponsored by ASTM Committee D-22 and the Environmental Information Association.

The purpose of the conference was to focus on recent advances in research on measurement methods for asbestos in bulk building materials, as well as ambient, indoor, and work place air, water, and settled dust. The program included discussion of measurement methods, monitoring strategies, data interpretation, and quality assurance for asbestos measurements. It was the intent of the program to bring the disciplines of analytical chemistry together with investigators who are assessing exposure to asbestos in the environment and to promote better understanding of their mutual interests, needs, and limitations. The papers presented at the conference have been subjected to peer review, and those accepted are published in this ASTM Special Technical Publication.

Asbestos is a useful material and has been used as a component of many building materials. However, when asbestos fibers become airborne and are inhaled they may produce adverse effects such as asbestosis, lung cancer, and mesothelioma. The U.S. Environmental Protection Agency, the Occupational Safety and Health Administration, and various state and local governments have issued regulations to control exposure to the asbestos fibers. These governmental units have also named analytical methods and procedures that must be used to be in compliance with the regulations. These compliance methods address monitoring asbestos in drinking water, building materials, and in workplace and ambient air.

There are also asbestos-monitoring interests where no government regulation has been promulgated. Such an interest is asbestos in settled dust. While government regulations generally address visible deposits of dust in areas where asbestos-containing materials have been identified, there have been no analytical methods for sampling and analysis of asbestos in this medium. Likewise, there are no regulatory monitoring or control strategies other than requiring that all visible dust should be cleaned. ASTM has addressed these needs by developing draft methods for asbestos in settled dust, and two have become ASTM standards (D 5755: Test Method for Microvacuum Sampling and Indirect Analysis of Dust by Transmission Electron Microscopy for Asbestos Structure Number Concentrations; and D 5756: Test Method for Microvacuum Sampling and Indirect Analysis of Dust by Transmission Electron Microscopy for Asbestos Mass Concentration).

Analytical methods are constantly being reviewed and revised by users to meet new or special analytical needs. ASTM methods are subject to this review process and are required to be re-approved every five years. Governmental compliance monitoring methods for asbestos have proved to be more difficult to amend. Although regulations require periodic review, technical improvements may not be adopted because they may increase the cost of the analysis and thus the burden to the public. While the government will accept results from a more stringent analytical procedure, they are reluctant to require procedures considered burdensome to the public. Where there is no standard method or governmental compliance monitoring procedure, the analytical needs are filled by the so-called "state of the art procedure." These procedures are commonly used by laboratories to meet the demanding analytical requirements of a wide variety of materials submitted for analysis.

While these "state of the art procedures" may ultimately become the practice of all and be incorporated into the regulations, their adoption may lag in meeting the immediate monitoring needs of the analyst. It is this need that the 1997 ASTM Boulder Conference addresses. Many asbestos-monitoring techniques have been developed for problem materials such as vinyl asbestos floor tiles, bulk samples with less than 10% asbestos content, and asbestos in settled dust. The goal of this conference was to provide a forum for these state-of-the-art improvements and to have them published for wider distribution and dissemination. This Special Technical Publication will provide documentation of this forum and serve as a guide for monitoring asbestos using improved analytical techniques. This publication will be especially useful to those unable to attend the conference and as a foundation for those who are continuing research to meet these analytical needs.

The Conference was organized into technical sessions dealing with four measurement areas: (1) Measurement Methods for Asbestos in Bulk Building Materials; (2) Measurement Methods for Asbestos in Ambient, Indoor, and Workplace Air; (3) Measurement Methods for Asbestos in Water; and (4) Measurement Methods for Asbestos in Settled Dust. Papers describing analytical methods, monitoring strategies, and quality assurance procedures were presented and discussed.

The session on Methods for Asbestos in Bulk Building Materials included discussions concerning polarized light microscopy (PLM), X-ray diffraction (XRD), and transmission electron microscopy (TEM) techniques for analysis of these materials. The performance of regulatory methods in the analysis of a variety of bulk building materials, soils, and paints was presented and discussed. Shortcomings of the regulatory procedures were highlighted, and research to develop improvements, especially for the 1% regulatory statute, was presented.

The session on Asbestos in Ambient, Indoor and Workplace Air included presentations on OSHA, EPA, and ISO methods for monitoring airborne asbestos by either phase contrast microscopy (PCM) or TEM. Interesting research on techniques for determining fiber length/diameter distributions and the depth of penetration of fibers into membrane filters were also presented.

The session on Measurement Methods for Asbestos in Water reviewed EPA and American Water Works Association methods for asbestos in drinking water and research on improved sample preparation techniques. These small fibers dictate the use of TEM for analysis. This session also includes the editor's choice for most interesting title in the conference, namely "Sludge, Crud and Fishguts: Creative Approaches to Non-Standard Asbestos Water Analysis." This title epitomizes the innovative spirit and talent that analysts must exercise in dealing with a wide variety of environmental monitoring needs.

The final session on Measurement Methods for Asbestos in Settled Dust was perhaps the most controversial session in the conference. Analytical methods employing TEM developed by ASTM Subcommittee D 22.07 for monitoring asbestos in settled dust and monitoring strategies and results were presented. Many asbestos in settled dust monitoring efforts have required litigation for final interpretation of datasets. Some of the presentations in this session exemplify the diversity of opinions in this area. Additional studies are needed in this field to determine the effect of human, mechanical, and natural activity on generating asbestos aerosols from settled dusts. Research is also needed to better define the quantity of airborne asbestos that constitutes an environmental exposure hazard.

The 1997 ASTM Boulder Conference on Advances in Environmental Measurement Methods for Asbestos served as a focal point for issues related to the needs for improved monitoring techniques for asbestos. This ASTM Special Technical Publication will serve as a documentaiton for our collective understanding of these issues as they were at the time of the conference. It is hoped that the papers published here will guide others in understanding these monitoring issues and lead to research for further improvements for us all.

Michael E. Beard
Consultant;
Raleigh, NC

Harry L. Rook
National Institute of
Standards and Technology
Gaithersburg, MD

Measurement Methods for Asbestos in Bulk Building Materials

Robert L. Perkins[1]

Analysis of Asbestos in Bulk Materials--1980 to 1997

REFERENCE: Perkins, R. L., **"Analysis of Asbestos in Bulk Materials—1980 to 1997,"** *Advances in Environmental Measurement Methods for Asbestos, ASTM STP 1342,* M. E. Beard and H. L. Rook, Eds., American Society for Testing and Materials, West Conshohocken, PA, 2000.

ABSTRACT: Federally sponsored asbestos proficiency testing programs have operated continuously in the United States since 1980. The U. S. Environmental Protection Agency published a test method for the analysis of asbestos in bulk materials in 1982, commonly referred to as the "Interim Method." Major revisions were made to the method and the revised version was published in 1993. Polarized light microscopy , supplemented by x-ray diffraction, is the primary analytical technique presented in the 1982 and 1993 test methods. The 1993 version of the test method also recommends additional analytical techniques such as gravimetric sample reduction, transmission electron microscopy, and the use of bulk calibration standards. Research Triangle Institute's more than 17 years' experience evaluating the test methods and characterizing thousands of proficiency testing samples for testing programs indicates that the available analytical techniques can provide very accurate results for qualitative analysis of asbestos-containing materials and reasonably accurate results for quantitation of asbestos concentrations. The quality of the results being produced by the asbestos laboratory community appears to be most influenced by the skill level of the analysts and the degree of employment of the available analytical techniques.

KEYWORDS: proficiency testing, test method, polarized light microscopy, quantitation, laboratory performance

---

Introduction

Federally sponsored proficiency testing (PT) programs for asbestos laboratories have operated continuously since 1980 when the U. S. Environmental Protection Agency's (EPA's) Bulk Sample Quality Assurance Program was initiated. The EPA program was replaced by the National Institute of Standards and Technology's (NIST's), National Voluntary Laboratory Accreditation Program (NVLAP) in 1989 as mandated by

[1]Manager, Earth and Mineral Sciences Department, Research Triangle Institute, Post Office Box 12194, Research Triangle Park, NC 27709.

the Asbestos Hazard Emergency Response Act (AHERA) [*1*].

There were approximately 100 laboratories enrolled in the initial round of the EPA program in 1980 and approximately 1 100 enrolled in the 18th and final round conducted in 1988. Enrollment in the NVLAP has ranged from 661 laboratories in the initial test round conducted in 1989 to a high of 707 laboratories in 1990. There are currently 350 laboratories enrolled in the program. In addition to the NVLAP, Research Triangle Institute (RTI) conducts two other PT programs for polarized light microscopy (PLM) asbestos laboratories, namely the American Industrial Hygiene Association (AIHA) program with 280 laboratories and the U.S. Navy program with 87 laboratories. There have been significant changes in the enrollment in these three programs (Figure 1) over the past several years.

The Test Method

Although PT of asbestos laboratories was initiated in 1980, the EPA test method, the so-called Interim Method [*2*] , was not published until 1982. This method designated PLM as the method of choice for analysis of asbestos in bulk building materials; the method also included a section on analysis of asbestos using x-ray diffraction (XRD) as a confirmatory method for identification and quantitation of asbestos in bulk material samples that have undergone prior analysis by PLM or other analytical methods.

The revised EPA test method was published in 1993 [*3*]. PLM, supplemented by XRD, is the primary analytical technique presented in the revised method. The revised method was expanded from the 1982 version to include additional analytical techniques such as gravimetric sample reduction, transmission electron microscopy (TEM), and the formulation and employment of bulk asbestos calibration standards.

Effectiveness of the EPA Test Method

As stated previously, the primary analytical technique for analysis of asbestos in bulk building materials is PLM. After nearly 20 years of employing this analytical technique for the analysis of asbestos, RTI is in a position to comment on the adequacy of the method and the ability of asbestos laboratories to successfully utilize the test method.

Qualitative Analysis

In addition to extensive evaluation by EPA, the test method has been utilized to characterize test samples used in the various PT programs for the past 17 years. Unlike analytical methods that rely on sophisticated equipment to provide analytical results, PLM depends greatly on the skill and experience of the analyst to provide accurate and complete results. Qualitative analysis of asbestos in bulk materials requires that the analyst be able to accurately determine the optical properties of fibrous particles. Accurate measurement of refractive indices (RIs), angles of extinction, birefringence, and so on is necessary to preclude the occurrence of false positives attributed to incorrect identification of asbestos look-alikes such as polyethylene fibers, wollastonite, and fibrous talc.

The combination of low asbestos concentration, small fiber size, and interfering

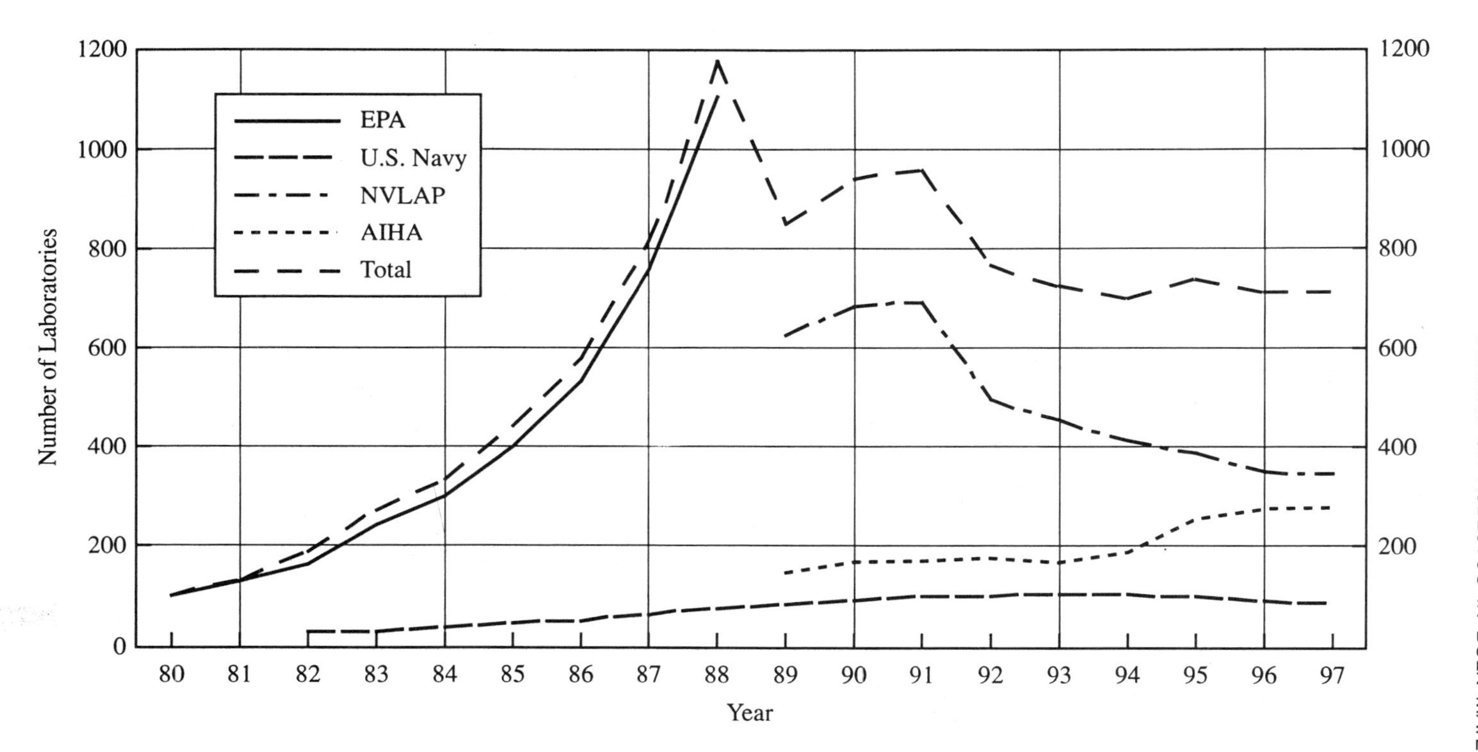

Figure 1 - *Average annual laboratory enrollment, by program.*

binder/matrix causes some bulk materials to be very difficult to analyze by PLM. Analysis of such samples may be facilitated by employment of additional analytical techniques such as gravimetric sample reduction by ashing and/or acid washing. Samples such as vinyl flooring materials, asphaltic roofing materials, and plasters may be gravimetrically reduced. This procedure removes the interfering matrix/binder and concentrates the asbestos, providing greater opportunity to detect and verify it.

The detection limit for PLM asbestos analysis is sample-dependent, but for the majority of bulk samples, this value is <1 %. RTI analysts have never failed to detect asbestos in a positive sample (with negative samples verified by TEM).

Quantitative Analysis

Although PLM is a very effective method for the qualitative analysis of bulk materials, quantitation of asbestos content using this technique is less certain. The only options available to the PLM analyst for determining the asbestos concentration are: 1) visual estimation and 2) point counting.

Visual estimation may be performed using a stereomicroscope at 10-40x magnification. The analyst estimates the relative volume proportions of asbestos and matrix components, resulting in an asbestos concentration expressed as a percent volume. Visual estimation may also be performed on slide mounts using PLM, resulting in an asbestos concentration expressed as a percent area.

Point counting using PLM is a systematic procedure that involves traversing a slide mount and recording the type of particle(s) directly under the intersection of the reticle cross lines or the points of the Chalkley point array. A minimum of 400 occupied points is recommended for each sample. The asbestos concentration determined by this technique will be a projected area percent concentration.

There are shortcomings to each of these techniques. Visual estimation, with stereomicroscopy or with PLM, is subject to analyst bias. Test results submitted by laboratories enrolled in the PT programs directed by RTI indicate a continuing tendency of laboratories to overestimate the asbestos content and also a continuing reliance by most laboratories on visual estimation for determining asbestos concentrations. Visual estimation is a viable quantitative technique only if: 1) asbestos fibers/bundles are visible by microscopy and 2) the analyst has been "calibrated" through the use of asbestos standards or reference materials containing known concentrations of asbestos. Not only should the analyst receive training with such materials, but reference materials should be included in the quality assurance/quality control (QA/QC) program as blind samples. This provides documentation of analyst bias and also information on the accuracy and precision of the analyst's quantitative results.

Results obtained by point counting should not be affected by analyst bias as greatly as results obtained by visual estimation if the technique is performed properly. As with quantitation by visual estimation, asbestos fibers/bundles must be visible,(i.e., the microscope set up should maximize fiber visibility). As is the requirement for all quantitative techniques using microscopy, the sample should be homogeneous to ensure that the small subsamples used for the quantitative procedure(s) are representative of the total sample. Assuming optimum conditions, an individual analyst's precision, and the precision between and among analysts, should be very good. As would be expected, precision and accuracy should increase with the increasing number of points counted.

Point counting precision and accuracy may be illustrated (Table 1) by examining the results produced from the point counting of formulated (by weight percent) asbestos calibration standards by two experienced analysts. The following observations may be made about these point-counting data: 1) the concentration of chrysotile determined by point-counting tended to be lower than the true weight-percent concentration; 2) the concentrations of amphibole asbestos types tended to be higher than the actual weight-percent concentrations; 3) point counting of samples containing less than 1 % chrysotile always resulted in mean values of less than 1 %, and 4) the point counting values for samples containing less than 1 % amosite were always 1 % or greater. These relationships are explained in the discussion that follows.

Quantitative results may also be biased by sample composition and relative particle thickness [*4*]. Building materials may contain a variety of components having a wide range of densities. For example, a sample could contain cellulose (specific gravity of 0.9), perlite (0.4), and chrysotile (2.6). The asbestos concentration determined for such a sample would be expected to be biased low (as compared to the true weight-percent concentration) because of discrepancies in the relative volumes of the components.

Visual area estimates and point counting are really measurements of the relative projected areas of particles as viewed on a microscope slide. Such estimates may be biased by differences in particle thicknesses. For example, if a sample contained relatively thick bundles of asbestos and a fine-grained matrix such as clays or calcite, the asbestos concentration measured by the projected area (and volume) would likely be underestimated. Conversely, if a sample contained thick "books" of mica and thin asbestos bundles/fibers, the asbestos content would likely be overestimated. It is apparent from these two simple examples that particle thickness is an important factor when relating area percent to volume percent and that the ideal situation would involve quantitation of materials having components of uniform particle thickness and similar densities. It is recommended that asbestos concentration be reported as volume percent, weight-percent, or area percent depending on the method of quantitation used. A weight percent concentration **cannot be determined** without knowing the relative densities and volumes of the sample components.

As stated previously, the employment of gravimetric sample reduction may greatly improve quantitative results. Difficult samples such as floor tiles, plasters, and roofing materials may be reduced greatly (greater than 85 % for some samples) by ashing and/or acid washing. Removal of interfering matrix material, resulting in a concentration of the asbestos, should greatly improve the accuracy of quantitative results.

There are some sample types for which the asbestos concentration cannot be determined adequately by PLM, (e.g., vinyl floor tiles). Although ashing followed by acid washing generally removes 80 % or more of the sample, the remaining residue usually contains a considerable amount of titanium dioxide (TiO_2), a pigment material that coats the asbestos fibers very effectively. This negates quantitation by PLM because few fibers are visible. For such samples, analysis by TEM or XRD is recommended.

Table 1 - *Results of point counting of asbestos standards*

Composition of Standard	Wt. % Asbestos	Point-Counting values (%)	
		Range	Mean
Chrysotile, gypsum, perlite	0.3	0-0.25	0.1
Chrysotile, gypsum, perlite	2.0	0.5-1.75	1.4
Chrysotile, gypsum, perlite	7.0	2.5-3.5	3.0
Amosite, mineral wool	1.5	5.0-5.75	5.4
Chrysotile, amosite, mineral wool	Chrysotile - 2.0 Amosite - 0.4	2.25-3.5 1.5-2.0	3.0 1.7
Chrysotile, gypsum, vermiculite	0.5	0-0.5	0.3
Chrysotile, gypsum, vermiculite	3.0	1.5-2.25	1.9
Amosite, gypsum, vermiculite	7.0	11.75-14.0	12.9
Amosite, crocidolite, vermiculite	Amosite - 3.0 Crocidolite - 3.0	6.5-7.5 1.0-2.75	7.1 2.0
Amosite, gypsum	0.5	0.75-1.25	1.0
Tremolite, calcium carbonate, perlite	5.0	8.0-8.25	8.1
Anthophyllite, gypsum, vermiculite	5.0	6.75-7.5	7.1

It can be stated with a great deal of certainty that unless a sample has been gravimetrically reduced, resulting in a residue composed completely or almost completely of asbestos, quantitative results may not reflect the true weight-percent concentration of asbestos. The technique of quantitative XRD provides a true weight-percent concentration. Concentration values determined by visual estimation and point counting are subject to the biases discussed previously.

Laboratory Performance

RTI has been evaluating the performance of PLM asbestos laboratories for more than 17 years. As would be expected, the laboratory performance in the early test rounds of the EPA testing program was not at the desired level. False negative, false positive, and identification errors were higher than would be expected given the capability of the test method. The high error rates are probably best explained by the proliferation of asbestos laboratories in the United States during the 1980's. As shown earlier, the EPA program grew from 100 laboratories to 1 100 laboratories in eight years. It is doubtful that all of these laboratories were staffed with experienced, qualified microscopists. As stated previously, the quality of analytical results determined with PLM is largely dependent on the experience and skill of the microscopist.

With the inception of the NIST testing program, NVLAP, the requirements for participating laboratories became more stringent. Laboratories are required to implement and document a plan for QA/QC and to submit to an on-site assessment every other year. PT is also much more rigorous than was the case with the EPA program. Laboratories are required to report all optical properties of the asbestos minerals and also to report asbestos concentrations. Test samples used in the NIST program are also more "challenging" than those that were generally used in the EPA testing program; samples having low asbestos concentrations, interfering binder/matrix, and/or unusual characteristics are commonly included in the NIST testing program.

Performance of laboratories participating in the NIST program has gradually improved with time. The false positive error rate has been reduced greatly, and the false negative rate has been at expected levels for the last several test rounds. This improvement is probably a result of the following: 1) reduction in the number of PLM asbestos laboratories operating in the United States; (many of the "marginal" laboratories have closed, resulting in an overall improvement in laboratory quality) and 2) the institution of more demanding requirements in the NIST program.

Error rates for quantitation of asbestos and determination of refractive indices (RIs) continue to be unacceptably high. As mentioned previously, quantitation of asbestos concentration by microscopic methods is subject to analyst, sample, and method bias, but analysis of test samples by RTI and NIST analysts indicates that the test method is capable of producing results more accurate than are generally reported by the laboratories. This discrepancy is probably best explained by a failure of the laboratories to "calibrate" their microscopists through use of quantitative standards, a failure to employ point counting, and by a failure to implement proper QA/QC procedures that would historically monitor the precision and accuracy of an analyst's quantitative results.

The high error rate for determination of RIs can only be attributed to a poor application of the techniques for determining these optical characteristics and an incomplete understanding of optical mineralogy. The two techniques commonly employed are dispersion-staining and Becke line, with most laboratories preferring the dispersion-staining technique. Both techniques produce acceptable results if: 1) the techniques are properly used, 2) the asbestos fibers are relatively free of masking binder/matrix material, and 3) the analyst has a clear understanding of the principles of optical mineralogy.

Harvey et al., [5] have historically documented the types, magnitudes, and changes in rates of the various types of errors committed by laboratories in each of the national PT programs conducted in the United States.

Conclusions

Although the number of PLM asbestos laboratories currently operating in the U.S. is down from its peak in the late 1980's, the quality of analyses being produced by the laboratories has improved. The EPA test method is quite capable of producing accurate qualitative results for almost all sample types, with the possible exception of some particularly challenging vinyl floor tiles. The PLM test method is less capable for the quantitation of asbestos concentrations, but utilization of additional and/or alternate analytical techniques can greatly improve the accuracy of these results.

References

[1] "Asbestos-Containing Materials in Schools: Final Rule and Notice," 40 *Code of Federal Regulations (CFR)* Part 763, October 1987.

[2] U.S. Environmental Protection Agency (EPA), "Interim Method for the Determination of Asbestos in Bulk Insulation Samples," EPA 600/M4-82-02D, Research Triangle Park, NC, December 1982.

[3] U.S. Environmental Protection Agency (EPA), "Method for the Determination of Asbestos in Bulk Building Materials," EPA/600/R-93/116, Washington, DC, July 1993.

[4] Perkins, R. L., "Estimating Asbestos Content of Bulk Samples," *National Asbestos Council Journal*, Spring 1991, pp. 27-31.

[5] Harvey, B. W., Ennis, J. T., Greene, L. C., and Leinbach, A. A., "Bulk Asbestos Laboratory Programs in the United States-Seventeen Years in Retrospect," *Advances in Environmental Measurement Methods for Asbestos*, ASTM STP 1342, M. E. Beard and H. L. Rook, Eds., American Society for Testing and Materials, 1998.

Jennifer R. Verkouteren[1], Eric B. Steel[2], Eric S. Windsor[1], Robert L. Perkins[3]

COMPARISON OF QUANTITATIVE TECHNIQUES FOR ANALYSIS OF BULK ASBESTOS PROFICIENCY TESTING MATERIALS

REFERENCE: Verkouteren, J.R., Steel, E.B., Windsor, E.S., and Perkins, R.L., **"Comparison of Quantitative Techniques for Bulk Asbestos Proficiency Testing Materials,"** *Advances in Environmental Measurement Methods for Asbestos, ASTM STP 1342,* M.E. Beard and H.L. Rook, Eds., American Society for Testing and Materials, Philadelphia, 2000.

ABSTRACT: The methods used by NIST for quantitative analysis of NVLAP bulk asbestos PTMs are described. Quantitative XRD using the method of standard additions is the most generally applicable tool used. PTMs are also analyzed gravimetrically using acid dissolution and ashing to remove the matrix quantitatively. Gravimetry is used for many samples, even when the matrix cannot be completely removed. Point counting is performed on all the PTMs, but the results are used primarily to adjust acceptance ranges and not to provide reference values. The general application of these analytical methods to the analysis of the PTMs is described, and the results from multiple techniques are compared for selected samples. The reliability of each method is described, along with precautions that must be exercised to obtain accurate results. Results from Research Triangle Institute using the same general methods on the same PTMs are compared with results from NIST.

KEYWORDS: bulk asbestos, x-ray diffraction, gravimetry, point count, NVLAP, standard additions

When the National Voluntary Laboratory Accreditation Program (NVLAP) for bulk asbestos began in 1988 very few validated methods for the quantitative analysis of bulk asbestos existed. The point count method in the Environmental Protection Agency (EPA) interim method [1] was required for the analysis of materials in schools, but the accuracy of the technique had not been tested. There was also uncertainty about the unit of measurement (area or volume) of the point count method. A quantitative x-ray diffraction (XRD) method was given in the EPA interim method, but it relied heavily on comparison to standards that were not available, and, again, the accuracy had not been

[1] Research Scientist, Chemical Science and Technology Laboratory, NIST, Gaithersburg, MD 20899.

[2] Leader, Microanalysis Research Group, NIST, Gaithersburg, MD 20899.

[3] Manager, Department of Earth and Mineral Sciences, RTI, Research Triangle Park, NC 27709.

tested. The asbestos analysis group at the National Institute of Standards and Technology (NIST) wanted two independent quantitative methods that could be used for a wide range of samples to determine reference values for the proficiency testing materials (PTMs) used in the accreditation program. We decided to concentrate on techniques of weight measurement, rather than area or volume measurement, since the procedure for making standards by gravimetry was straightforward.

To develop a quantitative XRD method NIST took the classic analytical chemistry approach of using the method of standard additions [2, 3]. The standard additions method is used on samples that have complex or widely variable matrices, where it would be difficult or impractical to prepare representative calibration materials [3]. The effects of sample matrix and texture on the asbestos peak intensities do not have to be modeled or characterized because the standard is added to each sample. Further advantages include the relative ease of data collection and analysis, as very few diffraction peaks need to be analyzed, and the calculations are done with standard spreadsheet programs. Other researchers developing quantitative XRD methods for bulk asbestos have used more complicated data analysis procedures {combined Rietveld and reference intensity ratio (RIR) method [4]} or have concentrated on specific sample types {gypsum-based bulk materials [5]}. Matrix removal methods, described briefly in the EPA interim method, were also developed as quantitative gravimetric tools. The description of gravimetric methods given in the revised EPA test method for bulk asbestos [6] is based largely on our investigations and information supplied to the EPA by Dr. Eric Chatfield.

This paper will discuss the application of the standard additions XRD method and the gravimetric method for the characterization of PTMs. The general reliability of each method is discussed along with precautions that must be exercised to obtain accurate results. The robustness of each technique is demonstrated by comparison of NIST's results with those obtained at Research Triangle Institute (RTI) using the same general methods. Point count results for the proficiency testing materials are used to describe the general reliability of the method as compared with standard additions XRD and gravimetry.

Quantitative XRD using Standard Additions

The standard additions method involves the addition of a known weight of a standard of the analyte in question into the sample [3]. A series of such additions is made, the instrumental response to the analyte is plotted against the amount of standard added, and a regression line is fit to the data. The x-intercept of the regression line gives the concentration of the analyte in the sample. For the analysis of chrysotile, the peak integral of the reflection at 12.2° 2θ is used, since it is generally free of interferences. An example is given in Figure 1 using data from a typical PTM.

The asbestos materials used for the standard additions are Standard Reference Materials (SRMs) 1866 (Common Commercial Asbestos) and 1867 (Bulk Asbestos - Uncommon). Each material is prepared using a cutting mill to reduce the fiber length without affecting the crystallinity. Although most quantitative XRD techniques call for reducing the grain size to a few micrometers, grinding can often be a problem for asbestos materials because they are susceptible to crystallinity changes [4, 5]. The PTMs are prepared using a cutting mill, or if there is hard, granular material present, by using a mortar and pestle.

Each PTM is generally an authentic building material removed during an abatement procedure which is subdivided and packed into 500 or more individual vials. Only handmixing is performed on the material prior to subdivision to maintain the integrity of the material's texture and simulate as closely as possible the real samples likely to be received by asbestos laboratories. A random sampling of the vials of each PTM are analyzed to test for variations among the individual vials. Each vial is subdivided into the number of aliquots needed for the standard additions, or if there is insufficient material, the additions can be made sequentially to the sample [3]. After the standard has been added to the sample, the mixture is agitated in a shaker and then packed into a standard XRD sample holder using a backloading technique.

The results for individual vials of a typical PTM are plotted in Figure 2, using the calculated concentrations determined by the linear fit to the standard additions data plotted in Figure 1. The spread in the data due to variations among vials is larger than the spread due to repeated analyses of a given vial. This is typically the case for samples that can be homogenized readily without compromising the asbestos components, as mentioned earlier. For PTMs that are more difficult to homogenize, such as samples with vermiculite or mica as major matrix components, the analytical variability can be as large as the variability among vials.

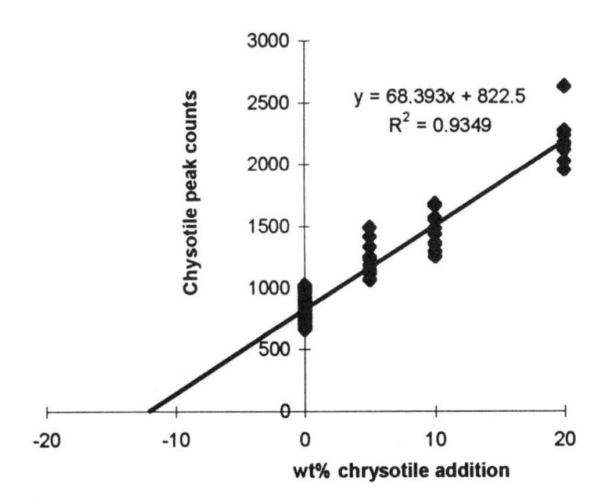

Figure 1. Standard additions of chrysotile to a PTM. The PTM contains 12.0 wt% chrysotile, as determined by the linear equation (x=12.0 when y=0).

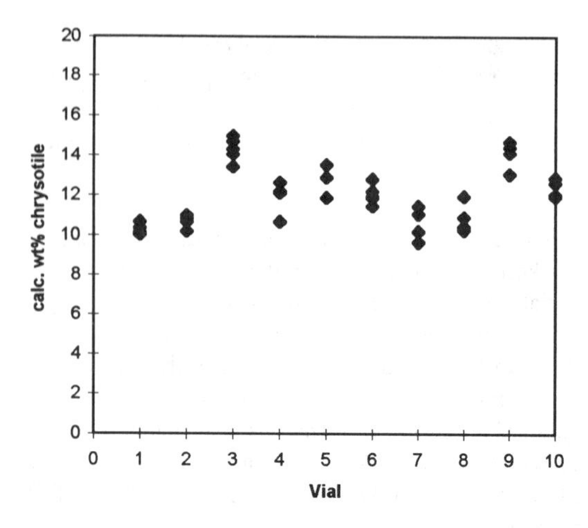

Figure 2. Concentrations of individual vials of the PTM in Figure 1. The relative standard deviation for the measurement of an individual vial averages 5.0%, and the relative standard deviation for all the measurements is 12.0%

The inability to grind the samples and the asbestos standards to a fine grain size can also create problems in homogenizing the mixtures. In some instances, the asbestos standard is of a sufficiently different texture from the sample that the two components do not mix well. If either the sample or the standard is over-represented on the surface of the prepared mount, the apparent instrumental response to the addition of the analyte can be affected, changing the calculated slope and concentration of asbestos in the sample. It is important to observe the physical appearance of the prepared XRD mount to check for homogeneity.

Another consideration in the application of the standard additions technique is the potential for absorption of the Cu x-ray beam by amosite, crocidolite, and actinolite, the Fe-bearing asbestos types. Absorption can cause loss of intensity and result in a non-linear response with standard additions. The non-linear response is readily observed in the data, and the solution is to perform the standard additions over a limited range where the response is linear.

One of the assumptions in the standard additions method is that the standard being added is representative of the unknown in terms of crystallinity, orientation, grain size, and elemental composition. Since each asbestos type can come from different mines and localities, it is difficult to know whether the standard is appropriate for a given sample. We have used chrysotile from different source localities as the standard for a given unknown, and have not detected any differences.

Gravimetry

Gravimetry is used to determine the concentration by weight of asbestos in bulk materials by removing matrix components with dilute HCl and/or by ashing. The sample is weighed prior to and then following matrix removal to determine the weight percent of the residue. Details of this procedure are given in the revised bulk method [6]. Many of the building materials we have analyzed contain some non-asbestos phases in the residue that cannot be removed with HCl or by ashing; however, the residue can be further characterized using polarized light microscopy (PLM) or other techniques to estimate the relative percentage of asbestos in the residue. Removing a significant amount of the matrix improves the precision of a point count analysis by increasing the number of asbestos points relative to the non-asbestos points.

Incomplete removal of the matrix produces results that are biased high relative to the true amount of asbestos in the sample. Loss or destruction of the asbestos components produces results that are biased low relative to the true amount of asbestos. Chrysotile can be leached by hot, concentrated HCl solutions [7]; thus it is recommended that cold, dilute HCl be used. The weight recovery and optical properties of chrysotile were tested after soaking for 15 minutes in cold solutions of full strength and dilute HCl; the chrysotile was found to be stable under both conditions. Full strength HCl is used only for materials that contain dolomite. Chrysotile starts to become amorphous at approximately 500 °C [8], so samples are not ashed above 450 °C. The amphibole asbestos types evolve a small amount of water (<2.5 wt%) below 500 °C and decompose above 600 °C [7]. Both amosite and crocidolite form oxyamphiboles at 300 to 350 °C [9] which results in a change in color and optical properties of the asbestos, but the mass and texture are unaffected. Some samples have combusted upon ashing, which results in temperatures high enough to destroy the asbestos components. Milling the samples prior to ashing can, in most cases, prevent combustion.

Mixtures of chrysotile and calcite were prepared to test the accuracy of the acid dissolution technique. SRM 1866 chrysotile was milled three times to produce very fine fibers to maximize the surface area (worst case for acid dissolution) and test the recovery procedures. An average loss of 3 wt% occurred for the milled chrysotile which probably represents original soluble material in the chrysotile standard. [SRM1866 is a standard for the optical properties of asbestos, and is not a quantitative standard. The chrysotile in SRM 1866 is greater than 80% pure, but contains some accessory phases.] Successive acid treatments of this material resulted in diminishing losses of 1 wt% or less. Gravimetric analysis of the mixtures of milled chrysotile and calcite containing from 0.1 wt% to 10 wt% chrysotile resulted in errors between ± 1% to 4% relative as shown in Table 1.

Floor tiles are particularly amenable to the use of gravimetric methods since they contain both ashable and acid-soluble components. The asbestos is typically 80 wt% or more of the residue left after ashing and acid dissolution, with additional phases such as

TiO_2 and quartz. We were concerned about ashing floor tiles since earlier work with differential thermal analysis (DTA) had indicated that the temperature of the tile could exceed 450 °C even at lower furnace temperatures due to exothermic reactions. To test whether the chrysotile was preserved during the ashing and acid dissolution procedures, the XRD peak integral of chrysotile in the original untreated tile was compared with the weight percent of the residue following the two procedures (Figure 3). There is an almost linear correlation between the amount of chrysotile in the original tile, as indicated by the XRD data, and the weight percent of the residue, indicating that the chrysotile was not lost during the gravimetric procedures.

Table 1. Calibration results for gravimetry.

Wt.% chrysotile[1], mg	Result[2], mg	1 σ, mg	Relative error, %
0.115	0.114	0.009	-0.9
1.015	1.036	0.067	2.1
5.005	4.805	0.074	-4.0
10.008	9.700	0.052	-3.1
100	96.420	1.296	-3.6

[1]formulated weight
[2]average of 5 measurements

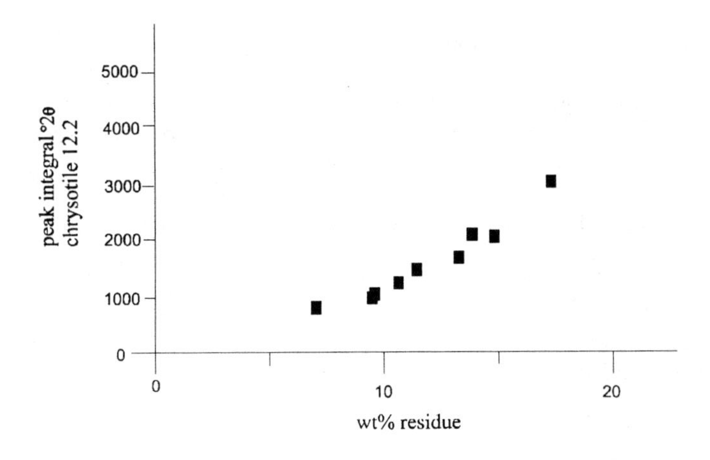

Figure 3. Comparison of gravimetric residues of 9 vinyl floor tiles with the XRD peak integral of chrysotile from the untreated tiles.

Comparison of Results

Between XRD standard additions and gravimetry

A comparison of results from gravimetry and quantitative XRD using standard additions is given in Table 2. In general, the results from gravimetry and quantitative XRD compare well, with a slight bias for the gravimetry results to be a little higher than the XRD results. This is explained by assuming that some non-asbestos components are still present in the asbestos residue. Those samples that contained two asbestos types in the residue were further analyzed by XRD or PLM to determine the relative amounts of the two types.

Table 2. Comparison of gravimetry and quantitative XRD for selected samples

Sample	Asbestos type	Mean Wt% by Gravimetry	Mean Wt% by XRD	1 σ Grav.	1 σ XRD
2A	Chrysotile	8.0	5.8	0.2	0.9
4A	Amosite	8.0	6.0	1.2	1.9
4C	Crocidolite	10.8[a]	9.2	1.4	2.1
	Chrysotile	4.7[a]	4.9	0.2	0.9
4D	Chrysotile	7.6	6.5	1.5	0.7
8A	Chrysotile	47.0	35.7	5.8	5.8
8C	Amosite	9.1[b]	13.1	0.9	4.0
	Chrysotile	0.8[b]	0.7	0.1	0.4

[a] The relative concentrations of crocidolite and chrysotile in the residue were determined by XRD using a calibration curve determined for synthetic binary mixtures.

[b] The amounts of amosite, chrysotile, and matrix phases in the residue were determined by point counting. The total residue was 13.0 wt% of the sample, of which the matrix phases were found to comprise 24% by point counting. Bias in the point count results may be responsible for the relatively low amount of amosite by gravimetry as compared with the standard additions XRD.

Between NIST and RTI

The gravimetric results from NIST and RTI for samples that have completely removable matrices are the same within a 5% relative error. The results from quantitative XRD using standard additions also agree well, as illustrated by the examples given in Table 3 for samples from recent proficiency testing rounds.

Table 3. Comparison of quantitative XRD using standard additions for NIST and RTI.

Sample	RTI mean (wt%)	RTI 95/95 tolerance	NIST mean (wt%)	NIST 1σ
13a (chrysotile)	18.8	7.0-30.6	22.5	3.3
13b (chrysotile)	8.9	4.4-10.7	9.2	0.7
14a (chrysotile)	3.7	0.7-6.6	3.4	0.8
15b (chrysotile)	13.3	10.0-16.6	9.6	2.2
15c (amosite)	1.7	0.3-3.1	2.0	0.4

In general the results from gravimetry and XRD agree relatively well, as do the results between the two laboratories, but there are cases when the results do not agree. In most cases the cause of the problem can be determined. For example, one of the laboratories originally obtained a result of 10.0 wt% chrysotile from standard additions XRD for sample 13a. The discrepancy between one result near 20 wt% and another near 10 wt% was resolved by both labs preparing new samples with standard additions to check the findings. The low result could not be reproduced and was determined to be a case of sample preparation error. By carefully preparing the samples, both labs arrived at approximately the same result. This sample did not have a soluble or ashable matrix, thus preventing the use of gravimetry as an independent technique, and the point count results were too variable to be helpful.

Between weight percent results and point counts

NIST employs the point count technique as outlined in the EPA interim method, collecting 400 positive points over 8 slides. If good slide preparations cannot be prepared from the sample in its original state due to particle size or heterogeneity, the sample is milled or ground in a mortar and pestle. Sample preparation has an effect on the results of the point count, which is why it is difficult to prepare projected area standards or to compare projected area percent with weight percent. In a point count, the thickness of each particle is not taken into account, and therefore a large difference in thickness between matrix and asbestos components causes a large difference between weight and projected area analysis [6, appendix C]. If the sample preparation affects the relative thickness of the matrix and asbestos components, the projected area results will change.

One of the proficiency testing materials was used to test the point count method, both within our laboratory and among the NVLAP laboratories. Slides were prepared from the sample as received (untreated) and after grinding. (The sample was an authentic building material containing 6 wt% chrysotile in a matrix of calcite, gypsum, portlandite, and other minor phases.) The averages of the point count results from the two different preparations were quite different, with an average for our results of 7.3% (projected area) for the ground material, and 3.3% for the unground material. The same general result was seen in the laboratory data as shown in Figure 4.

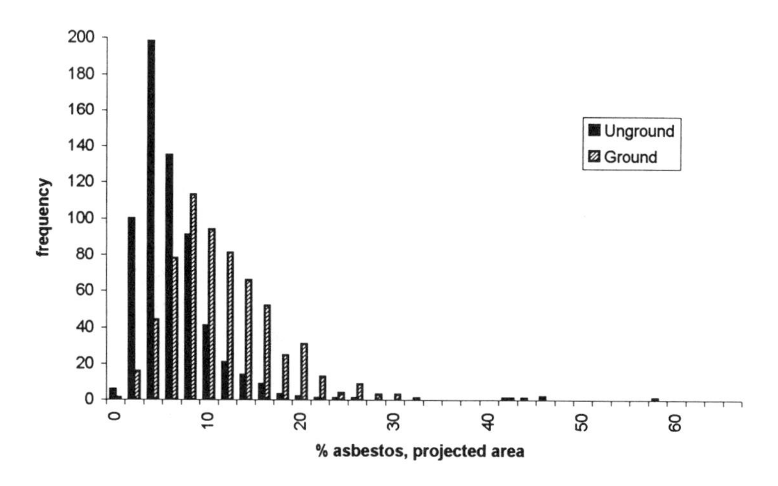

Figure 4. Laboratory point-count results for slides prepared from a PTM either with grinding, or without grinding. The PTM contains 6 wt% chrysotile.

Another source of bias in the point count results is the presence of matrix that coats the asbestos components, making them less visible under the optical microscope and more difficult to identify, resulting in a low concentration compared with weight percent determinations. This problem can sometimes be rectified by removal of the matrix using gravimetric procedures. In addition to variability in point count results due to sample preparation, thickness variations, and matrix interference, there are variations in results among different analysts that can be non-random. An analyst must make a decision at each point whether the material under the cross-hair intersection is asbestos or not, and some analysts become increasingly conservative in the criteria they use to make an identification of asbestos, while others become too liberal. This problem can be addressed using standard quality control procedures, as described in [10].

In some cases point count results agree with the weight percent results, as in the case for sample 15c where both laboratories had point count results that averaged approximately 3% amosite, compared to the results from XRD (Table 3) which average approximately 2 wt% amosite. On the other hand, the point count results can be quite biased, for a synthetic mixture of amphibole asbestos and calcite the point count results were biased high by a factor of 3. We use quantitative XRD and gravimetry to determine reference values for the PTMs, and use point count results as guidelines for determining the range of acceptable values for a given sample.

Conclusions

The weight percent of asbestos in bulk materials can be determined reliably using a combination of gravimetry and quantitative XRD using standard additions. Both techniques are widely applicable to bulk asbestos materials, although the XRD method is the most generally useful. Two independent laboratories (NIST and RTI) using the same methods on the same samples arrive at very similar results. The point count method produces results in different units (projected area) that are subject to more variability and bias. As such, we use the results from gravimetry and quantitative XRD produced by NIST and RTI to determine reference values for the weight percent of asbestos in the PTMs, and use point count results to adjust the grading criteria.

References

[1] U.S. Environmental Protection Agency Interim Method of the Determination of Asbestos in Bulk Insulation Samples: Polarized Light Microscopy, 40 CFR Ch. 1, Pt. 763, Subpt. F, Appendix A, 7/1/87 edition.

[2] Klug, H.P. and Alexander, L.E., 1974, *X-Ray Diffraction Procedures for Polycrystalline and Amorphous Materials 2nd Ed.*, John Wiley and Sons, New York, pp. 553-554.

[3] Skoog, D.A., 1985, *Principles of Instrumental Analysis 3rd Ed.*, Saunders College Publishing, New York, pp. 210-212.

[4] Gualtieri, A., and Artioli, G., 1995, "Quantitative Determination of Chrysotile Asbestos in Bulk Materials by Combined Rietveld and RIR Methods", Powder Diffraction 10 [4] pp. 269-277.

[5] Hu, R., Block, J. Hriljac, J.A., Eylem, C., and Petrakis, L, 1996, "Use of X-ray Powder Diffraction for Determining Low Levels of Chrysotile Asbestos in Gypsum-Based Bulk Materials: Sample Preparation", Anal. Chem. 68 [18] pp. 3112-3120.

[6]U.S. Environmental Protection Agency Method for the Determination of Asbestos in Bulk Building Materials, EPA/600/R-93/116 NTIS document PB93-218576, July 1993, R.L. Perkins and B.W. Harvey.

[7] Hodgson, A.A., 1979 "Chemistry and Physics of Asbestos" in Asbestos, Vol. 1, Properties, Applications, and Hazards, L. Michaels and S.S. Chissick, Ed., John Wiley and Sons, New York.

[8] Martinez, E., 1961 "The Effect of Particle Size on the Thermal Properties of Serpentine Minerals" Amer. Min. 46, 901-912.

[9] Hodgson, A.A., Freeman, A.G., and Taylor, H.F.W., 1965 "The Thermal Decomposition of Amosite" Mineralogical Magazine 35, 445-462.

[10] Verkouteren, J.R., and Duewer, D.L. 1997 "Guide for Quality Control on the Qualitative and Quantitative Analysis of Bulk Asbestos Samples: Version 1" NISTIR 5951, U.S. Department of Commerce, NTIS publication PB97-153530/AS, Springfield, VA 22161.

Peter M. Cooke[1]

A PERSONAL PERSPECTIVE ON TEACHING ASBESTOS ANALYSIS: LESSONS FROM THE CLASSROOM AND LABORATORY

REFERENCE: Cooke, P.M., **"A Personal Perspective on Teaching Asbestos Analysis: Lessons from the Classroom and Laboratory,"** *Advances in Environmental Measurement Methods for Asbestos, ASTM STP 1342,* M.E. Beard and H.L. Rook, eds., American Society for Testing and Materials, 2000.

ABSTRACT: In 1973, McCrone Research Institute taught a course called "Asbestos Identification Using Dispersion Staining" for the Environmental Protection Agency. This was the first asbestos identification course utilizing polarized light microscopy (PLM) and associated techniques taught in the United States (it was first taught at our facility in London in 1969). More than 582 classes and 7457 students later, that single course has grown into what is now a series of five PLM courses. Have we produced competent analysts? Have students' interests and capabilities changed over the years? Have the various accreditation programs and new regulations and standards affected what is taught in the classroom? Are we addressing the science and art of microscopy? This paper focuses on these questions, and on the answers that have been learned from both students in the classroom and practicing analysts at accredited laboratories.

KEYWORDS: polarized light microscopy (PLM), asbestos, microscopy education, McRI, NVLAP

Introduction

McCrone Research Institute (McRI) is a 37-year-old not-for-profit organization dedicated to teaching and research in chemical microscopy. As such, it offers a range of microscopy courses. Most, but not all, feature applied polarized light microscopy (PLM) and associated techniques. *Microscopical Identification of Asbestos* is one such course.

Offered in the United States for the first time in 1973, *Microscopical Identification of Asbestos* was originally entitled *Asbestos Identification Using Dispersion Staining*. It was offered primarily for professional microscopists interested in extending their capabilities into asbestos analysis. That introductory class has grown into what is today a series of five PLM courses that cover asbestos-related topics ranging from a basic introduction to the microscope and associated theory to quality control. The courses are entitled *Microscopical Identification of Asbestos*, *Advanced Asbestos Identification*, *Quantitative Methods in Asbestos Analysis*, *Special Asbestos Problems*, and *QA/QC in the Laboratory*. Related courses include *NIOSH 582 Fiber*

[1] Principal, Microscopy Instruction, Consultation & Analysis (MICA), 5807 N. Maplewood Avenue, Chicago, IL 60659. email:pmcooke@xsite.net

Counting using phase contrast microscopy (PCM) for airborne fibers, and those requiring the use of the transmission electron microscope (TEM).

The evolution and development of McRI asbestos courses was a response to the burgeoning asbestos industry, which was being driven largely by new regulations, primarily the Asbestos Hazard and Emergency Response Act (AHERA) of 1986 [*1*]. The passage of AHERA created the need for the analysis of hundreds of thousands of samples, which led to the establishment of many new laboratories. The demand for competent analysts grew at an unprecedented rate. This demand was reflected in the growing number of students enrolled in the McRI asbestos-related microscopy courses. Table 1 illustates the growth in the numbers of both classes and students at McRI and McCrone Scientific Ltd.

TABLE 1—*Asbestos Courses at McRI and McCrone Scientific Ltd. (London).*

Year	Total Number of Courses	Total Number of Students
McCrone (London)		
1969	1	7
1970	1	10
1971	2	18
1972-1996	28	175
McRI (U.S./Canada)		
1973	1	13
1976	4	57
1977	2	31
1978	2	26
1979	11	149
1980	4	47
1981	4	53
1982	5	66
1983	8	94
1984	12	159
1985	17	218
1986	20	351
1987	37	692
1988	67	1096
1989	73	1152
1990	64	746
1991	61	578
1992	42	433
1993	30	352
1994	32	356
1995	31	297
1996	24	272
Total	**582**	**7457**

The demand for training an increasing number of analysts brought about changes at McRI including the development of additional specialized courses and the need to increase the faculty. The McRI courses were further influenced by the development in 1987 of the environmental programs of the National Voluntary Laboratory Accreditation Program

(NVLAP) which consisted of the PLM bulk- and the TEM airborne-asbestos fiber analysis programs. McRI had been teaching the NIOSH 582 PCM course; McRI then added the TEM and the related Selected Area Electron Diffraction (SAED) courses. The advent of NVLAP brought students who not only needed to learn how to properly identify bulk asbestos and air samples, but also needed answers to questions such as: "What must my laboratory do to become accredited" (which over the years has changed to "...stay accredited")?

Personal Reflections on Teaching Asbestos Analysis

The microscopy courses taught at McRI are typically one week in duration. Originally the non-asbestos courses were two weeks long, but were eventually changed to week-long courses because companies simply would not absorb the perdiem costs for two weeks in addition to the course tuition.

In the early 1970s, before I began teaching at McRI, individuals who enrolled in *Asbestos Identification Using Dispersion Staining* were typically seasoned microscopists wanting to learn how to identify asbestos minerals and related building materials. These students were, for the most part, professional microscopists, mineralogists, and analysts with previous experience in the use of polarized light microscopy, hence the initial course offering was only three days in duration.

With the passage of AHERA, McRI experienced a significant change, both in the rapid growth of our student population and in the backgrounds of those students. Students had more varied backgrounds, from bachelor- to doctorate-level geologists and others with strong science credentials, to those with little formal science education. With the broader range in student skill levels, the introductory three-day course necessarily evolved into a five-day course.

The new student body presented unique challenges in the classroom. What was the best way to introduce practical, college-level material to students who didn't have adequate preparation? A review of crystallography and optics, for example, can be intimidating to some students, but such a review is critical for a thorough understanding of the origin of, differences in, and measurement of optical properties of the asbestos minerals and associated particles. In addition, the new vocabulary, including terms such as diffraction, refractive index, dispersion, retardation, interference, compensation, extinction, uniaxial, biaxial, etc., can be overwhelming without appropriate instruction.

Another classroom consideration is the student-to-microscope ratio. McRI firmly believes that for effective instruction, each student should have a microscope. I can't imagine the instruction being as effective if microscopes are shared.

The sequence of topics follows a logical path: introduction to physical and geometrical optics; microscope alignment; illumination; particle morphology and crystallography; use of plane polarized light for pleochroism, refractive index measurement by Becke lines and dispersion staining; use of crossed polars for isotropy determination, uniaxial and biaxial optics, birefringence, extinction characteristics and sign of elongation by compensation. (A detailed course outline can be found in Appendix I.)

Overall, I believe that the design of the McRI courses provides analysts with the potential to do good work. Of course, once back at their laboratories, we expect that beginning analysts will continue studying the course manual and evaluating their sample analysis results.

Successful completion of an introductory McRI asbestos identification course is based upon the following factors: 1) the ability to properly align, calibrate, and use a polarized light microscope; 2) passage of various quizzes; and 3) correct identification of the fibrous components of a requisite number of bulk insulation samples.

I use one of 3 different sets of 50 samples as unknowns in the introductory class; hundreds of additional samples are also available. All regulated asbestos minerals are frequently represented in the samples, as are a variety of other fibers and matrices that the student can expect to encounter. Samples are drawn primarily from a variety of "real world" materials, mine-grade minerals, and past proficiency rounds from various laboratory accreditation and proficiency testing programs. I also have fifteen introductory sets of five samples each. These permit all students to work on the same initial five samples concurrently, allowing a more expedient review of their first efforts at analysis. Those who are more experienced or proficient receive additional more difficult samples.

From a purely pedagogical perspective, the introductory *Microscopical Identification of Asbestos* and the *Advanced Asbestos Identification* courses have probably been the most rewarding for me to teach, in part because of the challenge to reach all levels of students with various backgrounds, and to observe first-hand the individual triumphs that come while learning the science (and art) associated with polarized light microscopy.

The array of student capabilities requires that I be particularly attuned to each student's growing skill level so I can use my time most efficiently to assist the greatest number of students. Deciding when and how to assist, review, and/or demonstrate requires paying close attention to individual needs and progress. Those who fall short of analyzing the requisite number of samples are offered the opportunity to analyze additional proficiency samples back at their laboratories in order to successfully complete the course. Some students choose to come in early or remain after class.

As in any endeavor, motivation is a key factor regardless of individual background or prior experience. Those who have a keen desire to learn, do so.

How competent are the analysts after this introductory class? Much depends on the students' expectations and preparation preceding the class. Students who arrive at McRI without adequate preparation may leave the introductory course not fully understanding exactly what they are doing or the underlying theory. In a practical sense, however, they are able to start "routine" sample analyses, identifying the more frequently encountered fibers including chrysotile, amosite, crocidolite, mineral wool, glass fibers, and cellulose associated with the more common matrices.

A number of laboratories offer a cursory introduction to the microscope and sample preparation and analysis before their employees attend a formal PLM course. These students tend to do noticeably better during the analysis of the unknown samples. Many laboratories I have visited as a NVLAP assessor use McRI's or another introductory course as a first step in training their analysts, after which comes further in-house training and/or supervision. The laboratories do not expect the student to immediately begin unsupervised sample analysis, nor do they expect the educational process to end there. Laboratories that permit unsupervised analysis and ignore the educational process are the ones that NVLAP assessors have identified as having quality assurance/quality control (QA/QC) inadequacies. Laboratories with only one analyst have to address this challenge by increasing their QA/QC activities and supplementing their analyst's educational training. In fact, some of the better laboratories I have audited have been "one-person" laboratories that have successfully addressed these issues. Potential analyst proficiency problems are not unique to the NVLAP PLM bulk asbestos program; assessors in the TEM NVLAP have encountered similar problems.

Many laboratories have or are required to provide some sort of in-house training regimen. As a NVLAP assessor and, perhaps more importantly, as a teacher, I have always found it extremely interesting to evaluate in-house training course outlines at individual laboratories. They range from sparsely written to quite detailed outlines replete with quizzes and a repository of routine-to-challenging samples for analysis. Many in-house training programs are rigorous and provide a thorough educational experience. I am disappointed when I find in-house training manuals that are inadequate or unrealistic. For instance, I once visited a laboratory whose outline for beginning analysts without any prior outside training stated that five minutes were to be devoted to learning refractive index measurement, and ten minutes were to be spent on observations made in crossed polars. Of course, the true quality of any training is a summation of the interaction of the particular analyst, his or her supervisor, and the laboratory.

Throughout the industry today, a significantly heightened emphasis on quantitation exists, in marked contrast to its importance pre-AHERA. Regulations dictate accountability to 1% levels, and in some states, to the 0.1% level. Laboratories expecting to be held to these standards need to know it can be accomplished, although it requires a more rigorous anaytical approach than the routine "visual"estimate. There simply is not sufficient time in the introductory class to cover quantitation rigorously; emphasis must be placed on qualitative analyses. It is suggested to students that they acquire additional training in quantitative methods.

My experience auditing laboratories accredited by NVLAP offers unique benefits to my students because the time I spend in the laboratories enables me to be more sensitive to the realities they have to contend with on a day-to-day basis. Keeping up with the latest regulations is challenging for both the laboratory and the McRI staff (and not as enjoyable or rewarding as teaching microscopy). Former and current students always pose questions about regulatory interpretation and application. If McRI has been kept informed, I provide copies of the regulations for the students. As teachers we have an obligation to try to keep up with the regulations, but our primary role is to teach microscopy. For a number of years I have found that setting up conference calls in the classroom via a speaker telephone between students and government or laboratory officials is the most efficient means for getting an interpretation of regulations, evaluations of national proficiency round performance, and alternate sample analysis or preparation recommendations. Individuals, including Mike Beard, formerly with the U.S. EPA, Bruce Harvey and Bob Perkins at Research Triangle Institute (RTI), and Eric Chatfield of Chatfield Technical Consulting, have been gracious in providing such useful and needed telephone consultation. This is typically done in our more advanced classes on an as-needed basis as students show interest. The classroom thus serves as an alternate forum to disseminate first-hand information on the ever-changing regulatory aspects of asbestos analysis.

McRI PLM Course Listing

Following are descriptions of the asbestos-related PLM courses offered at McRI in the U.S., and the year in which they were first offered.

Identification of Asbestos (1973)

First developed and taught in England by Dr. Walter C. McCrone in 1969, the content and subject sequence of this introductory course is designed for students with no formal training in asbestos analysis. The course outline can be found in the appendix.

Advanced Asbestos Identification (1983)

Initially this was a review course brought about by student interest in unique analytical problems. Originally three days in length, it was extended to five days primarily as a result of the following: increased concern over asbestos, which had extended the search from obviously friable building material to all other possible sources; the introduction of new fibrous mineral and non-mineral "look-a-like" substitutes; the revelation that asbestos minerals exposed to sufficient heat, fire, and chemicals alter physical, chemical, and optical properties; and the obligation to meet regulatory standards.

Most analysts appreciate —and need—the review of the material presented in the introductory class. Course content includes plotting individual dispersion-staining graphs of the asbestos minerals; alternate techniques for refractive index measurement, compensation, and birefringence measurements; more detailed recording of optical properties, including numerical values for refractive index (thus satisfying NVLAP requirements); means for refractive index liquid calibration; identification of fibrous synthetic, ceramic, natural, and other look-a-like materials; identification of heat- and chemically altered fibers; various sample preparation techniques (including quantitation and gravimetric procedures); and less common and more difficult sample matrices. Accordingly, the "unknowns" that students analyze are more difficult than those in the introductory class. Students are also encouraged to bring samples from their laboratories to the class.

Special Problems in Asbestos Analysis (1989)

This is an advanced PLM course for those who have successfully completed both the introductory and the advanced asbestos identification courses. It differs from those first two courses in that it is more structured to individual needs. In a sense, analysts get personal consulting. Some students are interested in further developing in-house training and request a thorough review of the underlying theory. While analyzing the more difficult samples, students are encouraged to push the limits of the instrumentation capabilities, pressing the science of microscopy nearer to the art of microscopy. Alternate instrumentation is also evaluated. Many students share similar concerns (difficult samples, regulations, and NVLAP or other accreditation requirements) and these we address together as a class.

Quantitative Methods in Asbestos Analysis (1991)

This is an advanced course in quantitative methods in asbestos analysis; students must have successfully completed the introductory course.

Examination of the performance of the laboratories on the test samples [*2*] distributed in the proficiency testing programs indicate that more emphasis needs to be placed on quantitation techniques. This emphasis was also driven, in part, by regulations, viz. the "1% dilemma" [*3*]. Laboratories now had additional reasons (other than being accurate) for being interested in quantitation techniques: to maintain accreditation and/or comply with new or existing regulations. Prospective students often ask, "Is this the point-counting course?"; in actuality, the course is devoted to various sample preparation techniques, gravimetry, sample reduction methods, and the preparation and use of quantitative reference samples [*4,5*]. Methods that are statistically significant and legally defensible [*6*] are taught. Issues affecting point-counting results are evaluated [*7,8,9,10*]. Alternate instrumental means of analyses are compared for increased accuracy and precision. Maintaining NVLAP accreditation, New York State Environmental Laboratory Approval Program (ELAP) approval, American Industrial Hygiene Association (AIHA) proficieny, and regulatory compliance have been primary reasons for continued laboratory interest in this course.

Advanced Asbestos QA/QC (1996)

Most NVLAP assessors have indicated that laboratories vary widely in their approach to the collection and relevant handling of QA/QC activities. This course was developed solely to meet the need for uniform QA/QC practices. It not only continues the laboratory's qualitative and quantitative improvement, but also tracks the data required by NVLAP in a statistically [*11,12*] relevant way.

Teaching Innovations (Teaching Microscopy Can Be Fun)

Early in my teaching career I remember agonizing over how much to cover in a one-week course. Dr. Walter McCrone advised me that within the structure of our week-long courses, it is just as important to know what to leave out of the lecture as what to leave in. He maintained that students need to spend time just looking through the microscope. His comments had an affect on my approach to teaching microscopy. I now spend a significant amount of time explaining what the students are seeing and asking questions of them *while the students are at the microscope* rather than in the lecture room.

A teaching aid that has been invaluable is one I have always associated with Dr. McCrone and McRI. Three cameras [*13*] are utilized to relay and project images alternately from a polarized light microscope, a stereomicroscope, or a desktop notebook; we who teach there casually refer to this as our "command post." We all use it for demonstration. When teaching the TEM or SAED courses, an McRI colleague, Bob Stevenson, also utilizes video projection through the side window of a JEOL 1200 TEM. The normally circular field-of-view is somewhat elongated, but it allows projection of bright-field and diffraction images. An excellent text [*14*] on video microscopy far beyond what is needed in the classroom exists, yet a number of chapters are very useful in understanding video projection of microscopical images.

Developing and utilizing innovative and informative teaching aids can be challenging, and when successful, their use can be downright fun. Using them in the classroom has been integral to my approach to teaching. Jan Hinch, of Leitz, is always coming up with new "gadgets" with which to teach. He once used two similar halves of clear oval pantyhose containers filled with colored water to make a demonstration model of a uniaxial indicatrix. I have also found the model to be useful, but have abandoned indicatrix theory in the introductory asbestos courses; I do not think that, in a practical sense, the students need it.

I now use a battery-operated soap-bubble-blowing gun to demonstrate (in Lawrence Welk-like fashion) Newtonian-sequence interference colors. Seeing those familiar colors first-hand makes it easier to then proceed with explanations of the interference or polarization colors associated with anisotropic media viewed between crossed polars and the varied uses of the Michel-Lévy chart.

I also use two large sheets of polarizing filters and an overhead projector to demonstrate the effects associated with examining particles viewed between crossed polars. It is an efficient way to be able to illustrate all six orders of interference colors from a quartz wedge at once, along with different compensators and various other anisotropic media. Students can then relate these colors to what they might be seeing projected via our "command post" microscope, or confirm their own microscopical observations. Using the overhead projector, polyester or acetate strips can be placed on top of each other in various orientations to further demonstrate retardation and compensation.

Optical-grade calcite rhombs and polarizing filters are helpful in demonstrating anisotropy, double refraction and the effect of using polarized light.

Large crystal models aid in demonstrating crystal optics to larger classes in addition to the wooden crystal models normally associated with "orientation" exercises in a crystallography class.

I use uranium glass [*15*] to aid in tracing light paths in oculars, objectives and substage condensers. I find the uranium glass to be a valuable resource in demonstrating angular aperture, from which I discuss and illustrate the importance of numerical aperture, and the effect that the opening and closing of the substage diaphragm has on microscopical image quality and resolution.

My colleague, John Delly, is constantly offering innovative classroom aids and means for improving the versatility of a microscope, like retrofitting illuminators to make them fully focusable and centerable [*16*] or making one's own dispersion staining (focal masking) objective [*17,18*]. He has also introduced the use of a high-magnification compound microscope as a low-magnification preparative microscope; the conversion of a pocket field microscope for on-site asbestos analysis; and a motorized substage polarizer, permitting rapid determination of optical properties of fibers without stage rotation [*19*]. The classroom use of the motorized polar is also an effective way to demonstrate optical properties and to also quickly get students attention.

Bob Stevenson also uses visual aids when teaching McRI's TEM and SAED courses. When covering diffraction he literally uses "fuzzy brown balls" to illustrate the three-dimensionality associated with reciprocal space points. Opening and closing a small umbrella serves to illustrate the effects that accelerating voltage has on Ewald's sphere. Students have a better appreciation for the integrity and brightness of spots associated with Laue zones as Ewald's sphere (the umbrella) makes contact through various parts of the "fuzzy brown balls." Of course all of this can then be explained with Fourier Transform equations, but given a choice, the diversity of students, and Bob's personality....

He also uses acetate sheets on which he has drawn a regular array of dots to represent reciprocal space points; individual layers are separated with transparent plastic blocks to illustrate the third dimension. It helps students to think of reciprocal space as "being there." Using the small umbrella, or by simply cupping his hands, Bob can illustrate the effect that stage tilting has as Ewald's sphere moves through reciprocal space. Students can more readily see the origin of the changing diffraction patterns and begin to recognize that as Ewald's sphere becomes tangent to those acetate planes, a zone axis has been located. The end result is that students have a better appreciation for reciprocal space and diffraction.

I have found visual aids like these to be valuable in conveying difficult concepts and in making learning more enjoyable.

Valuable Reference Materials

The Asbestos Particle Atlas [*20*] grew out of an EPA-sponsored project. The Atlas' text, as well as the arrangement and order of the photomicrographs that illustrate asbestos fibers and related material in the many types of illumination associated with the polarized light microscope are excellent. The Identification of Asbestos [*21*] is a manual, used in all of McRI's asbestos identification courses, which grew out of that initial atlas and years of classroom experience. There is no other text with as much asbestos-specific information explaining the approach and underlying theory of PLM examination of bulk materials.

Identification of Asbestos, An Audiovisual Training Program for Microscopists [*22*] consists of 35mm slides, cassette tapes, and a text of the narration. It is an excellent resource for individual study or classroom training following a formal course.

Another book that I have found useful for its coverage of many inorganic fibers, their mineralogy, crystal chemistry, and associated health effects is Asbestos and Other Fibrous Materials [*23*]. The Proceedings of Workshop on Asbestos: Definitions and Measurement Methods [*24*] contains much of the original analytical methodologies.

The revised EPA test method—Method for the Determination of Asbestos in Bulk Building Materials [*25*]—has much more useful information for the analyst and classroom than the Interim Method for the Determination of Asbestos in Bulk Insulation Samples [*26*].

THE MICROSCOPE, founded in 1937 (and published since 1963 by Microscope Publications, an affiliate of McRI), is another valuable resource. The asbestos-related contributions can be used to trace the historical development of the analysis of asbestos such as use of dispersion staining (Cherkosov's focal masking) as a rapid means for asbestos fiber identification; the change in mineral-fiber optical properties associated with exposure to heat, fire, and chemicals; various microscope innovations; the development of TEM protocols; the increased interest in quantitation techniques; and the development of protocols for meeting regulatory standards or NVLAP guidelines. In 1987, the journal began a regular column featuring contributions on asbestos-related analyses. (A compendium of the asbestos-related articles is found in Appendix II.) I have found these contributions, at one time or another, to be a resource not only in the classroom but also in the laboratory when representing NVLAP.

Many Bureau of Mines Report of Investigations publications have been useful teaching resources, unfortunately most are now out of print.

The handbooks Mineralological Techniques of Asbestos Determination [*27*] and Applications of Electron Micoscopy in the Earth Sciences [*28*] are collections of lecture notes from short courses sponsored by the Mineralogical Association of Canada that provide useful information for both the classroom and laboratory.

Other valuable articles on asbestos analysis can be been found in the National Asbestos Council (now known as the Environmental Information Association) journals [*29*].

There are many books, references, and texts that I use and recommend; far too many are out of print and thus are more difficult to find. Here are a few that are relatively easy to locate: Introduction to the Methods of Optical Crystallography [*30*]; Crystallography and Crystal Chemistry An Introduction [*31*]; An Introduction to the Rock-Forming Minerals [*32*]; and Optical Mineralogy: The Nonopaque Minerals [*33*]. I have found that many students respond well to Crystals and Light [*34*]; it is inexpensive and perhaps less intimidating than college-level texts.

Finally, I have found Teaching Microscopy [*35*] to be an interesting and enlightening account of how others who teach microscopy approach the subject.

Conclusions

Both practicing microscopy and teaching microscopy are challenging and fun. It is always gratifying to see former students taking such a keen interest as they extend their microscopical capabilities to other specialties. The research and teaching by the McRI staff have had a significant effect on analysts' approach to the identification of asbestos and other particles. In a small way, I trust that I have had an impact on the skills and understanding of the analysts. They have been a delight to teach. I thank them for the lessons that I, in turn, have learned from them.

References

[1] USEPA, Asbestos-Containing Materials in Schools; Final Rule and Notice, Federal Register Vol. 52 (210): 41826-41903 (October 30, 1987).

[2] Harvey, B.W., "Classification and Identification Error Tendencies in Bulk Insulation Proficiency Testing Materials." *The Microscope*, 37/4, 1989, pp. 393-402. (Also published in *American Environmental Laboratory*, Vol. 2 (2), 1990, pp. 8-14.)

[3] Perkins, R.L., Harvey, B.W. and Beard, M.E., "The One Percent Dilemma." *EIA Technical Journal*, Summer, 1994, pp. 5-10.

[4] Harvey, B.W., Perkins, R.L., Nickerson, J.G., Newland, A.J. and Beard, M.E., "Formulating Bulk Asbestos Standards," *Asbestos Issues*, Vol. 4(4), 1991, pp. 22-29.

[5] Harvey, B.W., Perkins, R.L., Nickerson, J.G., Newland, A.J. and Beard, M.E., "Feasibility Study for the Formulation of Asbestos Bulk Sample Calibration Standards," *Proceedings of the 1991 U.S. EPA/A&WMA International Symposium*, VIP-21, EPA/600/9-91/018, pp. 220-225, 1991.

[6] Chatfield, E.J., "Legally-Defensible Discrimination Between Asbestos-Containing and Non-Asbestos-Containing Materials at 1% and 0.1% Levels." Presented at Inter/Micro, 1994.

[7] Stewart, I., "Asbestos Content in Bulk Insulation Samples: Visual Estimates and Weight Composition," EPA/560/5-88-011, 1988.

[8] Perkins, R.L., "Point Counting Technique for Friable Asbestos-Containing Materials," *The Microscope*, 38/1, 1990, pp. 29-40.

[9] Perkins, R.L. and Beard, M.E., "Estimating Asbestos Content of Bulk Samples," *National Asbestos Council Journal,* Vol.9 (1), 1991, pp. 27-31.

[10] Perkins, R.L. and Beard, M.E., "Asbestos Bulk Samples Analysis: Visual Estimates and Point-Counting," *Proceedings of the 1991 U.S. EPA/A&WMA International Symposium*, VIP-21, EPA/600/9-91/018, pp. 214-219, 1991.

[11] Verkouteren, J.R. and Duewer, D.L., "Guide for Quality Control on the Qualitative and Quantitative Analysis of Bulk Asbestos Samples: Version 1," NISTIR 5951, 1997.

[12] Munch, C.P. and Jaber, J.D., "Application of Fundamental Statistical Technique to Asbestos Bulk Sample Quality Assurance," *The Microscope*, 45/1, 1997, pp. 19-26.

[13] McCrone, W.C. and Laughlin, G.J., "Video as a Teaching Aid," *The Microscope*, 38/2, 1990, pp. 135-139.

[14] Inoué, S., "Video Microscopy," Plenum Press, New York, 584 pp., 1982.

[15] Delly, J.G., "Uranium Glass," *The Microscope*, 38/1, 1990, pp. 109-116.

[16] Delly, J.G., "Fully-Adjustable Olympus BHSP Illuminator," *The Microscope*, 36/4, 1988, pp. 327-337.

[17] Delly, J.G., Sirovatka, J.C., "A Dedicated Central-Stop Dispersion Staining Objective," *The Microscope*, 36/3, 1988, pp. 205-212.

[18] Sirovatka, J.C., "A Dedicated Central-Stop Dispersion Staining Objective (Nikon),"*The Microscope*, 37/1, 1989, pp. 43-47.

[19] Delly, J.G., "Portable Microscope for On-Site Analysis and Counting of Asbestos Fibers," *The Microscope*, 34/4, 1986, pp. 331-340.

[20] McCrone, W.C., "The Asbestos Particle Atlas," Ann Arbor Science Publishers Inc./The Butterworth Group, Ann Arbor, Michigan, 120 pp., 1980.

[21] McCrone, W.C., "Asbestos Identification," McCrone Research Institute, Chicago, Illinois, 199 pp., 1987.

[22] McCrone, W.C., "Identification of Asbestos: An Audiovisual Training Manual," Brian Howard and Associates, Inc., Brooklyn, New York, 1983.

[23] Skinner, H.C.W., Ross, M. and Frondel, C., "Asbestos and Other Fibrous Materials," Oxford University Press, New York/Oxford, 204 pp., 1988.

[24] Gravatt, C.C., LaFleur, P.D. and Heinrich, K.F.J., eds., *Proceedings of Workshop on Asbestos: Definitions and Measurement Methods*, National Bureau of Standards Special Publication 506, 1977.

[25] Perkins, R.L. and Harvey, B.W., "Method for the Determination of Asbestos in Bulk Building Materials," EPA/600/R-93/116, 1993.

[26] "Interim Method for the Determination of Asbestos In Bulk Insulation Samples," EPA/600/M4-82-020, 1982.

[27] Ledoux, R.L., ed., "Mineralogical Techniques of Asbestos Determination: Short Course Handbook," Vol. 4, 279 pp., 1979.

[28] White, J.C., "Applications of Electron Microscopy in the Earth Sciences: Short Course Handbook," Vol. 11, 213 pp., 1985.

[29] *National Asbestos Council Journal*, vols. 1-15, 1983 - 1997.

[30] Bloss, D.F., "An Introduction to the Methods of Optical Crystallography," Holt, Rinehart and Winston, New York, 294 pp., 1961.

[31] Bloss, D.F., "Crystallography and Crystal Chemistry: An Introduction," Holt, Rinehart and Winston, New York, 545 pp., 1971.

[32] Deer, W.A., Howie, R.A., and Zussman, J., "An Introduction to the Rock-Forming Minerals," Longman Scientific & Technical, England, 696 pp., 1992.

[33] Phillips, W.R. and Griffen, D.T., "Optical Mineralogy: The Nonopaque Minerals,"_W.H.Freeman and Company, San Francisco, 677 pp., 1981.

[34] Wood, E.A., "Crystals and Light: An Introduction to Optical Crystallography," 2nd edition. Dover Publications, Inc., New York, 156 pp., 1977.

[35] Delly, J.G., ed. "Teaching Microscopy," Microscope Publications, Chicago, Illinois, 264 pp., 1994.

Introduction to Microscopical Identification of Asbestos

Course Outline

Monday a.m. *Lecture and Demonstration*: physical optics including reflection, refraction, Snell's Law, refractive index, dispersion, image formation using plane and concave mirrors, image formation with thin lenses; aberrations; geometrical optics including ray tracing and magnification.

Lecture and Demonstration: compound microscope and simple magnifier optics; oculars, objectives, condensers; microscope alignment; Köhler illumination.
Laboratory: microscope alignment, Köhler illumination.

Monday p.m. *Lecture and Laboratory*: Köhler illumination, Nelsonian illumination; diffuse illumination, tracing image-forming and illuminating ray paths; conjugate foci, resolution; diffraction; magnification; contrast; optical and mechanical tube length; numerical aperture; micrometry.
Lecture and Demonstration: asbestos terminology, morphology, mineralogy, and polymorphs; stereomicroscope alignment and optics.
Laboratory: asbestos minerals and associated fibers and building materials using the compound microscope and the stereomicroscope.
Assigned reading

Tuesday a.m. *Lecture and Laboratory*: Quiz on Köhler illumination. Review: Köhler illumination, microscope alignment, asbestos and other fibrous materials
Lecture and Demonstration: crystallography and optical crystallography, crystal systems; plane polarized light; pleochroism, isotropy, anisotropy, uniaxial optics; biaxial optics; refractive index measurement, immersion method.
Laboratory: pleochroism; refractive index measurement using Becke lines, relative refractive index measurement of asbestos minerals and other particles mounted in n_D =1.66 and Meltmounts n_D = 1.552, 1.605, 1.680.

Tuesday p.m. *Lecture and Demonstration*: central and annular-stop dispersion staining, origin of dispersion colors, dispersion and dispersion staining graphs, refractive index measurement.
Laboratory: refractive index measurement using dispersion staining.
Lecture and Laboratory: alignment of the stereomicroscope, sample preparation, mounting of standard reference fibers in Cargille HD RI liquids n_D = 1.550, 1.605, 1.680.
Assigned reading

Wednesday a.m. *Lecture and Laboratory*: Quiz on dispersion staining and refractive index measurement. Review of dispersion staining and fiber reference mounts.
Lecture and Demonstration: crossed polars, birefringence, Michel-Lévy chart, origin of polarization colors, isotropy, anisotropy, extinction characteristics.
Laboratory: microscope and polarizer/analyzer alignment, determination of isotropy, anisotropy, birefringence measurements.

Wednesday p.m. *Lecture and Laboratory*: crossed polars: extinction characteristics and measurements, reading a stage vernier, sign of elongation, use of compensators.
Lecture and Laboratory: integration of plane polarized light, crossed polars, crossed polars with the Red l compensator, dispersion staining. Determination of morphology, size and optical properties: pleochroism, relative refractive index, birefringence, extinction characteristics, sign of elongation, sample preparation techniques, sample quantitation techniques.
Quiz on microscope setup for analysis. Assigned reading.

Thursday a.m. *Laboratory*: analysis of unknowns.
Lecture and Laboratory: review sample preparation techniques, measurement of optical properties of asbestos and related particles, analysis of unknowns.

Thursday p.m. *Laboratory*: analysis of unknowns

Lecture and Laboratory: review quantitation techniques, analysis of unknowns.

Friday a.m. *Lecture and Laboratory*: Quiz on asbestos analysis. analysis of unknowns, review alternate sample preparation and analysis techniques, review NVLAP requirements.
Course review, evaluation, adjournment.

Compendium of Asbestos-Related Articles from *The Microscope*

Identification of Asbestos Fibers by Microscopical Dispersion Staining. Y. Julian; W.C. McCro
1970, 18/1, 1-10

Dispersion Staining of Fibers. L. Forlini; W.C. McCrone, 1971, 19/3, 243-254

Practical Aspects of Counting Asbestos on the Millipore IIMC. E.J. Jones, 1975, 23/2, 93-101

A New Dispersion Staining Objective. W.C. McCrone, 1975, 23/4, 221-226

Determination of n_D, n_F, n_C by Dispersion Staining. W.C. McCrone, 1975, 23/4, 213-220

An Obvious Illuminator for Dispersion Staining. O. Goldberg, 1976, 24/4, 291-294

An Alternative Dispersion Staining Technique. R.G. Speight, 1977, 25/4, 215-225

Identification of Asbestos by Polarized Light Microscopy. W.C. McCrone 1977, 25/4, 251-264

Dispersion Staining Colors. W.C. McCrone, 1978, 26/2, 109-120

OSHA Regulations & Methods for Asbestos. W. Dixon, 1978, 26/4, 183-186

Cargille PCB-Free Refractive Lasex Liquids. M. Liva; W. Sacher, 1979, 27/2, 87-100

Hexametaphosphate Pretreatment of Insulation Samples for Microscopical Identification of Fibro
Constituents. D.H. Taylor; J.S. Bloom, 1980, 28/1, 47-49.

Fiber Optics Illumination for Use in Dispersion Staining. R. Resua; N. Petraco, 1980, 28/2, 51-

Quantitative Determination of Chrysotile in Building Materials. H.W. Dunn; J.H. Steward, Jr.,
1981, 29/1, 39-46

Dispersion Staining Techniques High Numerical Aperture. D. Schmidling, 1981, 29/2, 121-126

Application of Water Immersion Microscopy to the Examination of Thermal Insulation Fibers.
R.G. Speight, 1982, 30/1, 5-9

Typical Asbestos Testing Procedure. F. Goldblatt, 1982, 30/3, 259-263

A Fluorescent Dye Binding Technique for Detection of Chrysotile Asbestos, F.R. Albright; D.V.
Schumacher; B.J. Felty; J.A. O'Donnell, 1982, 30/3, 267-280

Changes in Optical Properties of Chrysotile During Acid Leaching. J.R. Kessler, 1983, 31/2, 16
174

ıe Evaluation of Asbestos Contamination of Surfaces A New Approach with an Old Technique. R.G. Speight, 1983, 31/2, 175-186

ew Materials for the Microscopist. R. Sacher, 1985, 33/4, 241-246

:otch Magic Tape—An Aid to the Microscopist for Dust Examination. G. Nichols, 1985, 33/4, 247-254

outine Detection and Identification of Asbestos. W.C. McCrone, 1985, 33/4, 275-284

n Examination of the Influence of Microscope Characteristics on Asbestos Fiber Counting. R.G. Speight, 1986, 34/2, 93-106

ermanence of Membrane Filter Clearing and Mounting Methods for Asbestos Measurements. T. Sheaton-Taylor; T.L. Ogden, 1986, 34/3, 161-172

uparel and Its Use in Measurement of Asbestos. T.L. Ogden; D.J.M. Thompson; P.A. Ellwood, 1986, 34/3, 173-180

ortable Microscope for On-Site Analysis and Counting of Asbestos Fibers, J.G. Delly, 1986, 34/4, 331-340

sbestos in Water. J.R. Millette, 1986, 34/4, 371-374

alibration of the Electron Diffraction Camera Constant. J.R. Millette; J.J. Smith, 1987, 35/1. 107-117

lectron Diffraction of Asbestos. J.R. Millette, 1987, 35/2, 207-215

he Effects of Interocular Distance on Microscope Calibration. J.A. Armstrong, 1987, 35/3, 267-271

Reference List for Microscopical Methods for the Measurement of Asbestos. J.R. Millette; J.R. Krewer, 1987, 35/3, 311-318

licroscopy and the Asbestos Hazard Emergence Response Act (AHERA). J.R. Millette, 1988, 36/1, 71-77

uality Assurance in the Electron Microscopy Asbestos Laboratory. J.A. Krewer, 1988, 36/2, 153-159

Dedicated Central-Stop Dispersion Staining Objective. J.G. Delly; J. Sirovatka, 1988, 36/3, 205-212

nexpensive Orange Filter for Refractive Index Determinations. J.G. Delly, 1988, 36/3, 213-215

ounting Rules for TEM Analysis of Clearance Samples, J.R. Millette; S.B. Burris, 1988, 36/3, 273-280.

evelopment and Operation of an In-House Asbestos Program. M.P. Gerrity, 1988, 36/4, 309-318

ully-Adjustable Olympus BHSP Illuminator. J.G. Delly, 1988, 36/4, 327-337

The Effect of Heat on the Microscopical Properties of Asbestos. G.J. Laughlin; W.C. McCrone 1989, 37/1, 9-15

A Dedicated Central Stop Dispersion Staining Objective (Nikon). J.C. Sirovatka, 1989, 37/1, 4 47

Calculation of Refractive Indices from Dispersion Staining Data. W.C. McCrone, 1989, 37/1, 4 53

Design and Construction of an Asbestos Microscopy Laboratory Facility. V.H. Ainslie; S.M. Hays, 1989, 37/1, 77-88

A Data Sheet for Electron Diffraction Analysis of Amphibole Asbestos. J.J. Dodge, 1989, 37/2, 189-194

Making Cargille Liquid Preparations Permanent. R.D. Moss, 1989, 37/3, 223-231

Readily Obtainable Electron Diffraction Patterns in Monoclinic Amphibole Asbestos. S.B. Burri 1989, 37/3, 303-309

The Application of Point Counting Procedures for the Analysis of Bulk Asbests Samples. T.F. Bergin, 1989, 37/4, 377-387

Classification and Identification Error Tendencies in Bulk Insulation Proficiency Testing Materia B.W. Harvey, 37/4, 393-402

Discussion of the Asbestos Bulk Sample Analysis Quality Assurance Programs. W.C. McCrone 37/4, 403-409

Darkfield Microscopy. W.C. McCrone; R.D. Moss, 1990, 38/1, 1-8

The Importance of an In-House Training Program in an Asbestos Laboratory. D.C. Mefford, 1990, 38/1, 23-27

Point-Counting Technique for Friable Asbestos-Containing Materials. R.L. Perkins, 1990, 38/1 29-39

The Identification of Halotrichite. L. Miles; T. Hopen; R. Kuksuk, 1990, 38/1, 41-46

3M's "Post-it™" as an Aid in Particle Sorting. J.T. Fisk, 1990, 38/2, 197-198

Settled Dust Analysis Used in Assessment of Buildings Containing Asbestos. J.R. Millette; T. Kremer; R.K. Wheeles, 1990, 38/2, 215-220

Analytical Protocol for Determination of Asbestos Contamination of Clothing and Other Fabrics. E.J. Chatfield, 1990, 38/2, 221-222

Training Methodology for Area/Volume Percentage Determination in Asbestos Analysis. D. Mefford, 1990, 38/3, 281-288

Stepping on Asbestos Debris. J.R. Millette; W. Ewing; R. Brown, 1990, 38/3, 321-326

A Standard TEM Procedure for Identification and Quantitation of Asbestiform Minerals in Talc. Kremer, 1990, 38/4, 457-468

Modified Preparation Methods for the Analysis of Asbestos in Vinyl Floor Tiles. M.L. Harris; S.Y. Yu, 1991, 39/2, 109-120

Searching for Asbestos, Identifying the Matrix. J.R. Millette; W.R. Boltin; O.S. Crankshaw, 1991, 39/2, 131-134

Point Counting Simple Methods of Different Sized Spheres, Cubes, and Cylinders. J.R. Kessler, 1991, 39/3-4, 203-213

The Crystallography of Strained Crocidolite. T.F. Bergin, 1991, 39/3-4, 287-298

Observations on Studies Useful to Asbestos O&M Activities. R.C. Wilmoth; T.J. Powers; J.R. Millette, 1991, 39/3-4, 299-312

Calibration of Refractive Index Liquids Using Optical Glass Standards and Dispersion Staining Techniques. S.C. Su, 1992, 40/1, 95-108

The Effect of Color on PLM Measurements. T.F. Bergin, 1992, 40/4, 227-239

A Close Examination of the Surfaces of Asbestos Gasket Materials. J.R. Millette; R.S. Brown, 1992, 40/2, 131-135

Calibration of the Rotation Angle Between TEM Brightfield Images and Electron Diffraction Patterns—A Response with New Data. J.S. Smith; J.R. Millette, 1992, 40/2, 115-120

Permanent Preparations with Cargille Liquids—Another Look. W.C. McCrone; R.D. Moss, 1992, 40/2, 123-124

Demonstration of the Capability of Asbestos Analysis by Transmission Electron Microsocpy in the 1960s. J.R. Millette; W.E. Longo; J.L. Hubbard, 1993, 41/1, 15-18

Removal of Matrix Calcium Compounds from Asbestos Fibers in Bulk Insulation. S.Y. Yu; M.L. Harris; V. Llacer, 1993, 41/2-3, 45-49

Titanium Dioxide: Anatase or Rutile? J.R. Millette; T.J. Hopen; J.P. Bradley, 1993, 41/4, 147-153

Evaluation of Asbestos Quantitation Methods. R.D. Moss, 1994, 42/1, 7-14

The Effects of High Temperature Over Short Periods on Chrysotile and Amosite Varieties of Asbestos. P. Goymer; H.S. MacDonald, 1995, 43/4, 177-178

A Combination Calibration Grid for Transmission Electron Microscopes. P. Few; J.R. Millette, 1996, 44/4, 175-179

Application of Fundamental Statistical Technique to Asbestos Bulk Sample Quality Assurance. C.P. Munch; J.D. Jaber, 1997, 45/1, 19-26

Reference Methods for Asbestos Analysis. J.R. Millette, 1997, 45/2, 57-59

Bruce W. Harvey,[1] J. Todd Ennis,[1] Lisa C. Greene,[1] and Adrianne A. Leinbach[1]

BULK ASBESTOS LABORATORY PROGRAMS IN THE UNITED STATES—SEVENTEEN YEARS IN RETROSPECT

REFERENCE: Harvey, B. W., Ennis, J. T., Greene, L. C., and Leinbach, A. A., **"Bulk Asbestos Laboratory Programs in the United States—Seventeen Years in Retrospect,"** *Advances in Environmental Measurement Methods for Asbestos, ASTM STP 1342,* M. E. Beard and H. L. Rook, Eds., American Society for Testing and Materials, West Conshohocken, 2000.

ABSTRACT: Quality assurance and proficiency testing programs have been available for 17 years in the United States for laboratories analyzing for asbestos in bulk building materials by polarized light microscopy. After peaking in the late 1980's, total enrollment has dropped but stabilized in the last four years to 700-750 laboratories, and participation rates currently exceed 95 %. Almost 122 000 test samples have been sent worldwide by Research Triangle Institute to laboratories in four major programs. A review of nearly 113 000 analysis results reveal that a small number of bulk building materials cause the majority of difficulties in each of the programs, although error frequencies vary from program to program. Improvement has occurred in overall laboratory performance since the first programs began in the early 1980's, with a significant decline in qualitative errors and a much less dramatic decrease in errors associated with semiquantitative estimation of the amount of asbestos present.

KEYWORDS: asbestos, laboratories, polarized light microscopy, quality assurance, proficiency testing, enrollment, error rates, performance

---

Polarized light microscopy (PLM) is the definitive method for the analysis and identification of asbestos, rendering quick and conclusive results in the hands of the experienced analyst [*1*]. Over the past 17 years, Research Triangle Institute (RTI) has been involved with four quality assurance (QA) and proficiency testing (PT) programs for laboratories using PLM to analyze for asbestos in bulk building materials. RTI has, on a contract basis, provided a variety of technical support services to the various sponsoring federal agencies/departments and accreditation organizations for the four programs. That support includes maintenance of a bulk materials repository; selection, characterization, packaging, and distribution of test samples; receipt and evaluation of laboratory results of analysis; preparation of final reports to the sponsor and participants; and maintenance of a

[1] Senior research geologist, geologist, optical mineralogist, and research geologist, respectively, Earth and Mineral Sciences Department, Research Triangle Institute, Post Office Box 12194, Research Triangle Park, NC 27709-2194.

technical information service. By way of this broad and extended involvement, and with access to archived analysis results, RTI took a retrospective look at the four programs to: a) define trends in laboratory enrollment and participation; b) identify the bulk materials most likely to cause analysis problems; c) evaluate the analytical performance of bulk asbestos laboratories; and d) determine, to the extent possible, whether participation in these programs has resulted in improved analytical performance over time.

The Four Major Programs

Four QA or PT programs of a national scope, described below, have been offered by federal agencies/departments or by accreditation organizations for laboratories using PLM to analyze for asbestos in bulk building materials. Three of the four are being conducted currently.

The U.S. Environmental Protection Agency (EPA) Asbestos Bulk Sample Analysis Quality Assurance Program

The only nonactive program of the four, this program (later titled the Bulk Sample Analysis Interim Accreditation Program) was initiated in late 1979 in conjunction with the EPA's Asbestos-in-Schools Technical Assistance Program [*2*]. Program participants were mostly commercial laboratories, but noncommercial laboratories also participated. Eighteen semiannual test rounds were conducted, and laboratories' qualitative analysis results were evaluated. The analysis results were graded using consensus values from participating laboratories. The results from the last two rounds were used to establish interim accreditation for laboratories prior to the availability of the National Institute of Standards and Technology (NIST) National Voluntary Laboratory Accreditation Program (NVLAP) for bulk laboratories. Nearly 27 000 test samples were sent to participants, mostly in the United States.

The U.S. Navy Asbestos Identification Proficiency Testing Program

The smallest yet longest-running of the three current programs, this program was initiated in late 1982 as a result of Navy standards [*3*] regarding the health and safety of service and civilian personnel. The program provides quarterly testing of shipyard, medical, shipboard, and other Navy laboratories. Only laboratories' qualitative analysis results are evaluated, and laboratories earn a proficiency rating based on established evaluation criteria. The analysis results are graded using reference values from RTI and two external reference laboratories. More than 40 000 test samples have been distributed in 56 test rounds to laboratories worldwide.

The NIST Bulk NVLAP

The bulk NVLAP was initiated in 1989 by NIST as a direct result of passage of the Asbestos Hazard Emergency Response Act (AHERA) [*4*]. The program provides the sole vehicle for accreditation of laboratories performing analysis of bulk samples from schools.

Participants are, for the most part, the commercial (and some noncommercial) laboratories that participated in the EPA program. Laboratories' qualitative and semiquantitative analyses are evaluated, and the results are graded using reference values determined by RTI and NIST. The graded results are used by NIST in conjunction with on-site assessment results to determine each laboratory's accreditation status. Sixteen test rounds have been conducted, and nearly 31 000 test samples have been sent to participants in the United States and Canada.

The American Industrial Hygiene Association (AIHA) Bulk Asbestos Proficiency Analytical Testing Program

This program, initially titled the Bulk Asbestos Quality Assurance Program, was begun in 1989 by AIHA to provide proficiency testing for commercial and noncommercial laboratories not desiring or requiring NVLAP accreditation. Laboratories' qualitative and semiquantitative analysis results are evaluated, and the results are graded using reference values determined by RTI and two external reference laboratories. A proficiency rating is determined based on established evaluation criteria. Nearly 24 000 test samples have been sent in 30 quarterly test rounds to laboratories in the United States, Canada, Europe, and the Far East.

Trends in Laboratory Enrollment

Factors influencing laboratory enrollment in bulk asbestos programs include the timing and requirements of specific legislation and standards, market competition, and even federal budget constraints. These factors have all had considerable effect on the number of laboratories performing PLM analysis for asbestos. (Figure 1) shows the laboratory enrollment pattern for each program and the combined enrollment for all programs. From about 100 laboratories in the first EPA test round, total enrollment in all programs rose to almost 1 200 laboratories in 1988. It dropped dramatically as the bulk NVLAP began, rose to nearly 1 000 laboratories in 1991–1992 with the initiation of the AIHA program, and since 1993 has stabilized at 700 to 750 laboratories.

The dramatic increase in enrollment in the EPA program was a reflection of strong growth in all sectors of the emerging asbestos industry of the 1980's, which was, in turn, being driven in large part by the requirements of AHERA and by EPA's heightened emphasis on the safety of children in schools throughout the United States. The failure of approximately 500 laboratories from the EPA program to enroll in the bulk NVLAP was probably due in part to the fee-based structure of the latter and to the pressures of greater market competition. The slow, long-term decline in the bulk NVLAP may be indicative of completion of much of the work associated with the analysis of building materials in the schools and laboratory diversification into other available work. A strong upturn in AIHA program enrollment beginning in 1995 is attributed to recognition of that program by the Occupational Safety and Health Administration (OSHA) in its revised asbestos standards [5] and to a sharp increase in participation by laboratories outside the United States. Growth in the AIHA program has not come at the expense of the bulk NVLAP. Fairly steady enrollment figures in the Navy program are a reflection of the Navy's requirements

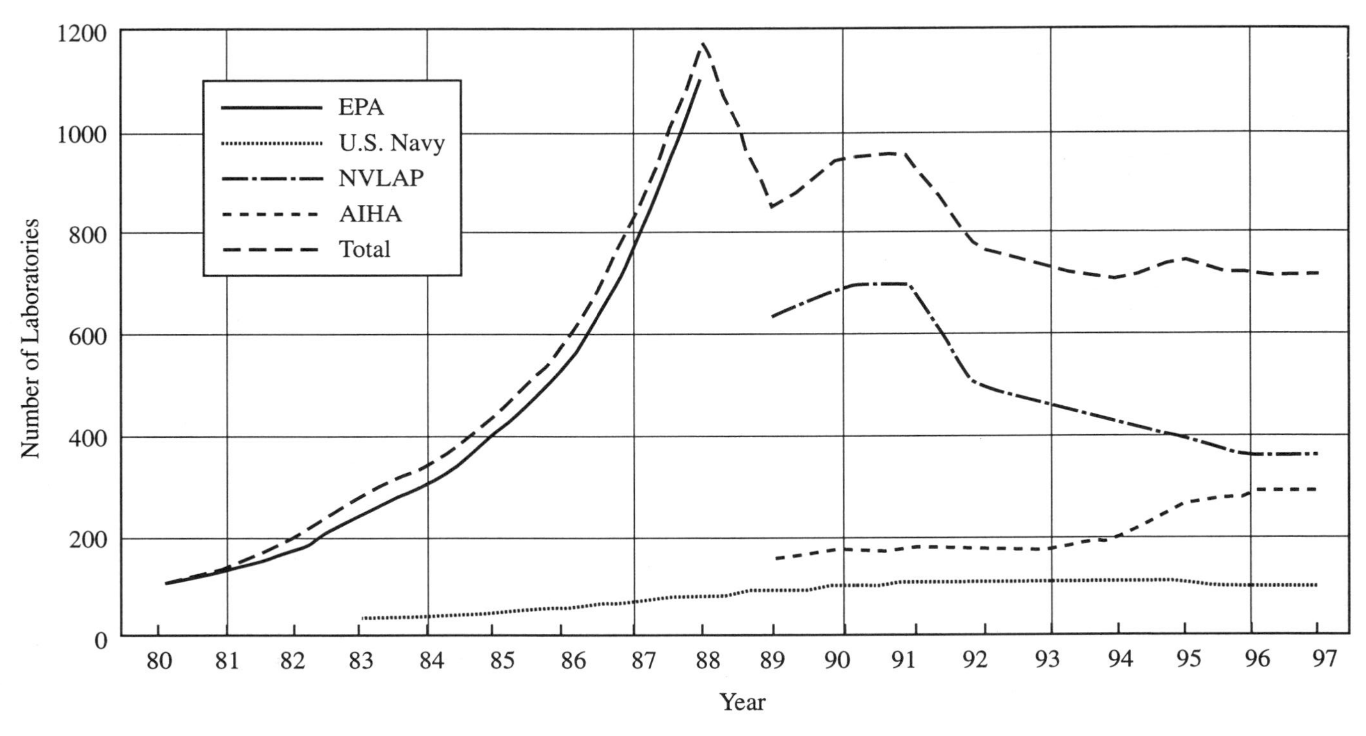

Figure 1 – *Average annual laboratory enrollment, by program.*

for participation of a small number of nonmarket-driven laboratories; a 15 % downturn in the last two years is the direct result of the closure of a number of shipyards and a reduction in the size of the Navy's fleet worldwide.

Laboratory Rates of Participation

Laboratory participation (the rate of return of analysis results) has been very strong in all programs. (Table 1) shows the rates of laboratory participation, by program. Participation in the last two years of each of the current programs has exceeded 96 %. In the EPA and AIHA programs, improved participation with time is apparently related to an increased benefit (the interim NVLAP accreditation offered in EPA Test Rounds 17 and 18 and the proficiency ratings offered since Test Round 22 of the AIHA program) made available to the laboratories. In the Navy program, an increased rate of participation since Test Round 42 is related to a heightened emphasis on that issue from higher Navy command. In the bulk NVLAP, where no significant changes in policy or increases in benefit to the laboratory have occurred, the higher participation rate in more recent years may reflect in part the tendency for attrition to remove laboratories less likely to participate.

Table 1 – *Rates of laboratory participation, by program.*

Program	Program format or policy change, or arbitrary time line division	Rate of participation, %
EPA	Rounds 1–16 (QA format)	85.3
	Rounds 17 and 18 (interim accreditation)	90.8
Navy	Rounds 1–42	92.0
	Rounds 43–56 (command emphasis on participation)	96.6
NVLAP	Rounds 1–8	94.5
	Rounds 9–16	97.2
AIHA	Rounds 1–21 (QA format)	93.4
	Rounds 22–30 (PT format)	96.2

Laboratory Performance

Qualitative Analysis

A review of laboratory results from the four programs was undertaken to determine the frequencies for types of qualitative and semiquantitative errors. Types of qualitative errors include sample classification errors (false negatives and false positives) and asbestos identification errors. The review also allowed for identification of those samples most likely to cause errors. An earlier summary [*6*] of errors presented limited data for the EPA and Navy programs through the 1989 test rounds.

This review covered all test rounds conducted through April 1997 – 18 rounds of the EPA program, 56 rounds of the Navy program, 15 rounds of the bulk NVLAP, and 30 rounds of the AIHA program. The programs differ in more regards (i.e., types of laboratories, analyst training and experience, quality systems requirements, analysis reporting requirements, data evaluation criteria, etc.) than they share similarities. The general level of analytical difficulty of samples in each program also varies but has increased dramatically with time. For these reasons, direct comparisons among the programs, especially as related to analytical performance, have been intentionally avoided.

Overall qualitative error rates – (Table 2) shows the overall rates, by program, for sample classification and asbestos identification errors. With the exception of the identification and false positive error rates in the EPA program, almost all error rates are in the 3.5 % to 5.5 % range. The uniformity in the false negative, false positive and overall classification (weighted average of false negative and false positive) error rates among the three current programs is no doubt somewhat coincidental, given the compound effects of the aforementioned program differences and the differences in general levels of difficulty of samples used in each program.

Table 2 – *Overall sample classification and asbestos identification error rates.*

Error type	EPA program error rate, %	Navy program error rate, %	NVLAP error rate, %	AIHA program error rate, %
Classification false negative	3.0	4.2	3.6	3.6
Classification false positive	8.8	5.6	4.3	4.4
Overall sample classification	5.1	4.6	3.9	3.7
Asbestos identification	6.4	4.9	1.9	4.2

Specific error types and the samples most likely to cause them – Results from each program were examined to identify the types of samples historically responsible for generating the highest frequencies of false negatives, false positives, and asbestos identification errors. A small number of samples, described in (Tables 3 through 5), were identified as causing the majority of analysis difficulties, with this pattern transcending all programs. The number in parentheses following each sample description is RTI's repository number for that sample. Where multiple samples are referenced, "various" appears.

(Table 3) lists the samples causing the highest frequencies of false negatives and the resulting error rates for each program, as appropriate. The table lists data for the samples causing the four highest error rates for each of the three current programs. The only sample from the EPA program for which data are given is also the only sample in the 18 rounds of that program that caused a false negative error rate exceeding 10 %. The average false negative error rate for the other 47 asbestos-containing samples used in the

Table 3 – *Samples most likely to cause false negative errors.*

Sample composition	EPA program error rate, %	Navy program error rate, %	NVLAP error rate, %	AIHA program error rate, %
2 % chrysotile, 48 % mica, and 50 % carbonate in spray-on (S68)	...	29.7	...	...
3 % chrysotile, 85 % mica, and 12 % carbonate in spray-on (T22)	11.2	...	...	...
2–3 % chrysotile, 20 % mica, and 78 % carbonate in spray-on (T29)	...	...	15.9	17.2
2 % chrysotile, 24 % organics, 3 % TiO_2, and 71 % carbonate in floor tile (T62)	...	...	...	14.5
2 % chrysotile, 9 % cellulose, 37 % mica, and 52 % carbonate in spray-on (T81)	...	33.3	...	15.0
2–3 % chrysotile, 20 % bleached cellulose, 12 % perlite, and 63 % carbonate in spray-on (U20)	...	49.1	14.6	33.9
1 % chrysotile, trace-level nonasbestiform tremolite/anthophyllite, 3 % TiO_2, and 96 % matrix in floor tile (U25)	...	...	39.7	...
2 % chrysotile, 21 % bleached cellulose, 10 % perlite, and 67 % carbonate in spray-on (U35)	...	34.4	...	...
5 % short-range chrysotile and 95 % carbonate in RTI formulation (V10)	...	...	21.9	...
All other positive samples	2.6	3.3	2.0	2.2

EPA program was only 2.6 %, reflecting the consistent use of test materials that offered little in the way of asbestos detection problems to participants.

Two types of asbestos-containing materials appear most likely to cause false negatives. The first are spray-on/textured coating materials containing very low percentages of chrysotile in combination with varying proportions of mica, cellulose, and perlite in a binder of gypsum or calcite; titanium oxide is often present. The second are vinyl floor tiles, also with low percentages of chrysotile, with matrices composed of varying proportions of organic polymers, calcium carbonate or sulfate, mica, cellulose, titanium oxide, and/or quartz. An individual sample, an RTI formulation containing extremely short-range (<50μm) chrysotile in a calcium carbonate binder, also caused an elevated number of false negatives in the bulk NVLAP. These three diverse material types are similar in that the asbestos is almost always chrysotile, the asbestos percentage is very low, and the fibrils are often very short and/or thin. Detection is often hampered by the presence of titanium oxide, a pigmenting material that tenaciously binds to the fibers and renders detection and the observation of optical properties very difficult.

The samples historically responsible for generating the highest false positive error rates, and the resulting error rates for each program, are listed in (Table 4). Polyethylene fibers, typically seen in test samples as a commercial spray-on insulation manufactured under the tradename Thorotex® (Thoro Systems Products, Miami, FL), and wollastonite have caused the highest frequencies of false positives. These materials have been previously identified [7] as asbestos look-alikes that might be encountered by analysts with some frequency. They have been responsible for 72 %, 50 %, 76 %, and 75 % of all false positives incurred in the EPA, Navy, bulk NVLAP, and AIHA programs, respectively.

Contained in (Table 5) are historical data for asbestos identification errors. These errors are incurred almost exclusively on samples containing the three less-common asbestos types – actinolite, anthophyllite, and tremolite. These asbestos types are

Table 4 – *Samples most likely to cause false positive errors.*

Sample composition	EPA program error rate, %	Navy program error rate, %	NVLAP error rate, %	AIHA program error rate, %
Polyethylene (pulped fibers) and silicate binder in spray-on (S34)	32.7	29.0	6.9	14.2
Wollastonite cleavage fragments; mine-grade and weight-percent formulations (various)	22.4	19.1	6.9	6.5
Tremolite cleavage fragments in floor tile (U24)	...	...	9.3	...
All other negative samples	1.5	3.1	1.5	1.5

Table 5 – *Samples most likely to cause asbestos identification errors.*

Sample composition	EPA program error rate, %	Navy program error rate, %	NVLAP error rate, %	AIHA program error rate, %
Anthophyllite; mine-grade and in various weight-percent formulations (various)	46.5	50.8	19.1	47.2
Tremolite/actinolite; mine-grade, in floor tiles, and in weight-percent formulations (various)	56.1	53.1	6.3	27.7
All other asbestos samples	3.8	3.2	1.5	2.8

responsible for 44 %, 34 %, 24 %, and 35 % of all asbestos identification errors incurred in the EPA, Navy, bulk NVLAP, and AIHA programs, respectively. It has been the policy in each program not to assign identification errors for failure to differentiate between actinolite and tremolite. There is likely a strong correlation between these high error rates and the frequency with which analysts have seen these three asbestos types. Sharply lower error rates in the bulk NVLAP are likely a reflection of program requirements that laboratories: a) report specific optical properties of all asbestos types for evaluation purposes and b) purchase (and assumedly use) NIST Standard Reference Material (SRM) 1867 containing the three uncommon asbestos types.

Semiquantitative Analysis

Quantitation of asbestos is an integral, yet controversial, aspect of bulk sample analysis. It is a required component of test sample analysis in most PT programs, yet it continues to present difficulties to laboratories [*8*].

Many semiquantitative techniques are available to the laboratory, including volume estimation by stereomicroscope and area estimation by point count or visual projection. Each technique has advantages and disadvantages [*9*], and each yields answers in units that may or may not be equivalent to those of another technique or to those stipulated in specific standards [*10*]. The EPA Interim Method [*11*] requires quantitation by point count or an "equivalent method" but never defines the latter. Not until the revised EPA Test Method [*12*] was published had specific recommendations been made for using gravimetric matrix reduction to refine visual estimates or point counts, or guidance on formulating and using asbestos weight-percent standards for analyst calibration. Analysts have typically learned estimation techniques from someone whose own estimates were poorly calibrated or not calibrated at all. The pervasive tendency for laboratories to overestimate the amount of asbestos in bulk samples has been documented previously [*10*].

The Navy program has never required quantitation of asbestos. Semiquantitative results were reported by laboratories in all 18 EPA rounds but were only evaluated in the two rounds (17 and 18) in which laboratories earned interim NVLAP accreditation. Because six of the seven asbestos-containing samples in those two rounds contained 30 % or more asbestos (levels above which significant bias generally does not occur), the evaluation of semiquantitative analysis performance is restricted in this paper to results from the bulk NVLAP and the AIHA program.

Shown in (Figure 2) are semiquantitative results data for all the asbestos-containing samples used in the bulk NVLAP. Data for the AIHA program are so strikingly similar as not to require inclusion in this paper. The data show the pervasive laboratory tendency to overestimate the amount of asbestos in a sample, where overestimation is defined as the consensus value of all reporting laboratories divided by the true content (as determined jointly by NIST and RTI using point counting and quantitative X-ray diffraction [XRD]). Positive bias is generally nonexistent in samples containing 20 % to 25 % or more asbestos but rises sharply as the asbestos percentage decreases. For those samples identified as "routine," in which the asbestos is easily detected, overestimation is generally two- to four-fold or higher. A small balance of "non-routine" samples have uncharacteristically low overestimation (or none at all), considering the amount of asbestos contained in them. These samples typically present significant detection difficulties because they contain very low asbestos percentages, the asbestos fibers are usually fine and/or short, the fibers are heavily masked by binder/matrix material, or a combination of the above. These samples are, for the most part, those spray-on insulations and floor tiles described in (Table 3) as being responsible for high incidences of false negatives.

Improvement in Laboratory Performance with Time

Qualitative Analysis

One would expect that overall analysis capabilities have improved over the last 17 years as a result of laboratories actively participating in QA or PT programs. (Table 6) provides error rate reduction data for the ten samples or sample types, from those difficult samples constituting (Tables 3 through 5), which have been used two or more times in a testing program (samples such as the polyethylene may have been used six or more times in a program). For samples used twice, each error rate is listed (first use followed by second use), and the change in error rate is indicated in parentheses. For samples used more than twice, the error rates listed are for the first half of the uses, and for the second half of the uses, respectively.

The data suggest that improvement, in some cases dramatic, has been made by laboratories in all four programs in their abilities to analyze those samples identified in this study as being especially difficult. The average decrease in error rate for the 23 sample/program combinations is 32.8 %, with improvement on specific samples in specific programs exceeding 60 % on nine occasions and ranging to a high of 85 %. In two of the three cases involving an increase in error rates (as in the one case involving no change), the error type was the misidentification of an uncommon asbestos type. The lack of

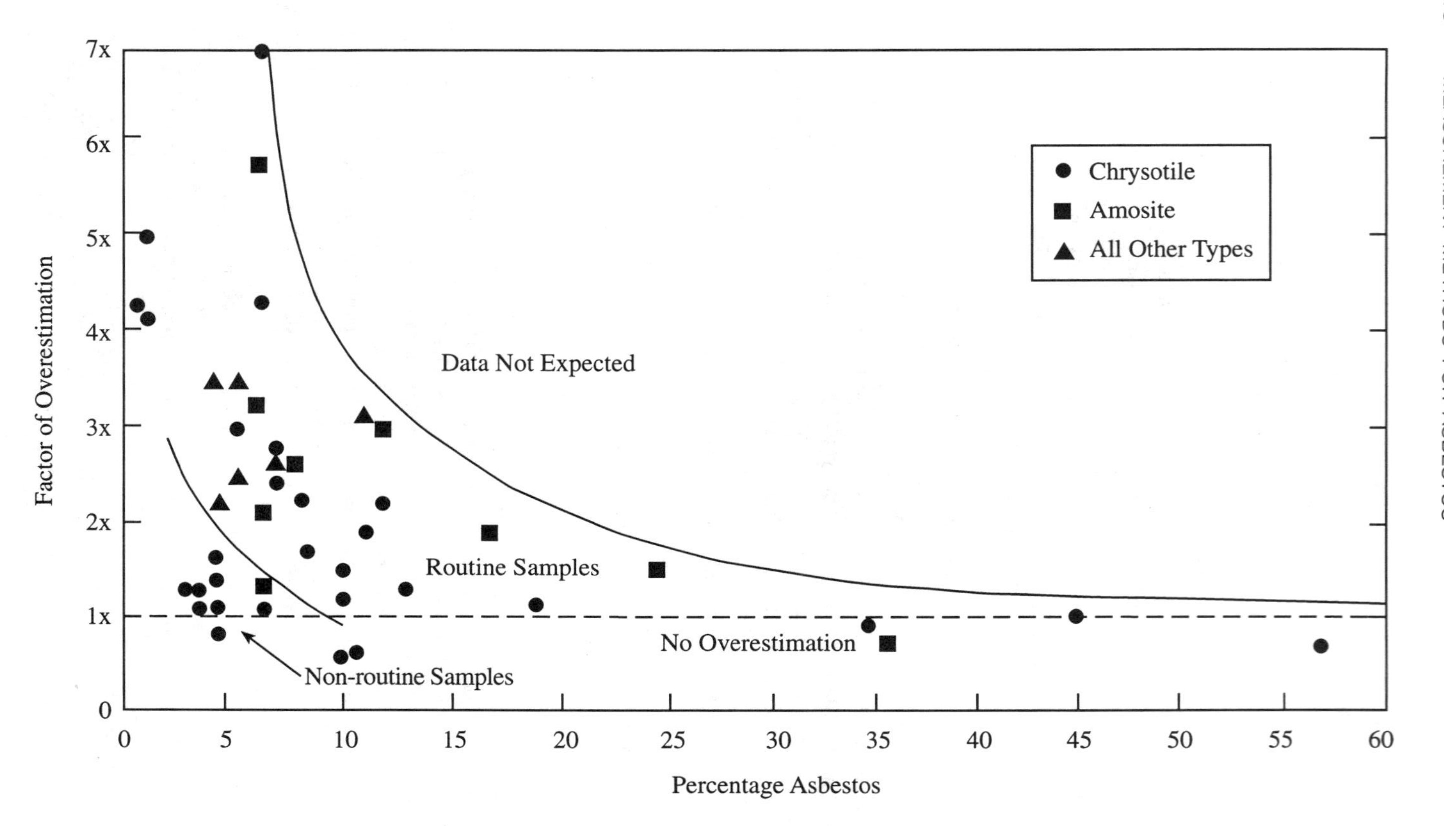

Figure 2 – *Overestimation on NVLAP semiquantitative results.*

Table 6 – *Error rate reduction on known difficult samples.*

Sample composition	EPA program error rates, %	Navy program error rates, %	NVLAP error rates, %	AIHA program error rates, %
False negatives:				
2 % chrysotile (S68)	...	47.1/17.5 (–62.8)	...	5.6/1.3 (–76.8)
3 % chrysotile (T22)	...	22.1/8.4 (–62.0)	...	9.4/4.0 (–57.4)
1–2 % chrysotile (T29)	...	...	15.9/2.4 (–84.9)	15.2/19.0 (+25.0)
2 % chrysotile (T62)	...	...	...	18.1/5.2 (–71.3)
2 % chrysotile (T81)	...	40.8/25.3 (–38.0)	...	14.2/14.1 (–0.7)
1–2 % chrysotile (U20)	...	60.1/22.6 (–62.4)	...	34.2/33.6 (–1.8)
False positives:				
Polyethylene (S34)	40.8/27.1 (–33.6)	32.2/26.2 (–18.6)	9.3/3.1 (–66.7)	17.8/10.8 (–39.3)
Wollastonite (various)	23.3/21.7 (–6.9)	19.6/18.8 (–4.1)	8.7/1.6 (–81.6)	9.3/1.7 (–81.7)
Asbestos identification:				
Anthophyllite (various)	44.6/47.7 (+7.0)	42.3/54.6 (+29.1)	...	32.7/21.0 (–35.8)
Tremolite (various)	...	53.1/53.1 (0.0)	...	...

improvement by EPA program laboratories on the identification of anthophyllite in samples may be attributed in part to a rapid increase in the number of new analysts late in the program, most of whom probably did not analyze the anthophyllite sample the first time it was used. The lack of improvement on the part of Navy laboratories may be attributed in part to the analysis requirements of the program; proficiency is based solely on a laboratory's ability to correctly classify a sample as asbestos-containing or not. Asbestos identification, while evaluated, is not part of the program's scoring structure.

Identifying the factors to which such broad-based, substantial error reduction can be attributed is difficult because the laboratories in any one program change, the incentive for successful participation may change, analysts within any one laboratory change, and the experience and training of any one analyst changes. A decrease in error rates in a particular program or on a particular sample may be a reflection of real improvement on the part of analysts in general, the result of a natural "winnowing out" of the weaker laboratories over time, or both.

Semiquantitative Analysis

The use of specific test samples has been repeated 5 times in the bulk NVLAP and 23 times in the AIHA program. (Table 7) shows semiquantitative data for each round of use for all 5 bulk NVLAP samples and for 5 samples selected randomly (for illustration purposes) from the 23 AIHA samples. For 9 of the 10 samples listed, the consensus quantitation mean was higher than the reference value, following the overestimation trend described earlier. On subsequent use, the consensus quantitation mean remained substantially higher than the reference value on 5 of the 10 samples listed.

In 26 of the 28 sets of paired measurements, the consensus mean of the values reported by laboratories in the subsequent use of a sample was lower than that of the initial use; however, student's t-tests on the 28 paired measurements revealed little statistically significant difference between the consensus means of the reported values from the initial and subsequent uses of any particular sample.

If it is therefore assumed that the consensus mean of the subsequent use of a sample has an equal chance of being higher or lower than that of an initial use, it is then indeed statistically significant that 26 of the 28 means on subsequent uses were lower. The statistical chance, calculated by sign test, that 26 of 28 subsequent means would be lower is less than 1 %; this suggests that there has been a real improvement in accuracy between initial and subsequent uses on the basis of the combined group of 28 sample uses. A similar comparison of coefficients of variation (CVs) showed 16 CVs increasing between initial and subsequent uses and 10 decreasing, suggesting that although accuracy has improved overall, precision has not.

Conclusion

Over the 17 years that QA and PT programs for laboratories analyzing for asbestos in bulk building materials have existed in the United States, total enrollment has been affected by legislation, market demand, and military cutbacks. Total enrollment in all programs has ranged from 100 to 1 200 laboratories and in recent years has stabilized at

Table 7 – *Semiquantitative results on selected bulk samples.*

Sample composition	Initial use		Subsequent use	
	Consensus mean, %	Coefficient of variation	Consensus mean, %	Coefficient of variation
6.8 % amosite (S01)	11.8	0.6	6.9	0.5
2–3 % chrysotile (T29)	4.3	0.9	3.3	0.8
9.8 % amosite and	34.6	0.4	19.3	0.6
3.4 % crocidolite (T44)	14.3	0.6	7.4	0.8
5.0 % chrysotile (T80)	4.6	0.9	5.2	0.9
2.3 % chrysotile (U20)	3.8	1.1	3.8	0.8
3.3 % chrysotile (S68)	12.0	0.8	11.0	0.7
4.5 % chrysotile (T30)	10.2	0.8	7.8	0.9
2.6 % chrysotile (T81)	6.0	0.9	3.2	0.9
15.5 % amosite (U08)	32.8	0.5	27.1	0.5
5.7 % chrysotile (U77)	12.5	0.7	10.5	0.8

700 to 750 laboratories. Laboratory participation of late has been strong, exceeding 96 % over the last two to three years.

Overall qualitative analysis error rates are in the 3.5 % to 5.5 % range. The majority of errors are being generated by a small number of specific sample types. Low asbestos percentage, thin/short asbestos fiber morphology, and/or binder interferences characterize the samples causing false negatives. Wollastonite and polyethylene have caused greater than two-thirds of all false positives, and major asbestos identification problems are associated with only the three uncommon asbestos types.

There is a pervasive tendency for laboratories to overestimate the amount of asbestos in proficiency test samples, and that bias increases sharply as the true asbestos content decreases. Overestimation is uncharacteristically low on samples where the detection of asbestos is difficult.

Sharp reductions have occurred in qualitative analysis error rates in all three current programs, with up to 80 % lower error rates on specific, difficult samples. Improvement in laboratories' semiquantitative analyses has occurred as well but less dramatically. Laboratory participation in bulk asbestos QA and PT programs, presumably coupled with greater analyst training and experience, has been instrumental in improving overall laboratory analysis capabilities.

References

[*1*] McCrone, W. C., *The Asbestos Particle Atlas*, Ann Arbor Science, Ann Arbor, Michigan, 1980, pp. 2-3.

[*2*] U.S. Environmental Protection Agency (EPA). *Asbestos-Containing Materials in School Buildings: A Guidance Document, Part 1*. Office of Toxic Substances C00090. Washington, DC, March 1979.

[*3*] Chief of Naval Operations. Navy Occupational Safety and Health (NAVOSH) Program, *Operations Navy Instructions 5100.23*, Washington, DC. May 1979.

[*4*] Asbestos-Containing Materials in Schools: Final Rule and Notice, 40 *Code of Federal Regulations* (CFR) Part 763, October 1987.

[*5*] Occupational Safety and Health Administration (OSHA). Occupational Exposure to Asbestos: Final Rule, 29 CFR 1910, 1915, and 1926, August 1994.

[*6*] Harvey, B. W., "Classification and Identification Error Tendencies in Bulk Insulation Proficiency Testing Materials," *The Microscope*, Vol. 37, Third/Fourth Quarter 1989, pp. 393-402.

[*7*] McCrone, W. C., "Discussion of Asbestos Bulk Sample Analysis Quality Assurance Programs," *The Microscope*, Vol. 37, Third/Fourth Quarter 1989, pp. 403-409.

[*8*] Perkins, R. L., "Point-Counting Technique for Friable Asbestos-Containing Materials," *The Microscope*, Vol. 38, First Quarter 1990, pp. 29-39.

[*9*] Perkins, R. L., "Estimating Asbestos Content of Bulk Samples," *National Asbestos Council Journal*, Spring 1991, pp. 27-31.

[*10*] Perkins, R. L., Harvey, B. W., and Beard, M. E., "The One Percent Dilemma," *Environmental Information Association Journal*, Summer 1994, pp. 5-10.

[*11*] U.S. Environmental Protection Agency (EPA). Interim Method for the Determination of Asbestos in Bulk Insulation Samples, EPA 600/M4-82-020. Research Triangle Park, NC, December 1982.

[*12*] U.S. Environmental Protection Agency (EPA). Method for the Determination of Asbestos in Bulk Building Materials, EPA/600/R-93/116. Washington, DC, July 1993.

Ann G. Wylie[1]

THE HABIT OF ASBESTIFORM AMPHIBOLES: IMPLICATIONS FOR THE ANALYSIS OF BULK SAMPLES

REFERENCE: Wylie, A. G., "**The Habit of Asbestiform Amphiboles: Implications for the Analysis of Bulk Samples**", *Advances in Environmental Measurement Methods for Asbestos, ASTM STP 1342*, M.E. Beard, H. L. Rooks, Eds., American Society for Testing and Materials, West Conshohocken, PA, 2000.

ABSTRACT: Evidence on the carcinogenicity of fibrous minerals supports the conclusion that amphiboles must form in an asbestiform habit in order to pose a risk to human health. Furthermore, the asbestiform habit controls many of the physical properties of asbestos. Because of the distinctive characteristics of the asbestiform habit, populations of asbestiform amphiboles can be distinguished from populations of amphibole cleavage fragments by light microscopy. Populations of asbestos fibers longer than 5 μm are characterized by fibers that occur in bundles, are often curved, and have very high aspect ratios (mean aspect ratio > 20:1 - 100:1) and narrow widths, usually less than 0.5 μm. It is inappropriate to apply a 3:1 aspect ratio criterion to identify amphibole asbestos. Other minerals that crystallize in a habit similar to asbestos do not necessarily pose the same risk because factors such as friability, biodurability, bioavailability and surface chemistry are important in determining carcinogenicity of mineral fibers.

KEY WORDS: asbestos, asbestiform, amphibole, mineralogical characteristics

Introduction

When the Occupational Safety and Health Administration (OSHA) issued the first permanent standard for exposure to asbestos dust in 1972 [*1*], the names of three minerals,

[1]Professor, Laboratory for Mineral Deposits Research, Department of Geology, University of Maryland, College Park, MD 20742

tremolite, actinolite, and anthophyllite, were included without the specification that they must be asbestiform. In 1992, the Occupational Safety and Health Administration revised its asbestos regulations so that only the asbestiform varieties of these minerals were covered. [*2*] During the 20 years separating these rulings, research was conducted that addressed the following issues: 1) what are the properties of amphibole-asbestos, 2) which of these properties make them carcinogens, 3) what is the carcinogenic potential of amphiboles that are not asbestiform, and 4) how can asbestiform amphiboles be differentiated from other amphibole habits. In this paper, I will summarize the mineralogical characteristics of asbestiform amphiboles with particular emphasis on differentiating among the habits of amphibole in the analysis of bulk samples.

The Mineralogical Properties of Amphibole Asbestos

A mineral name specifies two things about the naturally occurring crystalline solid it identifies: the relative proportions of the major elements that make it up and the arrangement, or structure, of the atoms. Associated with the chemical composition and atomic structure are physical properties by which the mineral can be identified. These include specific gravity (or density), indices of refraction and other optical properties, hardness, color, luster, cleavage, streak, solubility, electrical conductivity, and so forth.

Sometimes a mineral may form with a habit of growth or an unusual color that distinguishes it from the more common occurrences, and a varietal name is used to designate this form. Well known examples can be found in minerals that are prized as gems where color and clarity are the properties that determine the variety. For example, emerald and aquamarine are varieties of the mineral beryl, and ruby and sapphire are varieties of the mineral corundum. Because varietal names are usually assigned specifically on the basis of external appearance, varieties can not be distinguished from other habits of the same mineral by bulk techniques of analysis such as x-ray diffraction. Furthermore, if the appearance of a mineral specimen is severely altered by grinding to a fine powder, it is usually impossible to assign a varietal name, and it may be appropriate to conclude that it is no longer a particular variety at all. For example, an emerald crushed to a fine power can not be distinguished from a powdered aquamarine, although both powders remain easily identifiable as the mineral beryl. In the same way, because asbestos is a varietal name, if asbestos is reduced to a powder such that its distinguishing physical properties are destroyed, it is no longer asbestos.

The amphiboles regulated in their asbestiform variety

are usually listed as amosite, crocidolite, tremolite-asbestos, actinolite-asbestos, and anthophyllite-asbestos. These are the asbestiform varieties of the minerals grunerite, riebeckite, tremolite, actinolite, and anthophyllite, respectively. (Amphibole nomenclature is detailed in [*3*].) Amphiboles may be found in nature in compact masses; in stubby or acicular prismatic crystals, either singly or intergrown; in fine, brittle needles which, when separated by open space, are referred to as byssolite [*4*]; or as long, strong, flexible, fibers, which are usually composed of very thin single crystals called fibrils, and which are capable of being woven, a habit that is referred to as asbestos. Amphiboles may be 'fibrous', i.e., they give the appearance of being composed of fibers, whether or not they are actually made up of individually separable fibers. While all asbestiform amphiboles are fibrous, not all fibrous amphiboles are asbestiform.

In modern usage, only grunerite-asbestos and riebeckite-asbestos go by varietal names, amosite and crocidolite respectively. While crocidolite is generally recognized by mineralogists as a varietal name for riebeckite-asbestos, (although technically discarded in the official amphibole terminology [*3*]) the term 'amosite' is not, because it is a commercial term for a mineral product (Asbestos Mines of South Africa). However, AMOSite is composed mainly of the mineral grunerite in its asbestiform habit, and amosite remains the varietal name for grunerite-asbestos for regulatory purposes, despite attempts by mineralogists to discredit it. Tremolite, actinolite, and anthophyllite are mineral names and as such do not specify a particular habit or mineral variety. To specify the asbestiform variety, tremolite-asbestos, actinolite-asbestos and anthophyllite-asbestos must be designated. For the latter three, the term 'asbestos' was omitted in OSHA regulations prior to 1992, (although it has always been used by EPA), and unfortunately today it is still occasionally disregarded.

The varieties of amphibole known as asbestos possess a set of physical properties that make them commercially valuable, properties which have been described extensively in the literature [*5-11*]. For thousands of years asbestos has been prized for the fact that it can be woven into a cloth that will not burn. Necessary for this application are long, thin, flexible silicate fibers that possess high tensile strength. In amphibole-asbestos, the flexibility and tensile strength are directly relatable to the dimensional properties and the conditions of growth. Tensile strength may be ten times greater than the massive varieties [*12*]. Walker and Zoltai [*12*] attribute the high tensile strength of asbestos to both the very small widths of fibrils and the small number of flaws on fibrillar surfaces. However, heating asbestos fibers to temperatures too low to produce structural modifications lowers their tensile strength, suggesting that hydrogen bonds between fibrils, which are broken as water is lost during heating, may also contribute to tensile strength [*11*]. Further, it has been suggested that planar defects parallel to the fiber axis enhance tensile strength by mitigating the propagation of cracks and by providing sites of interplanar slip [*13,14*]. The flexibility of asbestos can be explained by both the small width and extreme aspect ratios of fibrils and by the fact that fibrils in bundles may slip past one another. The observation that anthophyllite-asbestos from the Paakila district in Finland has much lower flexibility and larger fibrillar widths than crocidolite from both the Cape Province in South Africa and the Hammersely Range in Western Australia is consistent with the conclusion that flexibility is inversely proportional to fibril width. Although amphibole-asbestos is cited for its chemical and electrical resistance, it has not been demonstrated that these properties are different from the nonasbestiform varieties.

Amphibole-asbestos fibrils range in width from about 1 to 0.01 μm. In some deposits the range in width is small; in others, it is large and may vary from one part of the deposit to another. Individual fibrils and bundles of fibrils may attain lengths of hundreds to thousands of times their widths. The geologic conditions that favor crystallization in the asbestiform habit include a water-rich fluid, probably at geologically low temperature, that is saturated or supersaturated with respect to the amphibole mineral. These conditions result in rapid nucleation at multiple sites and growth confined almost exclusively to one crystallographic direction (c-axis) usually without preferred orientation in the other two crystallographic directions [*15-18*]. The presence of some simple or complex ion (or ions) dissolved in this fluid may inhibit growth perpendicular to the fiber axis, effectively forming a barrier so that adjoining fibers remain separated. Thus, in asbestos, the individual fibrils are separate crystals, and they can be easily separated with hand pressure; their surfaces are surfaces of growth rather than breakage; and they are extremely thin and therefore can easily become airborne and be inhaled.

In most commercial deposits of amphibole-asbestos, the long axis of the fibrils are parallel to each other and perpendicular to the wall of the veins in which they are found. These are referred to as cross-fiber deposits. However, asbestos is also found in bundles that are parallel to the vein walls; the bundles may be interwoven, or they may occur in aggregates radiating out from an apex with shorter fibrils and fiber bundles filling in the spaces. Sometimes, asbestos occurs in mass fiber deposits in which the fibrils and fiber bundles occur without a predominant orientation. The Calidria chrysotile-asbestos deposit is such a mass fiber deposit. Mass fiber occurrences of amphibole sometimes result in a tough, compact rock from which the fibers cannot be separated; these samples are not asbestos. The best example of this is the variety of actinolite known as jade.

Fibrillar surfaces are growth surfaces. In amphibole asbestos, they are faces in the [001] zone and are most commonly {100}, {110} ({210} for the orthorhombic anthophyllite-asbestos), and {010} [*15-17*]. For most samples of amphibole-asbestos, {100} is prominently developed, resulting in flat ribbon-like fibrils [*19,20*]. The ribbon-shape was confirmed for amosite and crocidolite by experiments with TEM and SEM that directly measured both width and thickness [*21*]. It is not uncommon to find sheet silicates such as talc, chlorite, and serpentine forming thin epitaxial layers on the outside of fibrils and occasionally in the spaces between them [*16,17*].

Rapid growth from supersaturated, low temperature, water-rich fluids in which there may be fluctuations in temperature, pressure, pH and concentration of dissolved ions, are the conditions that favor the development of defects in the structure of minerals. Three types of defects are common in amphibole asbestos: Wadsley defects parallel to (010) resulting from chain width errors, twinning on (100), and stacking faults parallel to (100) [*7, 22,23*]. Structural defects produce planes of weakness called parting (as opposed to cleavage which is weakness inherent in a "perfect" structure), and some separation of fibrils during comminution may actually be parting along these defect surfaces. However, (100) twinning, stacking faults, and Wadsley defects are not confined to the asbestiform habit of amphiboles; they may be well developed in some nonasbestiform amphibole specimens although they do tend to enhance elongation [*19,24*]. For example, byssolite fibers of tremolite and actinolite often exhibit optical properties and/or physical shapes that are consistent with extensive twinning on {100} [Verkouteren and Wylie, in preparation,*18*].

The fibrillar structure of the asbestiform habit results in anomalous optical properties [*18*]. With the exception of anthophyllite, asbestiform amphiboles are monoclinic. However, in cross-polarized light, asbestos fibers generally display parallel extinction, instead of the expected inclined extinction. Crocidolite and amosite were once thought to be orthorhombic like anthophyllite because of this property. In tremolite-asbestos and actinolite-asbestos, oblique extinction can sometimes be seen in the larger fibers that are not bundles. However, oblique extinction is not observed in crocidolite or amosite.

The Carcinogenicity of Amphibole Asbestos and the Asbestiform Habit

Although this paper is not intended to review the vast literature on the carcinogenicity of asbestos, it may be helpful to the reader to review some of the major studies that have address the role of fiber size. The most influential work was the animal experiments of Merle Stanton and co-workers published in 1981[*25*]. From this work emerged the "Stanton Hypothesis": the carcinogenicity of inorganic particulate depends on dimension and durability rather than on physicochemical properties. It has been extended to include the following *corollaries*: the fibers of narrow width are the most carcinogenic, and carcinogenicity is proportional to the number of long, thin fibers in a material. Stanton et al. found that populations with abundant fibers longer than 8 μm and narrower than 0.25 μm were most closely linked to pleural tumor response in rats and that this response was independent of fiber type. The human experience with asbestos appears to support the conclusion that narrow widths contribute significantly to the carcinogenic potential of asbestos. Crocidolite from both the Cape Province of South Africa and from Western Australia has a the mean fiber diameter of less than 0.1 μm. Mesothelioma may account for as much as 18% of the proportional mortality for crocidolite workers [*26*]. In contrast, in the Transvaal, where both crocidolite and amosite with mean diameters of 0.21 and 0.24 μm respectively are mined, mesotheliomas are rare [*27,28*]. In Paakila, Finland, where the mean diameter is about 0.6 μm and widths less than 0.1 μm are quite rare [*29*], the mortality from mesothelioma among miners is less than 1% [*30*]. However, anthophyllite-asbestos is almost always intergrown with talc [*16,17,31*], and it is possible that surface characteristics associated with mineralogical variability may also affect the carcinogenic potential of mineral fiber. Since 1981, many reviews of both animal and human data relating to asbestos exposure have been written and little has changed with respect to the dimensional hypothesis [*32-38*]: the narrower the amphibole-asbestos is, the more likely it is to produce mesothelioma. Furthermore, and perhaps most importantly, although numerous mining populations have been studied, no association between mesothelioma and the inhalation of amphibole has been demonstrated unless the amphibole is asbestiform [*2, 39-43*]. Furthermore, when amphibole-asbestos is crushed extensively and the asbestiform habit is wholly or partially destroyed, animal experiments indicate that its carcinogenic potential is reduced or eliminated entirely [*25,37*].

Amphibole-asbestos is also known to cause lung cancer and asbestosis. There appears to be the same relationship between the potential to produce lung cancer and fiber size that there is for mesothelioma: the potency of airborne fibers to produce lung cancer appears to be lower where airborne fibers are relatively coarse [*2*]. Furthermore, there is no compelling evidence that lung cancer and asbestosis can be attributed to the inhalation of amphibole unless that amphibole is asbestiform [*2, 38-43*].

The question of why the inhalation of amphibole-asbestos can result in lung cancer and mesothelioma has never fully been answered [*44,45*]. The possibility that the surface chemistry of asbestos is important remains and, in fact, is supported by the results of numerous studies [reviewed in *44-47*]. Nonetheless, most attention has been focused on the properties of asbestos and other mineral particles that can be most easily measured, i.e., their dimensions. As analysts, we too focus on morphology since it is morphology that defines the asbestiform habit. However, it is important to remember that there may be many properties that accompany the asbestiform habit that we do not measure. Therefore, we must be sure that when we analyze for asbestos, we apply morphological criteria that are specific for asbestos.

The Definition of Asbestos Fiber

For air-monitoring purposes in occupational settings, the Occupational Safety and Health Administration defined an asbestos fiber as any particle of asbestos longer than 5 μm with an aspect ratio of 3 or greater [*48*]. This morphological standard originated in England as the result of an air-monitoring program in an asbestos textile factory. The 5 μm length was chosen as a lower limit because the optical microscope was the instrument being used for monitoring, and studies by Addingley [*49*] and Lynch et al. [*50*] had shown that counting of particles less than 5 μm in length leads to imprecise results. The choice of an aspect ratio of 3 was arbitrary [*51*]. The dimensional definition became fully entrenched in the United States 1973, when, in the Reserve Mine Case, the US District Court accepted as a definition of a fiber any mineral particle that is at least 3 times longer that it is wide [*52*].

Today, the simplistic "longer than 5, greater than 3:1" cannot be applied as a <u>definition</u> for an asbestos fiber. OSHA's Final Rule dealing with actinolite, tremolite and anthophyllite is quite clear on this subject. In removing the nonasbestiform amphiboles from the asbestos standard, OSHA effectively removed the old definition of a fiber:

> "OSHA does not believe that the current record provides an evidentiary basis to determine "the appropriate aspect ratio and length" for determining pathogenicity. Even if dimensional cut-

> offs were known for asbestos fibers, additional data do not support a standard for all ATA (i.e., actinolite, tremolite and anthophyllite) minerals based on fiber dimension alone" [*2*].

Since most habits of amphibole cleave into fragments that when longer than 5 μm are also elongated, a 3:1 aspect ratio is not specific for the asbestiform habit. While OSHA believes that fiber dimension is the most significant indicator of fiber pathology, they state "..the evidence which is available more likely associates fibers with dimension common to asbestos populations with disease causing potential" [*2*]. By this statement, OSHA effectively directs us as analysts to look for *populations* of fibers with dimensions of asbestos, and only the asbestiform habit will produce a population of particles with the appropriate dimensions. Therefore, we are really analyzing for the *habit* of amphibole-asbestos, not for a particular particle size and shape.

It may be possible to continue to use the criteria of longer than 5 μm and 3:1 or greater aspect ratio for counting airborne particles in an atmosphere known to contain asbestos, such as during or after an abatement procedure. However, the analyst must be sure that it is only asbestos that meets these criteria. If other elongated minerals are airborne, the analysis will be in error.

The Analysis of Bulk Samples

When an analyst receives a sample and is asked to determine if asbestos is present, and if so, how much, he/she must apply a set of criteria to answer this question, and these criteria must be applied to populations of particles, not to particles individually. The first question to be answered is "Is there an asbestiform mineral in the sample?". In other words, is there a population of fibers longer than 5 μm that displays the following characteristics:

1) Mean aspect ratios of individual fibers ranging from 20:1 to 100:1,
2) Bundles of fibers often displaying splayed ends,
3) Very thin fibers, mean widths generally less than 0.5 μm, and
4) Curvature in the longest fibers usually common.

These characteristics of asbestos have been documented for amphiboles implicated in human disease [*5,9-11,13, 15-19,22,32,53-55*]. It is important to note that bundles and matted masses sometimes do not have the extremely high aspect ratios characteristic of the fibrils that make them up. Whether elongated or not, bundles and matted masses of fibers that meet the criteria should be considered asbestiform.

There is often expressed a concern that an analysis that relies on optical microscopy might result in a false negative because the width of asbestos fibrils is below the resolution of the microscope. There are two reasons why this is unlikely to be the case. First, several studies have shown that fibers of amosite and crocidolite are visible by light microscopy, notably phase contrast, if their width is greater than about 0.1 to 0.15 μm [56-58] even though 0.15 μm is below the resolution of the microscope. Resolution and visibility are different: resolution is a mathematical approximation of the minimum distance by which two points can be separated and still be seen as separate. It is a function of the optics of the microscope. Visibility of course depends on size but equally important is the contrast in index of refraction with the surrounding medium [*59*]. The second reason that asbestos is unlikely to go undetected is that those fibers that are truly invisible in general make up only a small portion of the mass of amphibole-asbestos although they are very abundant and, in fact, may make up the majority of the asbestos fiber population. Modeling the distribution of the mass of asbestos fibers allows an estimate of the percentage of mass that might be invisible by optical microscopy [7, *60*]. The worst case is presented by crocidolite from the Cape Province and Australia, material that is characterized by very small fibrils. If crocidolite is well dispersed, as much as 70% of the mass could be 'invisible' (width less than 0.1 μm). However, in most bulk samples, crocidolite is not uniformly dispersed and the fraction of mass that is visible is likely to be much higher than 30%. For other types of amphibole-asbestos, the model predicts that more than 90% of the mass would be visible. In my experience, asbestos as a naturally occurring trace *contaminant* in industrial mineral products is most likely to be tremolite-asbestos, actinolite-asbestos or anthophyllite-asbestos. (Because of the restricted geologic environments in which they occur, crocidolite and amosite are almost never encountered.) Modeling of the dimensional data for these more common varieties suggests that over 98% of the mass should be visible by optical microscopy. Therefore, it is highly likely that amphibole-asbestos would be detected by optical microscopy in a bulk sample if it is present in an amount greater than about 0.1 wt.%.

Once asbestiform fibers have been found, the analyst must determine which mineral they are because amphiboles and serpentine are not the only minerals to crystallize in an asbestiform habit. For example, nemalite is the asbestiform variety of brucite ($MgOH_2$) and agalite and fibrous talc have been used to designate the asbestiform variety of talc.

Other somewhat common minerals that may develop the asbestiform habit include sepiolite, palygorskite, erionite, and tourmaline. Evidence on the carcinogenicity of asbestiform minerals that are not asbestos is mixed, but there is no compelling evidence that all asbestiform minerals are carcinogenic. Different minerals have different biodurability (survival *in vivo*), surface chemistry, friability, especially friability *in vivo*, and bioavailability, differences that influence their biological activity (reviewed in [*61*]. For example, xonotlite, $Ca_6Si_6O_{17}(OH)_2$, commonly found in an asbestiform habit, is soluble in water.

Mineral identification by optical microscopy requires the use of a petrographic (polarizing light) microscope. Normally, the most reliable property for the identification of a mineral by optical microscopy is the magnitude of the indices of refraction. Many misidentifications of minerals that are fibrous and/or asbestiform could have been avoided by carefully measuring indices of refraction. Minerals may show dispersion staining colors of light yellow or light blue when they are in oils that have indices of refraction that are quite different from those of the mineral, especially when the index of refraction is greater than 1.600. These colors do not indicate a "match", a fact that can be easily confirmed by examining the colors and relative intensities of Becke lines. For mineral identification, at least two (preferably principal) indices of refraction (n_D) should be measured with a precision of at least 0.005 or better. If a mineral possesses other characteristic properties, such as the negative elongation and distinctive blue color of crocidolite, it may be acceptable to measure only one principal index of refraction. It is also extremely important to measure indices of refraction at extinction positions and to relate these directions to morphology. Most texts in optical mineralogy provide diagrams that illustrate the relationships between principal vibration directions and morphology. In the case of fiber bundles, the indices of refraction are not necessarily those given in optical mineralogy texts [Verkouteren and Wylie, in preparation]. For example, the fibrillar structure may result in only two indices of refraction when there should be three. In this case, the maximum and minimum indices of refraction may be slightly less and slightly greater than the reference magnitudes of γ and α respectively.

Optical properties other than indices of refraction, such as color, sign of elongation and birefringence, are usually very helpful in identifying minerals. For fibrous minerals, properties that are derived from an analysis of an interference figure (i.e., 2V, orientation of the optic plane, optic sign and dispersion of the optic axes) are generally not useful because interference figures are almost impossible to obtain on particles smaller than several micrometers. Furthermore, since asbestos fibers wider than 1 μm are composed of bundles of fibrils, any interference figure would arise from the bundle effect, not the mineral structure.

Care must also be taken in the use of extinction angle in the mineral identification process. First, monoclinic amphiboles that are characterized by oblique extinction will exhibit the characteristic extinction angle only when they lie parallel to (010), a somewhat rare condition for both fibrous and nonfibrous amphiboles. Monoclinic amphiboles that lie parallel to (100) will display parallel extinction and (100) parting may be very common in some specimens. Furthermore, bundles of randomly aligned fibers, the hallmark of asbestos, will show parallel extinction in all orientations [*18*].

There are many analytical techniques other than optical microscopy that can be used to identify minerals. For example, qualitative or quantitative energy dispersive x-ray analysis provides information on the chemical composition and x-ray and electron diffraction provide structural information. These techniques can be used if the optical properties are not sufficient for mineral identification. While these methods are more expensive and time-consuming, they are not as expensive and time-consuming as defending an incorrect analysis.

Conclusions

When OSHA "deregulated" nonasbestiform amphiboles in 1992 [*2*], they placed the burden of identifying asbestos and discriminating it from other habits of amphibole on the analytical community. It may have been simple to look only for 3:1 particles, but it was (and still is) an insufficient criterion by which to identify asbestos. Now we must apply a set of criteria to a *population* of particles. It is appropriate to apply these criteria to a population of particles that is defined by a 3:1 aspect ratio and a 5 μm lower limit of length as long as we recognize that this population discriminator does not define asbestos nor does it define hazardous particles. Within this population, we must identify the particles as amphibole (or chrysotile), and we must find compelling evidence for the asbestiform habit; otherwise, the mineral is simply not asbestos.

References

[1] Occupational Safety and Health Administration, 29 CFR 1910.1001 37 Federal Register, June 7, 1972, p. 11318.

[2] Occupational Safety and Health Administration, Occupational Exposure to Asbestos, Tremolite, Anthophyllite and Actinolite: Final Rule, amendment to

29 CFR 1910.1001 and 29 CFR 1928.56, Federal Register, June 4, 1992.

[3] Leake,B.E., "Nomenclature of Amphiboles," *American Mineralogist*, Vol.63, 1978, pp.1023-1053.

[4] Dana, E.S., and Ford, W.E., *A Textbook of Mineralogy,* J. Wiley and Sons, New York, 1932.

[5] Steele, E. and Wylie, A., "Mineralogical Characteristics of Asbestos Fibers," *Geology of Asbestos Deposits,* P.H. Riordon, ed., Society of Mining Engineers, 1981, pp.93-100.

[6] Zoltai, T., "Amphibole Asbestos Mineralogy," *Amphiboles and Other Hydrous Pyriboles,* D. Veblen, ed., Mineralogical Society of America, 1981, pp. 237-278.

[7] Veblen, D. and Wylie, A., "Mineralogy of Amphiboles and 1:1 Layer Silicates," *Health Effects of Mineral Dusts: Reviews in Mineralogy Vol.28*, Guthrie, G.D. and Mossman, B.T., eds., Mineralogical Society of America, 1993, pp. 61-138.

[8] Skinner,H.C.W., Ross,M., and Frondel, C., *Asbestos and Other Fibrous Materials*, Oxford University Press, 1988.

[9] Wylie, A.G., "Discriminating Amphibole Cleavage Fragments from Asbestos: Rationale and Methodology," *Proceedings of the VIIth International Pneumoconiosis Conference Part II*, Department of Health and Human Services publication No. 90-108, 1990, pp.1065-1069.

[10] Wylie, A.G., "Relationship between the Growth Habit of Asbestos and the Dimensions of Asbestos Fibers," *Mining Engineering,* 1988, pp.1036-1040.

[11] Hodgson, A.A., "Fibrous Silicates," Lecture Series No.4, Royal Institute of Chemistry, London, 1965.

[12] Walker, J.S. and Zoltai, T., "A Comparison of Asbestos Fibers with Synthetic Crystals Known as 'Whiskers'," *Annals of the New York Academy of Sciences,* Vol. 330, 1979. pp.687-704.

[13] Ahn, J.H. and Buseck, P.R., "Microstructures and Fiber Formation Mechanisms of Crocidolite Asbestos," *American Mineralogist*, Vol. 76, 1991, pp.1467-1478.

[14] Whittaker, E.J.W., "Mineralogy, Chemistry and

Crystallography of Amphibole Asbestos," *Mineralogical Techniques of Asbestos Determination*, R.L. Ledoux, ed., Mineralogical Association of Canada, 1979, pp.1-34.

[15] Alario Franco, M., Hutchison, J.L., Jefferson, D.A., and Thomas, J.M., "Structural Imperfections and Morphology of Crocidolite (Blue Asbestos)," *Nature*, Vol. 266, 1977, pp.520-521.

[16] Cressey, B.A., Whittaker, E.S.W., and Hutchison, J.L., "Morphology and Alteration of Asbestiform Grunerite and Anthophyllite," *Mineralogical Magazine*, Vol.46, 1982, pp.77-87.

[17] Veblen, D.R., "Anthophyllite Asbestos: Microstructures, Intergrown Sheet Silicates and Mechanisms of Fiber Formation," *American Mineralogist*, Vol.65, 1980, pp.1075-1086.

[18] Wylie, A.G., "Optical Properties of the Fibrous Amphiboles," *Annals of the New York Academy of Science*, Vol.330, 1979, pp.611-619.

[19] Dorling, M., and Zussman, J., "Characteristics of Asbestiform and Non-asbestiform Calcic Amphiboles," *Lithos*, Vol.20, 1987, pp.469-489.

[20] Lee, R.J. and Fisher, R.M., "Fibrous and Nonfibrous Amphiboles in the Electron Microscope," *Annals of the New York Academy of Sciences*, Vol. 330, 1979, pp.645-660.

[21] Wylie, A.G., Shedd, K.B. and Taylor, M.E., "Measurement of the Thickness of Amphibole Asbestos Fibers with the Scanning Electron Microscope and the Transmission Electron Microscope," *Microbeam Analysis*, Heinrich, K.F., ed., 1982, pp.181-187.

[22] Hutchison, J.L., Irusteta, M.C., and Whittaker, E.J.W., "High Resolution Electron Microscopy and Diffraction Studies of Fibrous Amphiboles," *Acta Crystallographica*, Vol. A31, 1975, pp.794-801.

[23] Chisholm, J.E., "Planar Defects in Fibrous Amphiboles," *Journal of Material Science*, Vol.8, 1973, pp.475-483.

[24] Ahn, J.H,, Moonsup, C., Jenkins, D.M., and Busek, P.R., "Structural Defects in Synthetic Tremolitic Amphiboles," *American Mineralogist*, Vol. 76, 1991, pp.1811-1823.

[25] Stanton, M.F., Layard, M., Tegeris, A., Miller, E., May, M., Morgan, E., and Smith, A., "Relation of Particle Dimensions to Carcinogenicity in Amphibole Asbestoses and Other Fibrous Minerals, " *Journal of the National Cancer Institute*, Vol.67, 1981, pp.965-975.

[26] McDonald, A.D. and McDonald, J.C., "Epidemiology of Malignant Mesothelioma," *Asbestos Related Malignancy*, New York Grune and Stratton, 1986, pp.57-79.

[27] Harington, J.C., Gilson, J.C., and Wagner, J.C., "Asbestos and Mesothelioma in Man," *Nature*, Vol. 232, 1971, pp.54-55.

[28] Timbrell, V., Griffiths, D.M., and Pooley, F.D., "Possible Biological Importance of Fibre Diameter of South African Amphiboles, *Nature*, Vol. 232, 1971, pp.55-56.

[29] Timbrell, V., "Review of the Significance of Fibre Size in Fibre-related Lung Disease: A Centrifuge Cell for Preparing Accurate Microscope-evaluation Specimens from Slurries Used in Inoculation Studies," *Annals of Occupational Hygiene*, Vol.33, 1989, pp.483-505.

[30] Karjalainen, A., Merrman, L.O., and Pukkala, E., "Four Cases of Mesothelioma among Finnish Anthophyllite Miners," *Occupational and Environmental Medicine*, Vol. 51, 1994, pp.212-215.

[31] Watson, M.B, "The Effect of Intergrowths on the Properties of Anthophyllite," Master's Thesis, University of Maryland, 1999.

[32] Wylie, A.G., Bailey, K.F., Kelse, J.W., and Lee, R.J., "The Importance of Width in Asbestos Fiber Carcinogenicity and its Implications for Public Policy, "*American Industrial Hygiene Association Journal*, Vol. 54, 1993, pp.239-252.

[33] Lippmann, M., "Asbestos Exposure Indices," *Environmental Research*, Vol. 29, 1988, pp.86-106.

[34] Pott, F., "Some Aspects on the Dosimetry of the Carcinogenic Potency of Asbestos and Other Fibrous Dusts," *Staub Reinhalt Luft*, Vol.38, 1978, pp.486-490.

[35] Harington, J.S., "Fiber Carcinogenesis: Epidemiologic Observations and the Stanton Hypothesis," *Journal of the National Cancer Institute*, Vol. 67, 1981, pp.977-989.

[36] Wagner, J.C., Berry, G., and Timbrell, V., "Mesothelioma in Rats after Inoculation with Asbestos and Other Materials," *British Journal of Cancer*, Vol. 28, 1973, pp.175-185.

[37] Wagner, J.C, "Biological Effects of Short Fibers," *Proceedings of the VIIth International Pneumoconioses Conference Part II*, Department of Health and Human Services Publication No. 990-108, 1990, pp.835-839.

[38] Berman, D.W., Crump, K.S., Chatfield, E.J., Davis, J.M.G., and Jones, A.D., "The Sizes, Shapes and Mineralogy of Asbestos Structures that Induce Lung Tumors and Mesothelioma in AF/HAN Rats Following Inhalation," *Risk Analysis*, Vol. 15, 1995, pp.181-195.

[39] Steenland, K., and Brown, D., "Mortality Study of Gold Miners Exposed to Silica and Nonasbestiform Amphibole Minerals: An Update with 14 More Years of Follow-Up," *American Journal of Industrial Medicine*, Vol. 27, 1995, pp. 217-229.

[40] Higgins, I.T.T., Glassman, J.H., Mary, S.O., and Cornell, R.G., "Mortality of Reserve Mining Company Employees in Relation to Taconite Dust Exposure, *American Journal of Epidemiology*, Vol. 118, 1983, pp.710-719.

[41] Brown, D.P., Kaplan, S.D., Zumwalde, R.D., Kaplowitz, M., and Archer, V.E., "Retrospective Cohort Mortality Study of Underground Gold Mine Workers," *Silica, Silicosis and Cancer - Controversy in Occupational_ Medicine, Cancer Research Monograph*, Vol.2, Goldsmith D.G., Win, D.M. and Shy, C.M., eds., Praeger Publishers, 1986, pp.335-350.

[42] Cooper, W.C., Wong, O., and Graeber, R., "Mortality of Workers in Two Minnesota Taconite Mining and Milling Operations," *Journal of Occupational Medicine*, Vol. 30, 1988, pp.507-511.

[43] Cooper, W.C., Wong, O., Trent, L.S., and Harris, F., "An Updated Study of Taconite Miners and Millers Exposed to Silica and Non-asbestiform Amphiboles," *Journal of Occupational Medicine*, Vol. 34, 1992, pp.1173-1180.

[44] Mossman, B.T., Bignon, J., Corn, M., Seaton, A., and Gee, J.B.L., "Asbestos: Scientific Developments and Implications for Public Policy," *Science*, Vol. 247, 1990, pp.294-301.

[45] Health Effects Institute-Asbestos Research, *Asbestos in Public and Commercial Buildings: A Literature Review and Synthesis of Current Knowledge*, 1991

[46] Guthrie, G.D. and Mossman, B.T. eds., *Health Effects of Mineral Dusts*, Mineralogical Society of America, Reviews in Mineralogy, Vol. 28, 1993.

[47] Mossman, B.T. and Begin, R., eds., *Effects of Mineral Dusts on Cells*, Springer-Verlag, 1989.

[48] Leidel, N.A., Bayer, S.G., Zumwalde, R.D., and Busch, K.A., "USPHS/NIOSH Membrane Filter Method for Evaluating Airborne Asbestos Fibers," U.S. Department of Health Education and Welfare, NIOSH, Technical Report No.79-127, 1979.

[49] Addingley, C.F., "Asbestos Dust and Its Measurements," *Annals of Occupational Hygiene*, Vol. 9, 1966, pp.73-82.

[50] Lynch, J.R., Ayer, H.E., and Johnson, D.L., "The Interrelationships of Selected Asbestos Exposure Indices," *American Industrial Hygiene Association Journal*, Vol. 31, 1970, pp.598-604.

[51] Cossette, M. and Winer, A.A., "The Standard for Occupational Exposure to Asbestos Being Considered by ASTM Committee E-34," *Workshop on Asbestos: Definition and Measurement Methods*, NBS Special Publication 506, 1978, pp.381-386.

[52] U.S District Court, District of Minnesota, 5th Division, Supplemental Memorandum (No. 5-72, Civil 19, Appendix 5, Judge Miles Lord, May 11, 1974.

[53] Siegrist, H.G. and Wylie, A.G., "Characterizing and Discriminating the Shape of Asbestos Particles," *Environmental Research*, Vol. 23, 1980, pp.348-361.

[54] Campbell, W., Huggins, C., and Wylie, A.G., "Chemical and Physical Characterization of Amosite, Chrysotile, Crocidolite, and Non-fibrous Tremolite for Oral Ingestion Studies by NIEHS," *Bureau of Mines Report of Investigation #8452*, 1980.

[55] Wylie, A.G., Virta, R., and Russek, E., "Characterizing and Discriminating Airborne Amphibole Cleavage Fragments and Amosite Fibers: Implications for the NIOSH Method," *American Industrial Hygiene Association Journal*, Vol. 46, 1985, pp.197-201.

[56] Kenny, L.C., Rood, A.P., Blight, B.J.N, "A Direct Measurement of the Visibility of Amosite Asbestos Fibers by Phase Contrast Optical Microscopy," *Annals of Occupational Hygiene*, Vol. 31, 1987, pp.261-264.

[57] Pang, T.W.S., Schonfeld, F.A., and Patel, K., "The Precision and Accuracy of a Method for the Analysis of Amosite Asbestos," *American Industrial Hygiene Association Journal*, Vol. 49, 1988, pp.351-356.

[58] Rooker, S.J., Vaughan, N.P., and LeGuen, J.M., "On the Visibility of Fibers by Phase Contrast Microscopy," *American Industrial Hygiene Association Journal*, Vol. 43, 1982, pp.505-509.

[59] Van Duijun, C., "Visibility and Resolution of Microptical Detail," *The Microscope*, Vol. 11, 1956, pp.196-230, 237-309.

[60] Wylie, A.G., "Modeling Asbestos Populations: A Fractal Approach," *Canadian Mineralogist*, Vol. 30, 1993, pp.437-446.

[61] Wylie, A.G., "Factors Affecting Risk from Biologically Active Minerals," Society for Mining, Metallurgy and Exploration Symposium: Mineral Dusts" Their Characterization and Toxicology, Washington D.C., September, 1996.

D. Wayne Berman[1]

ASBESTOS MEASUREMENT IN SOILS AND BULK MATERIALS: SENSITIVITY, PRECISION, AND INTERPRETATION -- YOU *CAN* HAVE IT ALL.

REFERENCE: Berman, D.W., **Asbestos Measurement in Soils and Bulk Materials: Sensitivity, Precision, and Interpretation—You *Can* Have It All,"** *Advances in Environmental Measurement Methods for Asbestos, ASTM STP 1342,* M.E. Beard, H.L. Rook, Eds., American Society for Testing and Materials, 2000.

ABSTRACT: The Interim Superfund Method for the Determination of Releasable Asbestos in Soils and Bulk Materials [*1*] provides asbestos measurements that can be related directly to risk. Results from a pilot study indicate that target sensitivities of 5×10^7 total asb s/g and 3×10^6 long asb s/g are easily achievable using the method and this more than covers the range of interest. Results also indicate that the precision of the method is adequate to distinguish differences in concentrations required for supporting risk management decisions. Relative percent differences (RPD's) of 50% across split samples should be routinely achievable using the method. A controlled field study also confirms that source concentrations that are measured using the Superfund Method, which can be reported for any structure size range of interest, can be coupled directly to appropriate emission and dispersion models to accurately predict airborne exposure concentrations. Adjustable parameters (i.e. fudge factors) are not required.

KEYWORDS: asbestos, measurement, analytical method, soils, bulk materials, risk assessment, site assessment, exposure, sensitivity, precision, emissions, dispersion, modeling, dust

---

The Interim Superfund Method for the Determination of Releasable Asbestos in Soils and Bulk Materials [*1*] is designed to provide measurements of the concentration of asbestos in soils and bulk materials that can be related directly to risk. Because a defined, representative fraction of the releasable asbestos that is present in a sample is deposited directly on mixed cellulose ester filters, the sample can be prepared using either a direct or an indirect transfer technique. Because the subsequent analysis is by transmission electron microscopy (TEM) and incorporates flexible counting rules, any of the size ranges of asbestos structures that have been variously associated with health effects can be

---

[1]President, Aeolus Environmental Services, 751 Taft St., Albany, CA 94706.

quantified. Because the method incorporates a gentle tumbling action to separate asbestos from the remainder of the sample, which is otherwise unmodified, the method is expected to faithfully preserve the distribution of the sizes of the releasable asbestos structures within the sample.

The Superfund Method can also be applied to a broad variety of bulk materials of potential interest. Any friable or unconsolidated material from which a representative sample weighing between 40 and 80 g can be extracted can be analyzed using the method. Consolidated, non-friable materials can also be analyzed using the method, but additional preparation steps need to be formalized. Such options are not addressed further here.

The sensitivity and level of precision that are achievable using the Superfund Method have been previously determined [*2*] and are summarized below. The objective of this study is to test the hypothesis that emission and dispersion models that have been developed for dust can be adapted generically for use with bulk measurements from the Superfund Method to predict airborne asbestos exposure concentrations. The performance of the Superfund Method toward this end is also compared to that of other methods (i.e. polarized light microscopy based methods) that are commonly used to determine the concentration of asbestos in bulk materials. Thus, sensitivity, precision, and prediction for interpretation are addressed below.

Background

Sensitivity

The Superfund Method has already been shown capable of achieving adequate sensitivity to support risk assessment. This was demonstrated during a pilot study [*2*] that was conducted to compare the performance of two candidate techniques being considered for development as the Superfund Method.

In an earlier study [*3*], concentrations of 5×10^7 total asb s/g sample and 3×10^6 long asb s/g were set as target sensitivities that would allow adequate determination of the full range of concentrations that potentially pose health risks. These levels also incorporate margins of safety to allow for anticipated variability and uncertainty without diminishing the performance of the method. Note that long asbestos structures are defined as those longer than 5 μm.

Results of the pilot study demonstrate that the target sensitivities set for the Superfund Method are easily achievable [*2*]. If samples are prepared carefully, only 4 grid openings need to be scanned at a TEM magnification of 20,000 to achieve the target sensitivity of 5×10^7 s_{tot}/g_{smpl}. Similarly, only 61 grid openings need to be scanned at a TEM magnification of 10,000 to achieve the target sensitivity for long structures of 3×10^6 s_{long}/g_{smpl}.

Importantly, the discussion concerning sensitivities presented here is based on samples analyzed following preparation by an indirect transfer technique. This was the only option available for the Superfund Method at the time that the pilot study was conducted. It has since been modified, however, to allow collection of filters that can be analyzed following preparation by standard, direct transfer techniques [*1*]. Although data are not currently available for confirmation, it is believed that comparable sensitivities are achievable with this method even for directly prepared samples, although a greater number of grid openings may need to be scanned than what is reported in the last paragraph.

Precision

The Superfund Method has also been shown capable of delivering a level of precision that is adequate for distinguishing acceptable concentrations from potentially hazardous concentrations with a resolution that is considered acceptable for supporting risk management decisions [*2*]. During the pilot study, the counting of 50 structures produced an average relative percent difference (RPD) across duplicate sample pairs of 47%. Of 40 paired counts (five separate counts of each of a pair of filters from each of four sets of paired samples by each of two laboratories), 14 (35%) exceeded an RPD of 50% and 6 (15%) exceed an RPD of 100%. However, 8 of the 14 RPD's exceeding 50% derive from a single set of paired samples (i.e. those from the Diamond 20 site) so it is conceivable that special problems developed during the preparation of this one pair of samples. Excluding the Diamond 20 set of samples, the mean of the remaining RPD's drops to 33% with only 20% (6 of 30) exceeding an RPD of 50% and only 7% (2 of 30) exceeding an RPD of 100%.

That a problem may have occurred with the preparation of the Diamond 20 samples during the pilot study is further reinforced, if one looks at the results of the duplicate pair of samples that were also generated from the Diamond 20 site during a later Diamond 20 study [*4*]. The RPD's measured for this pair of samples (from multiple analyses) are all smaller than 50% (ranging from 16% to 34%).

Based on the combined results of the pilot study and the later Diamond 20 study, a target RPD of 50% appears to be generally achievable, if at least 50 structures are counted during analysis. Because the RPD should be inversely proportional to the square root of the number of structures counted, an RPD of less than 100% should be easily achievable even when only 10 structures are counted. Note that even an RPD of 100% represents a difference in concentrations of only a factor of 3, which should provide adequate precision for most risk assessment applications.

The results described above are based on data generated following preparation by an indirect transfer technique. It is anticipated that the achievable precision will be somewhat worse for samples prepared by direct transfer. Unfortunately, insufficient data are available at this time to judge how much precision might suffer for samples prepared using a direct transfer technique. However, the potential degradation in precision is not expected to be substantial primarily because filters to be used for direct preparation collect dust under well-controlled conditions (i.e. from air that has been isokinetically sampled from the inside of an elutriator [*1*]).

Prediction (and interpretation)

In a risk assessment, inhalation exposure is typically evaluated using coupled emission and dispersion models that are appropriate for the particular circumstances under which such exposure is expected to occur. Exposure levels are thus estimated by measuring the concentration of the releasable form of the contaminant of interest in the bulk phase and then using the measured concentrations as inputs for the selected models.

Once airborne, asbestos structures are expected to behave very much like generic, fine particles. This is because dispersion in air is driven primarily by air currents and turbulence that are not specific to the particular type of particles being dispersed [*5*]. However, emission models are strongly dependent on the relationship between what is measured in the bulk phase and the exposure concentrations of ultimate interest.

Empirical studies relating bulk asbestos concentrations to air concentrations have not been conducted widely so that emission models that are specific to asbestos are not generally available. However, a broad range of models are available for "nuisance" and "respirable" dust . Nuisance dust is commonly defined as particles exhibiting aerodynamic-equivalent diameters of 10 μm and smaller and is abbreviated as PM_{10} [*6*]. Respirable dust is typically defined as particles with aerodynamic-equivalent diameters less than or equal to 3 or 3.5 μm [*6*]. Because respirable dust represents the set of particles that can reach and settle in the deep lung following inhalation, it is believed that this size range of particles possesses the greatest potential to cause adverse health effects. Correspondingly, asbestos structures that have been associated with disease induction exhibit aerodynamic equivalent diameters within the respirable range. The aerodynamic equivalent diameter of a particle is defined as the diameter of a sphere of unit density that exhibits the same settling velocity as the particle of interest [*7*].

Emission models for nuisance and respirable dust are available in the literature, e.g. [*7,8*], for a wide variety of conditions of potential relevance to asbestos. These include, for example, models for estimating emissions due to direct wind entrainment from paved and unpaved surfaces, due to vehicular traffic on paved and unpaved surfaces, due to agricultural tilling of fields, and due to excavation, loading, and dumping of unconsolidated bulk materials. Many of these models have even been validated.

Ideally, this large body of literature can be adapted to model asbestos. This has proven problematic, however, both because the dimensionality of dust emission models has not always been preserved when adapting such models for asbestos and because, heretofore, a soil bulk method capable of providing the right kind of input has not been available.

In this study, a generalized approach is proposed in which the dimensionality of the underlying dust model is carefully preserved as the model is modified to allow prediction of asbestos. The approach is then applied specifically to a model that is used to predict (dust and asbestos) concentrations downwind of an unpaved road. This is so the validity of the approach can be evaluated using data from the Diamond 20 study [*4*].

Methods

Predictive models were adapted for this study from published emission and dispersion models for dust [*7,8*]. Data employed in the evaluation were obtained from a controlled study of road emissions conducted at the Diamond 20 site near Copperopolis, California [*4,9*]. Analyses were conducted using standard linear regression techniques [*10*].

Model Development

The Copeland model [*8*] was selected as the starting point for development of a model to evaluate asbestos release and transport in this study. Because the data to be evaluated come from a controlled study of emissions from a roadway, the selected model is an emission model for dust generated on unpaved roads. The selected dust model was then coupled with a Gaussian line-source dispersion model in the manner described by Stenner et al. [*11*]:

$$C_{dust} = 1.7kn\left(\frac{s}{12}\right)\left(\frac{V}{48}\right)\left(\frac{W}{2.7}\right)^{0.7}\left(\frac{w}{4}\right)^{0.5}\left[\frac{(365-p)}{365}\right]\left[\frac{2}{(2\Pi)^{0.5}\sigma_z U}\right] \qquad (1)$$

where:

- C_{dust} is the dust concentration estimated for a defined distance downwind of a fixed road length of interest in g/m^3;
- k is the aerodynamic particle size multiplier (dimensionless);
- s is the silt content of the soil in wt %;
- V is the velocity of the vehicle in km/hr;
- W is the weight of the vehicle in Mg (megagrams);
- w is the number of wheels on the vehicle (dimensionless);
- p is the number of days per year when precipitation exceeds 0.254 mm;
- n is the frequency with which vehicles pass the road length of interest in veh/sec;
- σ_z is the vertical dispersion parameter (m); and
- U is the wind speed (m/sec).

Before attempting to modify this equation further, a dimensional analysis was conducted to assure that later modifications would preserve the overall dimensional form of the equation. The emission factor in Equation 1, "1.7k," consists of an emission coefficient multiplied by k (the particle size multiplier). This product has units of kg/veh*km (or g/veh*m) so that when multiplied by n (veh/sec) and the dispersion term (the last bracket on the right), which has units of sec/m^2, the resulting units on the right side of the equation match those of the output, C_{dust}: g *(dust)*/m^3 (air). The remaining variables in the equation all appear as parts of dimensionless ratios that each represent the manner in which emissions change from the empirically derived 1.7k as the value of each such variable is changed from a reference value represented by the denominator of the ratio (e.g. s/12 for silt content with 12% indicating the reference value).

As noted by Stenner et al. [*11*], Equation 1 is designed to provide estimates of long-term average concentrations because it contains a term representing the long-term average moisture condition for the road (the second bracket from the right). Typically, it is also modified by adding a term representing the fraction of the time that wind blows in the direction of interest. However, we are interested in a real-time model (rather than a model for predicting long-term averages). Therefore Equation 1 must be altered by substituting a real-time moisture content term so that it can be applied to evaluate short-term field studies, as is the case here. Thus, a factor representing real-time moisture content, F_m, is substituted as follows:

$$C_{dust} = 1.7kn\left(\frac{s}{12}\right)\left(\frac{V}{48}\right)\left(\frac{W}{2.7}\right)^{0.7}\left(\frac{w}{4}\right)^{0.5}\left[F_m\right]\left[\frac{2}{(2\Pi)^{0.5}\sigma_z U}\right] \quad (2)$$

The precise functional form of the moisture content factor, F_m, is not known except that, to preserve the dimensionality of the overall equation, it is likely to be some power function of a dimensionless ratio. As indicated later, however, lack of knowledge concerning the precise functional form of this factor does not affect our ability to employ this equation in this study. Equation 2 is the model employed to predict downwind dust concentrations in this study.

The simplest and most direct way to modify Equation 2 so that it can be used to predict asbestos rather than dust concentrations is simply to multiply the right side of the equation by the ratio of releasable asbestos structures/releasable dust that are available in the solid matrix of the roadway from which emissions are occurring. Thus:

$$C_{asb} = 1.7kn\left(\frac{s}{12}\right)\left(\frac{V}{48}\right)\left(\frac{W}{2.7}\right)^{0.7}\left(\frac{w}{4}\right)^{0.5}\left[F_m\right]\left[\frac{2}{(2\Pi)^{0.5}\sigma_z U}\right]R_{a/d} \quad (3)$$

where:

C_{asb} is the asbestos concentration estimated for a defined distance downwind of a fixed road length of interest in s/m^3;

$R_{a/d}$ is the *measured* ratio of releasable asbestos structures to releasable dust in the road matrix; and

all other parameters have been previously defined.

Notice that the form of the factor used to convert Equation 3 from a dust model to an asbestos model preserves the exact nature of the all of the dimensions of the equation. Measured ratios of releasable asbestos to releasable (PM_{10}) dust (i.e $R_{a/d}$'s) represent one of the two kinds of measurements that can be derived as outputs from the Superfund Method. Therefore, such measurements can be input directly into this model without the need for any adjustable factors.

The form of the modification incorporated into Equation 3 can be applied generically to any dust emission model in which the output is an airborne concentration of dust (of a defined size range). This remains true as long as the dust in the denominator of the measured ratio, $R_{a/d}$, matches the size range of dust that is modeled. Note that the operating conditions currently recommended for preparing samples in the Superfund

Method provides measures of $R_{a/d}$ where the denominator is the PM_{10} fraction of dust, so it is easiest to couple this method to models that are used to predict PM_{10} concentrations. However, adapting the method for use with other dust fractions requires only that the air flow rate be properly adjusted during preparation of samples [*1*].

To optimize use of Equation 3, the intended purpose for predicting exposure concentrations needs to be considered so that the range of asbestos structure sizes to be reported can be specified and matched with those measured using the Superfund Method. The purpose for predicting exposure concentrations also needs to be considered to define the preparation technique (i.e. direct transfer or indirect transfer) that would otherwise be used to prepare TEM specimens for analysis (were they not to be predicted). The procedures employed for preparing samples derived using the Superfund Method should then be matched accordingly. This is because it has been shown that choice of the TEM specimen preparation technique substantially affects the magnitude of the resulting measurement [*12*].

Finally, because there is also a desire to contrast use of measurements derived from the Superfund Method with those derived from PLM-based methods, the following adaptations were incorporated into the model to allow use of PLM measurements as inputs for predicting airborne exposure:

$$C_{asb} = 1.7kn\left(\frac{s}{12}\right)\left(\frac{V}{48}\right)\left(\frac{\cdot W}{2.7}\right)^{0.7}\left(\frac{w}{4}\right)^{0.5}[F_m]\left[\frac{2}{(2\Pi)^{0.5}\sigma_z U}\right]\left(\frac{AC}{100}\right)\left(\frac{\%f}{100}\right)F_{i/d}F_{s/wt} \quad (4)$$

where:

AC is the weight percent of asbestos estimated using a PLM-based method (wt %);

%f is the weight fraction of fines (sent to the laboratory for analysis) in the total weight of samples collected in the field (wt %);

$F_{i/d}$ is a factor indicating the ratio of asbestos structures of a specific size range of interest that would be measured if prepared by an indirect transfer technique versus what would be measured if prepared by a direct transfer technique (dimensionless);

$F_{s/wt}$ is a factor indicting the number of asbestos structures of a specific size range of interest per unit weight of asbestos measured by a PLM-based method (s/wt); and

all other factors have been previously defined.

A comparison of Equations 3 and 4 highlights some of the differences between soil or bulk measurements derived from the Superfund Method and those derived from a PLM-based method. While preparation techniques and the size range of interest can be matched to those desired for the output exposure estimates when using the Superfund Method, such options are not available for PLM-based methods, due to their inherent limitations. Consequently, adjustable parameters do not need to be added when exposure is derived from models using Superfund Method measurements as inputs. However, at least two adjustable parameters must be incorporated into such models when bulk asbestos measurements are derived from PLM-based methods. These are $F_{i/d}$ and $F_{s/wt}$. In many cases (as in the Diamond 20 study discussed below), it is also necessary to assume that the "area percent" asbestos measured using PLM is equivalent to "weight percent" asbestos because formal gravimetric analysis is not typically conducted in tandem with the PLM measurement.

The factor, $F_{i/d}$ is required in this equation because the air measurements from the Diamond 20 study (to which predicted values are compared) were prepared using an indirect transfer technique [*4*]. However, PLM measurements are generally converted to measurements that are assumed to derive from directly prepared samples (i.e. the value of the factor $F_{s/wt}$ is traditionally selected assuming conversion to measurements derived from directly prepared samples). The factor $F_{s/wt}$ is the number of structures (of a defined size range) per unit mass of asbestos and is required to convert the weight percent of asbestos estimated during PLM analysis to a fiber concentration. The derivation of values for these parameters is described further in the Results and Discussion Section. Note that omitting these factors destroys the dimensional relationship established in the equation so that results would not be interpretable.

When using PLM-based measurements (as opposed to Superfund Method measurements), it is also necessary to assume that the fraction of asbestos measured in the bulk material is equivalent to the fraction of releasable asbestos in the releasable dust contained within the matrix. This assumption may or may not be reasonable, depending on the nature of the matrix. The validity of this assumption is explored further in the Results and Discussion Section.

The factor (%f) in Equation 4 is required because bulk samples collected in the Diamond 20 study were size separated in the field and only the fines (i.e. those smaller than 3/8th in) were shipped to the laboratory for analysis. However, values for this factor are measured directly in the field (i.e. they are not adjustable).

Sources of Data

In the Diamond 20 study [*9*], two roads were selected for study because they exhibit the required 300 ft stretch of straight road, prevailing winds that blow primarily perpendicular to the road, and the required clearance upwind and downwind of the road. During the study, airborne PM_{10} dust and asbestos measurements were collected from co-located stations set up 25 ft upwind, 25 ft downwind, 75 downwind, and 150 ft downwind of the road. Sampling stations were also set up at remote locations to evaluate local background.

Samples were collected during several runs (each of two or three hours duration) during which a single vehicle was driven along a study road at regular intervals and a constant speed of 30 mph. For different runs, intervals were set equal to one of three frequencies (0, 15, and 30 passes per hour). Runs at each frequency were repeated several times at each road over the span of several days. Meteorological monitoring (including temperature, humidity, wind direction, and wind speed) was also conducted during each run.

Prior to conducting the controlled runs, road material was collected from multiple random locations, composited, and subjected to analysis using the Superfund Method and by a modification to the EPA Interim Method for the Determination of Asbestos in Bulk Insulation Samples [*13*], which employs PLM. The method was modified to apply to road material by incorporating a procedure for homogenizing collected samples and sieving to isolate a fine fraction in the field [*9*]. Additional samples were also collected and analyzed for silt and moisture content.

Values for input parameters that are derived from the Diamond 20 study and that are used in this study as input values for the models described above are summarized in Table 1.

Note that, although multiple size fractions of asbestos were tracked during the Diamond 20 study [*4*], the following evaluation focuses on only two of these: the phase contrast microscopy equivalent (PCME) fraction and total long structures (Long S). The former is defined as the set of asbestos structures that are longer than 5 μm, thicker than approximately 0.3 μm, exhibit an aspect (length to width) ratio greater than 3, and that exhibit largely parallel sides. The latter size fraction, Long S, is the set of all parent asbestos structures (i.e. fibers, bundles, clusters, and/or matrices) that are longer than 5 μm and exhibit an aspect ratio greater than 3 and all components of non-conforming complex structures (i.e. clusters or matrices) that themselves exhibit lengths greater than 5 μm and an aspect ratio greater than 3. More detailed descriptions of the morphological forms of asbestos structures that are found in nature and the manner in which they are characterized can be found in the literature, e.g. [*12*].

Results and Discussion

Evaluating The Dust Model

Before considering whether the modified Copeland model can be adapted so that it can be applied to asbestos (Equation 3), the validity of the underlying dust model (Equation 2) was first evaluated. Predicted PM_{10} dust concentrations that were derived using Equation 2 with the input values in Table 1 are compared to PM_{10} dust concentrations measured at corresponding locations downwind of study roads at the Diamond 20 site. Figure 1 is a plot of the measured dust concentrations versus the modeled dust concentrations.

It is immediately obvious from Figure 1 that, with the exception of a single data point, a linear relationship exists between measured and modeled dust. Given that the measured dust concentration reported for point R2-15-3B [*4*: Appendix B] lies so far from the trend for the rest of the data that it is clearly an outlier, it was omitted from the regression analysis of the trend for the rest of these data. The remaining data set contains 29 pairs of measured and modeled concentrations.

Actually, two trend lines are presented in Figure 1: a solid line for Road 1 of the Diamond 20 study (the triangles in the figure) and a dashed line for Road 2 (the circles in the figure). As is clear from the figure, however, the two trend lines are virtually identical. The scatter of the data from the two roads also appear to exhibit similar variation and the data sets exhibit extensive overlap. This visual evidence is confirmed by regression in

TABLE 1 -- *Input Values Used for Dust and Asbestos Models*[a]

Variable	Symbol	Units	Values: Road 1	Values: Road 2	Source
Particle Size Multiplier	k	unitless	0.45	0.45	[*8*]
Silt Content	s	wt %	19.4	11.2	Measured
Vehicle Velocity	V	km/hr	50	50	Specified
Vehicle Weight	W	Mg	1.6	1.6	Specified
Number of Wheels	w	unitless	4	4	Specified
Frequency of Passes	N	p/sec			Specified
(15/hr)			0.0042	0.0042	
(5/hr)			0.0014	0.0014	
(0/hr)			0	0	
Wind Speed	U	m/sec	1.9	1.4	Measured[b]
Dispersion Coefficients	σ_z	m			[*5*]
(at 7.58 m)			1.62	1.62	
(at 22.7 m)			3.68	3.68	
(at 45.5 m)			6.84	6.84	
Moisture Content	M	wt %	0.8	0.7	Measured
Moisture Factor	Fm	unitless	1.27	1.27	Estimated
ASBESTOS CONCENTRATIONS					
from Superfund Method	$R_{a/d}$	s/gPM_{10}			
(PCME)			1.9e+09	3.6e+09	Measured
(Long S)			3.0e+09	7.4e+09	Measured
from PLM-based Method	AC	wt%			
(PCME)			7.5	10	Measured
Conversion Factor	Fs/wt	s/gsmpl	3.0e+10	3.0e+10	Estimated[c]
Conversion Factor	%fines	wt%	66	71	Measured
Conversion Factor	Fi/d	unitless	9	9	Estimated[d]

[a] Values from [*4,9*].

[b] Average wind speeds shown, actual wind speeds used were run specific.

[c] Frequently quoted value, no specific source referenced.

[d] Central value of measurements from five paired samples [*4*]

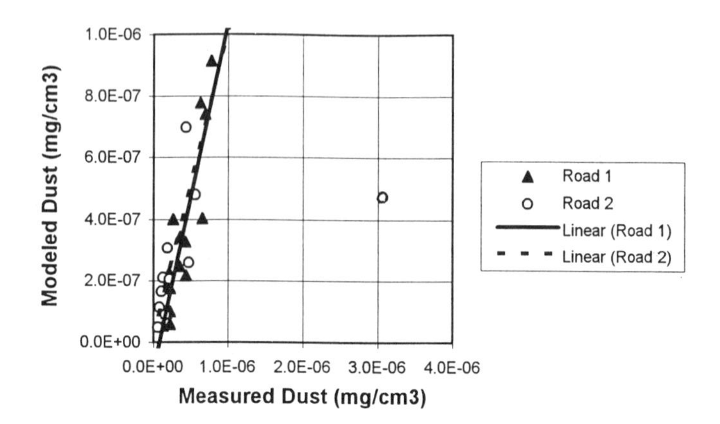

FIGURE 1 — *Measured vs. Modeled Dust Concentrations (Modified Copeland Model)*

which the estimated slopes for Road 1 (0.73) and for Road 2 (0.85) can in no way be distinguished ($p=0.38$). R^2 values for each road are also similar: 813 (Road 1) versus 0.848 (Road 2).

With the single point (R2-15-3B) omitted, the best-fit line describing the relationship between measured and modeled dust concentrations (for the combined data from both roads) exhibits a slope of 0.78 and a Y-intercept of 8.8×10^{-8}. The slope is significantly greater than zero ($p < 0.000001$) indicating a probability of less than one in a million that the observed relationship is due to chance. Importantly, this slope is also significantly different from 1 ($p = 0.0047$). The Y-intercept is indistinguishable from zero ($p = 0.36$). The correlation coefficient for the fit of this line is 0.90 (R^2 is 0.81). By convention, p-values less than 0.05, which represent no more than a 5% probability that the indicated difference is due to chance, are considered to be statistically significant. Such values are therefore assumed to indicate that the tested differences are real. Larger p-values suggest that the functions being compared are indistinguishable.

That the slope of the trend in Figure 1 is different from one indicates that the measured and modeled dust concentrations plotted in Figure 1 differ by a multiplicative constant. This suggests, among other possibilities, that the F_m factor introduced in Equation 2 (to represent the effect of moisture content) may not be one for the conditions studied. Setting this factor equal to 1.27 increases the slope of the best-fit line in Figure 1 to one, but does not otherwise alter the intercept or the fit of the line. Since this optimizes the relationship between measured and modeled dust, while our primary focus here is to evaluate asbestos, this modification is incorporated in all of the following analyses.

As indicated previously, we do not know the correct functional form for F_m in Equation 2. However, setting F_m to 1.27 to optimize the fit between measured and modeled dust serves only to remove extraneous sources of variation that in no way alters the nature of the relationships between measured and modeled asbestos (or between measured asbestos and measured dust). Preserving these latter relationships assures that the evaluation presented in this study remains applicable to any dust model that might be adapted for predicting airborne asbestos concentrations in the manner described in Equation 3.

The linear relationship presented in Figure 1 with a Y-intercept of zero (but with F_m adjusted to produce a slope of one) indicates that corresponding measured and modeled dust concentrations are equivalent. Therefore, the modified dust model represented by Equation 2 is confirmed. Note, even with F_m left unadjusted, the differences between measured and modeled dust concentrations average no more than 30%, which still signifies reasonable agreement.

Comparing Airborne Dust to Airborne Asbestos Measurements

With the relationship between modeled and measured dust established, the relationship between measured asbestos and measured dust was evaluated next. Figure 2 is a plot of PCME asbestos concentrations and PM_{10} dust concentrations measured at corresponding locations downwind of the study roads at the Diamond 20 site. As in Figure 1, triangles indicate concentrations measured downwind of Road 1 and circles indicate concentrations downwind of Road 2. The data point identified as R2-15-3B is omitted from this figure because the dust concentration associated with this sample was shown to be anomalous (Figure 1).

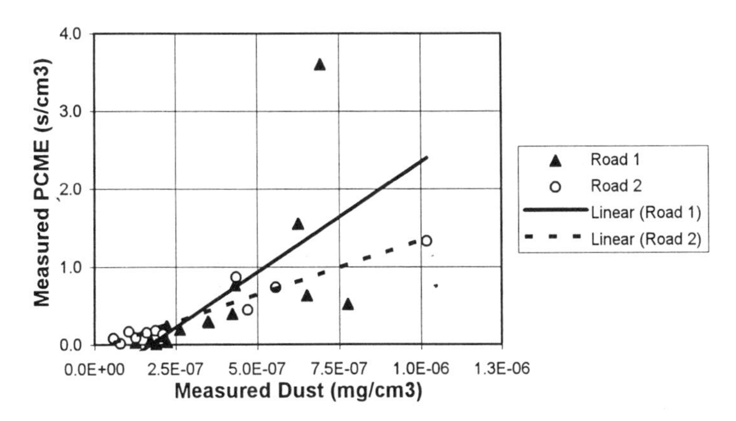

FIGURE 2 — *Measured Asbestos vs. Measured Dust Downwind of Diamond 20 Roads*

It is apparent from Figure 2 that the relationship between asbestos and dust concentrations measured downwind of the two study roads are different. Not only do the data from Road 1 exhibit greater overall scatter than the data from Road 2, but the best-fit trend lines for the two data sets appear to exhibit different slopes. Analysis indicates, however, that the difference between the slope for Road 1 (2.8×10^6) and that for Road 2 (1.4×10^6) is not significant, at least at the 5% level of significance ($p = 0.078$). However, the roads can be distinguished at the 10% level of significance. While this suggests that it is more likely than not that the two slopes do in fact differ, the confidence that can be placed in this conclusion is somewhat lower than one might otherwise hope.

Although not depicted, a similar trend is observed when measured Long S concentrations (the other size range of asbestos structures defined in the Methods Section) are compared to measured dust concentrations. These findings contrast with the very close agreement between measured and modeled dust depicted in Figure 1, where the trends from the two roads are entirely indistinguishable.

The relative scatter in the data from the two roads is strikingly different. While the correlation coefficient for the trend in the ratio of asbestos to dust downwind of Road 2 is 0.958 ($R^2 = 0.917$, n = 11), the correlation coefficient for Road 1 is only 0.677 ($R^2 = 0.458$, n = 18). Thus, while variation in dust concentrations accounts for more than 90% of the variation in asbestos concentrations downwind of Road 2, it explains less than 50% of the variation in asbestos concentrations observed downwind of Road 1. The source of the additional scatter observed among Road 1 data is not at all clear.

Because dust concentrations observed downwind of the two roads are comparable [*9*], any difference in the slopes of the ratios of asbestos to dust that are observed downwind of these roads, must be due to differences in downwind asbestos concentrations. The implication from Figure 2 is that Road 1 is emitting approximately twice the number of asbestos structures per unit mass of dust that Road 2 is emitting when vehicles traverse the roads. As indicated above, however, this difference is small and is barely distinguishable statistically, at least with the available data.

If the apparent difference in the magnitudes of the ratios of asbestos to dust concentrations measured downwind of Road 1 and Road 2 are real, this further suggests that the concentration of releasable asbestos in Road 1 bulk material is approximately twice the concentration of releasable asbestos in Road 2 bulk material. However, such differences were not observed in samples collected from the bulk material of each road. If anything, both the Superfund Method measurements and the PLM-based measurements suggest the opposite (i.e. that Road 2 concentrations are twice Road 1 concentrations). As indicated below, however, these differences are also too small to distinguish statistically so that none of these findings can be considered inconsistent. The finding that differences between Road 1 and Road 2 asbestos concentrations are too small to be readily distinguished using current air or soil measurement techniques is consistent with the results of an earlier Analysis of Variance (ANOVA) that was conducted on the Diamond 20 data [*4*].

Evaluating Bulk Asbestos Measurements

As shown above, the data set from Diamond 20 is not sufficiently rich to evaluate the ability of bulk methods to distinguish *relative* asbestos concentrations across different sources. However, the Diamond 20 data can still be used to compare the relative performance of the Superfund Method and PLM-based methods in terms of their respective abilities to provide measurements that display the correct *absolute* values.

Values of the ratio of asbestos structures to PM_{10} dust ($R_{a/d}$'s) found in air downwind of the roads studied at Diamond 20 are presented in Table 2. As the column headings indicate, the $R_{a/d}$'s presented are derived, respectively, from air measurements, from Superfund Method measurements, or from PLM-based measurements. Values are presented both for the PCME and for the Long S size fractions of asbestos. Units for all of the $R_{a/d}$'s presented are s/g (dust).

TABLE 2 -- *Estimates of the Ratio of Airborne Asbestos to Dust ($R_{a/d}$'s)*

Asbestos Size Fraction	Study Road	Air Data	Superfund Method	PLM Method
PCME	Road 1	2.8e+09	1.9e+09	1.3e+10
	Road 2	1.4e+09	3.6e+09	1.9e+10
	Mean	2.1e+09	2.7e+09	1.6e+10
Long S	Road 1	4.0e+09	3.0e+09	--
	Road 2	1.9e+09	7.4e+09	--
	Mean	2.9e+09	5.2e+09	--
Long/PCME	Road 1	1.4	1.6	--
	Road 2	1.4	2.1	--
	Mean	1.4	1.9	--

The $R_{a/d}$'s presented in the column under the heading "Air Data" are derived from the slopes of the best-fit trend lines for plots of measured asbestos concentrations versus measured dust concentrations. An example of such a plot is presented in Figure 2. Note, however, that the units of the slopes discussed in the last section above are s/mg, so that the values presented in Table 2 have been adjusted.

$R_{a/d}$'s based on the Superfund Method are presented in the next column of Table 2. These are also derived directly from measurement with no "adjustments." A comparison of the $R_{a/d}$'s presented in the "Air Data" Column and the "Superfund Method" Column indicates that the values are approximately equal. Corresponding $R_{a/d}$'s in these two columns generally differ by less than a factor of two. The greatest difference is observed for $R_{a/d}$'s representing the ratio of Long S to dust downwind of Road 2. These values differ by a factor of 3.9.

Unlike the $R_{a/d}$'s that represent slopes from regressions of air data, the Superfund Method derived $R_{a/d}$'s represent single measurements. Therefore, because the magnitude of these differences is similar to the magnitude of the differences for paired Superfund Method-analyses of duplicate samples that were observed during the pilot study [*2*], such differences are not statistically significant. Thus, the air measurements and the Superfund Measurements from which these $R_{a/d}$'s were derived should be considered to be consistent.

In contrast to those derived from air data or the Superfund Method, $R_{a/d}$'s derived from the PLM-based method are not measured directly. Rather, they are estimated using a series of conversion factors. Unfortunately, current knowledge is inadequate to define a "best" method for completing such a conversion. Therefore, options are discussed below.

For the values derived from PLM measurements that are presented in Table 2, it was assumed that the ratio of the number of asbestos structures to the mass of dust found downwind of a source remains equal to the number of asbestos structures per mass of sampled matrix. Thus, the values in the table are calculated from the following equation:

$$R_{a/d} = \left(\frac{AC}{100}\right)\left(\frac{\%f}{100}\right)F_{i/d}F_{s/wt} \quad (5)$$

where all terms are defined as described for Equation 4.

As indicated previously, AC (the weight percent asbestos) is assumed equivalent to the area percent asbestos that is the actual output measurement of the PLM method employed. Also, as indicated previously, $F_{s/wt}$ is an adjustable parameter that is not measured but is based on empirical relationships that are not well characterized. The value for Fs/wt that is generally quoted is 3×10^{10} (Table 1).

Although there was an attempt to quantify $F_{i/d}$ during the Diamond 20 study [*4*], no significant trend was observed among the five pairs of samples analyzed. Thus, because observed values for $F_{i/d}$ ranged between 3 and 16, a central value of 8 was chosen as the "best estimate."

Although Equation 5 represents the most commonly accepted procedure for estimating $R_{a/d}$'s from PLM data, the assumption underlying this equation may not be valid. For example, Superfund Method measurements indicate that $R_{a/d}$'s are equal to the quotient of the releasable asbestos concentrations measured in bulk samples and the concentration of releasable dust found in the bulk samples. If this were the case for PLM data, $R_{a/d}$'s would need to be derived using Equation 6:

$$R_{a/d} = \left(\frac{1}{F_{dust}}\right)\left(\frac{AC}{100}\right)\left(\frac{\%f}{100}\right)F_{i/d}F_{s/wt} \quad (6)$$

where:

F_{dust} is the concentration of dust in the bulk PLM sample; and all of the other variables are defined as described for Equation 4.

Use of Equation 6 to derive the PLM-based $R_{a/d}$'s presented in Table 2 would increase the values presented by a factor of 50, which would make the apparent lack of agreement between measured values and PLM-based values even worse.

Admittedly, PLM measurements ideally represent total asbestos in the bulk samples analyzed versus the *releasable fraction* of asbestos that is measured by the Superfund Method. Therefore, it is unclear that Equation 6 is any more appropriate for PLM than Equation 5. Perhaps the main conclusion to draw from this exercise is that, at this point in time, it is unclear how to derive input estimates for emission and dispersion models from PLM-based measurements. Note also that values are not presented in the PLM column for Long S because a different (and currently undefined) value would be required for $F_{s/wt}$ to estimate $R_{a/d}$'s that are appropriate for Long S.

Comparing Results With Similar Studies

It is instructive to compare this study with the two other recent studies in which asbestos measurements collected downwind of a serpentine-covered road were correlated with predictions from an asbestos-adapted dust model. As in this study, the modified Copeland Model that was introduced by Stenner et al. [*11*] served as the starting point for model development in these other studies [*14,15*]. However, the additional modifications proposed by Horie et al. [*14*] and adopted by Lynch [*15*] do not appear to preserve the dimensionality of the original model.

To better account for variation with vehicle speed, Horie et al. modified the Stenner et al. model by squaring the numerator of the velocity term in the equation. However, this serves to add an additional m/sec to the dimensions of the right side of the equation that is *not* also matched on the left. They also introduce a moisture content term of the form: $0.012/m^{0.6}$, where m is the moisture content in weight percent. The units on this term are either $(wt\%)^{-0.6}$ or $(wt\%)^{+0.4}$ depending on whether the 0.012 in the numerator of this term is considered a reference value for moisture content, which would itself have units of wt%. The additional units introduced on the right side of the equation by this term also remain unmatched on the left side of the equation. The model ultimately recommended by Horie et al. [*14*] has the form:

$$C_{PLM} = 1.7kn\left(\frac{s}{12}\right)\left(\frac{V^2}{48}\right)\left(\frac{W}{2.7}\right)^{0.7}\left(\frac{w}{4}\right)^{0.5}\left[\frac{2}{(2\Pi)^{0.5}\sigma_z U}\right]\left(\frac{AC}{100}\right)F_{s/wt}\left(\frac{0.012}{m^{0.6}}\right) \qquad (7)$$

where all of the terms have been previously defined (see Equation 4).

When referring to their study, note that Horie et al. [*14*] use the term "CF" for the name of the $F_{s/wt}$ factor in Equation 7. However, given that the dimensionality on the right and left sides of Equation 7 do not match, it is difficult to interpret the findings of their study.

Comparing Equation 7 with Equation 4 (the equation employed to relate air measurements and PLM measurements in this study), it is also apparent that the $F_{i/d}$ and %f/100 terms are missing in Equation 7. However, this is appropriate for the Horie et al. study [*14*]. Air samples in their study were prepared using a direct transfer technique so the $F_{i/d}$ term is reduced to one and can be eliminated. Also, it is assumed that bulk samples subjected to PLM analysis in their study were not first size selected in the field, so the %f/100 term is superfluous. Only the adjustable parameter, "CF" (i.e. $F_{s/wt}$) remains.

Because one of the roads studied by Horie et al. [*14*] was also sampled and analyzed using the Superfund Method (during the pilot study), it was hoped that downwind air concentrations observed in their study and predictions based on the Superfund Method could be compared. As already mentioned, however, TEM specimens analyzed in the Horie et al. study were prepared by a direct transfer technique while TEM specimens analyzed as part of the Superfund Method in the pilot study [2] were prepared by an indirect transfer technique. Thus, formal comparison is precluded. Still, qualitative trends can be evaluated.

Comparing Superfund Method-based predictions and air measurements from the Horie et al. study, trends in predicted and measured concentrations appear to be qualitatively consistent. Concentrations predicted using the pilot study data are about 10 to 15 times those measured for long structures and about 30 times those measured for total structures. It is typical for TEM analyses of indirectly prepared samples to produce results that are larger than directly prepared samples and multiples of 30 are common for total structures, e.g. [*12*]. Interestingly, a multiple of approximately 10 for long structures is what was also observed when comparing analyses of indirectly and directly prepared samples, respectively, in the Diamond 20 study [*4*]. Although qualitative, this suggests consistency between Superfund Method-based predictions and air measurements reported in the Horie et al. study [*14*].

Because Lynch [*15*] used the same model recommended by Horie et al. (Equation 7), interpretation of her results is also complicated by the lack of internal consistency among the dimensions of the terms in Equation 7. Further problems are also apparent because Lynch evaluated the Diamond 20 data, which (as discussed previously) requires that PLM-based measurements be multiplied by the $F_{i/d}$ and %f/100 factors from Equation 4 (in addition to "CF") before using such measurements as inputs for predicting downwind air concentrations.

Although Lynch [*15*] also attempted to compare predications based on Superfund Method measurements to the air data from Diamond 20, comparison with her analysis is complicated by use of mis-matched units on the inputs to the model. Lynch used measures of the concentration of releasable asbestos in total sample mass as the input to Equation 7 (rather than asbestos in dust as used here). Because there are no terms in Equation 7 with units of g of total soil mass that can cancel the denominator of the asbestos concentrations used by Lynch, it is not clear how to interpret her results. Due to these problems, the Lynch study [*15*] is not evaluated further here.

Conclusions

Given that modeled dust was shown to track measured dust (Figure 1) and that Superfund Method-derived $R_{a/d}$'s reasonably track the observed ratio of measured asbestos to measured dust (Table 2), the adapted model presented in Equation 3 can be expected to provide reasonably accurate predictions of airborne asbestos concentrations downwind of a road source. Direct comparison between measured and predicted airborne asbestos concentrations confirms this to be the case; a plot very similar to that depicted in Figure 2 results. Thus, when Equation 3 is used with $R_{a/d}$ estimates derived from a *single* Superfund Method measurement, the resulting predictions of airborne asbestos concentrations should be good to within a factor of 3 or 4.

If there is a need to increase precision further, $R_{a/d}$'s should be derived from Superfund Method analyses of a set of samples that are each representative of the entire source area of interest. Such samples might each be derived, for example, as composites of an independent set of samples collected from randomly (or systematically) selected sampling locations over the entire source area.

Only a small number of Superfund Method measurements will generally be required. For example, if as few as three or four duplicate measurements had been collected from each of the roads studied at Diamond 20, method measurements and any resulting airborne predictions would have achieved sufficient precision to accurately distinguish the apparent small differences that exist between Roads 1 and 2.

Importantly, the modifications proposed in Equation 3 do not depend on features that are particular to asbestos roads, they suggest a general approach for adapting dust methods to asbestos. The accuracy of predictions from such models would then be a combined function of the number of representative Superfund Method measurements employed to establish the input value for $R_{a/d}$ and the accuracy of the underlying dust model.

In contrast, it is unclear at this time how PLM-based measurements can be used as inputs for emission and dispersion models to predict airborne asbestos exposure concentrations. Before PLM-based measurements can be applied for this purpose, additional studies will be required to better define the values of the various adjustable (and measured) parameters that are required to convert PLM measurements to the appropriate units for use as inputs to such models.

Acknowledgments

I would like to thank Bruce Allen (ICF Kaiser Engineers, RTP, NC) for his assistance with the regressions and other statistical analyses conducted in this study. Thanks are also extended to Anthony Kolk (EMS Laboratories, Pasadena, CA), Andrew Gray (ICF Kaiser Engineers, San Raphael, CA), and Chatten Cowherd (Midwest Research Institute, Kansas City, MO) for their stimulating discussions during completion of this manuscript.

References

[*1*] Berman, D.W. and Kolk, A.J, 1995, *Superfund Method for the Determination of Asbestos in Soils and Bulk Materials*. Prepared for the Office of Solid Waste and Emergency Response under U.S. EPA Contract No. 68-W9-0059, Work Assignment No. 59-06-D800. Accepted for U.S. EPA publication as an interim method, July.

[*2*] Berman, D.W., Kolk, A.J., Krewer, J.A., and Corbin, K., 1994, "Comparing Two Alternate Methods for Determining Asbestos in Soils and Bulk Materials." Presented at the Eleventh Annual Conference and Exposition of the Environmental Information Association, San Diego, California, March.

[*3*] Berman, D.W. 1990, *Discussion Document: Development of a Superfund Method for the Determination of Asbestos in Soils and Bulk Materials*, prepared for the Environmental Services Branch of the U.S. EPA, Region 9, San Francisco,

California under U.S. EPA Contract No. 68-W9-0059, Work Assignment No. 59-06-D800, September.

[4] ICF Technology, Inc. 1994, *Evaluation of Risks Posed to Residents and Visitors of Diamond 20 Who Are Exposed to Airborne Asbestos Derived from Serpentine Covered Roadways*, prepared for U.S. EPA, Region 9, San Francisco, California under Contract No. 68-W9-0059, Work Assignment No. 59-06-D800., May.

[5] Turner, D.B., 1970, *Workbook of Atmospheric Dispersion Estimates,* U.S. Department of Health, Education, and Welfare, Cincinnati, OH, PB-191 482.

[6] American Conference of Governmental and Industrial Hygienists, 1993, *Threshold Limit Values for Chemical Substances and Physical Agents,* ISBN: 1-882417-03-8, ACGIH, Cincinnati, OH.

[7] Cowherd, C., Muleski, G.E., Englehart, P.J., and Gillette, D.A., 1985, *Rapid Assessment of Exposure to Particulate Emissions From Surface Contamination Sites,* prepared for the Office of Health and Environmental Assessment, U.S. EPA, Washington, D.C., EPA/600/8-85/002, February.

[8] U.S. Environmental Protection Agency, 1985, *Compilation of Air Pollution Emission Factors, Volume I: Stationary Point and Area Sources,* AP42, Fourth Edition, Office of Air Quality Planning and Standards, Research Triangle Park, N.C.

[9] Ecology and Environment, Inc., 1993, *Diamond 20 Asbestos, Phase II Site Assessment, Copperopolis, California,* Prepared for The U.S. EPA, Emergency Response Section (H-8-3), 75 Hawthorne St., San Francisco, CA, December.

[10] Bickel, P.J. and Doksum, K.A., 1977, *Mathematical Statistics: Basic Ideas and Selected Topics*, Holden-Day, Inc., San Francisco

[11] Stenner, R.D., Droppo, J.G., Peloquin, R.A., Bienert, R.W., and VanHouten, N.C., 1990, *Guidance Manual on the Estimation of Airborne Asbestos Concentrations as a Function of Distance from a Contaminated Surface Area for Area Suspension Evaluations*, prepared by Pacific Northwest Laboratory, Battelle Memorial Institute, prepared for the U.S. EPA under a Related Services Agreement with the U.S. DOE, Contract DE-AC06-76RLO 1830.

[12] Berman, D.W. and Chatfield, E.J., 1990, *Interim Superfund Method for the Determination of Asbestos in Ambient Air. Part 2: Technical Background Document*, Office of Solid Waste and Remedial Response, U.S. EPA, Washington, D.C., EPA/540/2-90/005b, May.

[*13*] U.S. Environmental Protection Agency, 1981, *Interim Method for the Determination of Asbestos in Bulk Insulation Samples,* EPA 600/M4-81-020.

[*14*] Horie, Y., Sidawi, S., and Tranby, C., 1992, *Development of a Technique to Estimate Ambient Asbestos Downwind from Serpentine-Covered Roadways*, A California Air Resources Board Final Report prepared under Contract No. A032-147.

[*15*] Lynch, K.P., 1995, *A Modeling Approach to Estimate Community Exposures to Airborne Asbestos Concentrations From Reentrained Road Dust,* Thesis for a Master of Science in Environmental Management, University of San Francisco, San Francisco, California.

Eric J. Chatfield[1]

A Validated Method for Gravimetric Determination of Low Concentrations of Asbestos in Bulk Materials

REFERENCE: Chatfield, E. J., **"A Validated Method for Gavimetric Determination of Low Concentrations of Asbestos in Bulk Materials,"** *Advances in Environmental Measurement Methods for Asbestos, ASTM STP 1342,* M. E. Beard and H. L. Rook, Eds., American Society for Testing and Materials, West Conshohocken, PA, 2000.

ABSTRACT: A method for determination of weight concentrations of asbestos in the 0.01% to 10% range is described. The method is based on gravimetric matrix reduction, incorporating oxidation of organic materials, dissolution of acid-soluble components, and removal of large particle size materials by sedimentation or flotation, followed by analysis of the remaining material using a well-defined point counting procedure optimized for determination of mass concentration. The method has been validated by an inter-laboratory study with the Optical Microscope Laboratory of the Hong Kong Environmental Protection Department using laboratory-prepared standards with known weight concentrations of asbestos ranging from 0.1% to 6% in a wide variety of matrices. The results of analyses by both laboratories exhibit a linear relationship with the known weight concentration of asbestos, with a correlation coefficient of 0.97. The method provides a legally-defensible basis for determining whether the concentration of asbestos measured in a material is lower than or higher than a legislated standard of either 0.1% or 1% asbestos by weight.

KEYWORDS: Asbestos, bulk, analysis, matrix reduction, weight concentration, gravimetric, point counting, legislation, legal, regulation

Introduction and Background

In many jurisdictions, an asbestos-containing material (ACM) is defined as any material containing more than 1% asbestos in terms of dry weight percentage. Some jurisdictions specify a limiting concentration of 0.1% asbestos, and others regard the presence of any detectable asbestos as sufficient to classify the material legally as an asbestos-containing material.

Polarized light microscopy (PLM) carried out on an untreated sample, and using visual estimation of the asbestos concentration, is satisfactory for determination of the

---

[1]President, Chatfield Technical Consulting Limited, 2071 Dickson Road, Mississauga, Ontario, Canada L5B 1Y8

asbestos in materials such as friable fireproofing and thermal insulation. When asbestos is present in these types of materials, it has usually been added during manufacture of the material, the concentration is usually substantially higher than 1%, and the experienced PLM analyst has no difficulty in concluding that the concentration is higher than a 1% or 0.1% standard. When no asbestos is detected by PLM, in fireproofing and thermal insulation, the experienced PLM analyst can be confident that a "none detected" result assures that the asbestos concentration is lower than a 1% or 0.1% standard. In practice, further attention to quantification of the asbestos concentration in these types of materials is of no interest from a regulatory or decision-making point of view, because they either contain a substantial concentration of asbestos which obviously exceeds the standard, or they obviously contain little or no asbestos. The regulatory requirement is therefore completely satisfied by identification of the asbestos and a visual estimate of the concentration present.

Quantification of the asbestos concentration, and the accuracy with which this can be achieved, becomes significant for materials in which the asbestos concentration is lower than approximately 10%, particularly if the matrix material compromises detection and identification of asbestos. Plasters, joint compounds, texture coats, ceiling tiles, floor tiles, asphaltic materials, and some cements fall into this category. For these types of samples, and in the range of asbestos concentration normally associated with these materials, PLM analysis alone is not sufficiently reliable, either to detect asbestos or to determine whether the concentration is above or below 1% or 0.1%. The U.S. Environmental Protection Agency (EPA) requires the use of point counting in the concentration range below 10% (or an assumption that the material contains more than 1% asbestos) [*1*]. However, the point counting method specified by EPA is neither well-defined nor does it have sufficient precision in the vicinity of 1% asbestos to achieve the required discrimination between regulated and unregulated materials at 1%. The EPA point counting method requires that, on appropriately-prepared slides, a total of 400 randomly-spaced, non-empty points be classified as to whether they fall on asbestos or non-asbestos particles. When asbestos fibers occupy 1% of the area of all particles on the slides, a mean of 4 asbestos points is predicted for the 400-point count. The best possible situation is that all particles, including the asbestos fibers, are deposited on the slides randomly according to a Poisson distribution, and therefore, for an area concentration of exactly 1%, all that can be stated is that 95% of repeat measurements would be predicted to lie between 1.09 points and 10.24 points, corresponding to 0.27% to 2.6% asbestos [*2*]. There is a significant probability that any single measurement could yield a result above or below the 1% regulatory limit for asbestos concentration.

Conventional point counting determines the relative projected areas occupied by different particle species on a microscope slide. The technique is useful for determination of the relative abundance of different phases or mineral species in polished rock sections of constant thickness. Provided that the grain size is larger than the thickness of the rock section, measurements of the relative areas of the various species are equivalent to measurements of the relative volumes. If the densities of the various species are known, the relative weights in the thin section can be calculated. The application of point counting, without defined counting criteria, to determination of the proportion of asbestos in a mixture of particles of different thicknesses and different densities is a mis-use of the method, because the result clearly depends on the types of particles and the extent of any sample preparation. For example, the relative areas occupied by different particle species vary with increased comminution, leading to differences in the reported asbestos concentration, even though the weight concentration of asbestos should have remained constant. There may also be extreme differences in density between the various particle species. Some materials such as perlite, vermiculite, cellulose or plastic foam, which have very low densities as a consequence of air pockets in their structures will change in density as well as relative area when they are processed for measurement. Given that, for the same weight concentration, a reported result in terms of percent area of asbestos depends strongly on both the nature of the sample and the sample preparation procedure, it is clear that percent area measurements **derived by any means** are not suitable for determination of compliance with regulatory standards expressed in terms of weight percent, although in some circumstances they can be indicative. Regulatory standards expressed in terms of percent area of asbestos are unenforceable, because the same weight concentration of asbestos can give rise to a range of percent area measurements, depending of the nature of the material and the sample treatment.

In the published methods [*3,4*] for determination of asbestos in bulk building material samples, EPA does not address the question of how compliance or non-compliance with a 1% regulation is to be demonstrated, other than reporting of a mean concentration based on detection of 4 or fewer asbestos points in a count of 400 non-empty points. For regulatory purposes, it is usually considered satisfactory if a compliance level is selected such that 95% of repeat measurements are predicted to comply with the regulation. Figure 1 shows the mean asbestos concentration and the 95% one-sided Poisson confidence limits as functions of the number of asbestos points

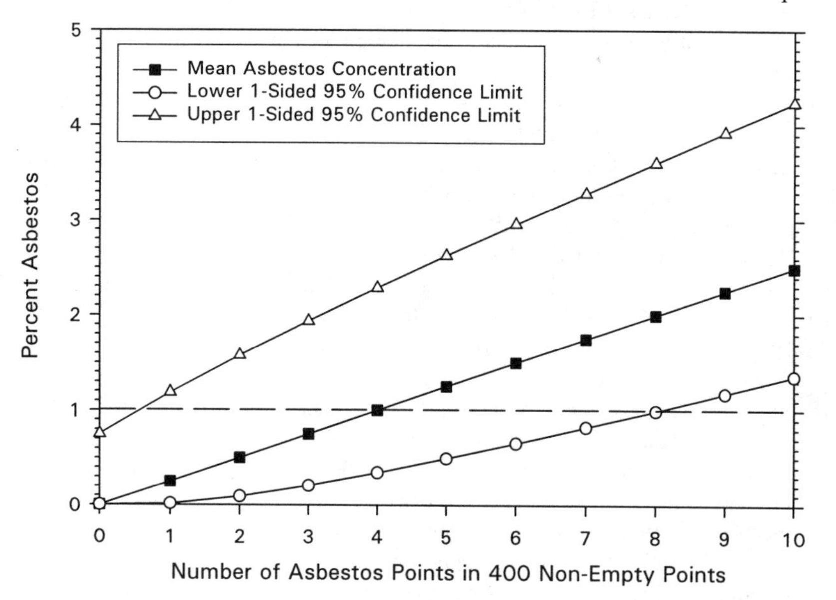

Figure 1--*One-sided 95% confidence limits for a 400-point count*

detected in a count of 400 non-empty points. The regulator should assume the lower 95% confidence limit value, to ensure that 95% of repeat measurements are predicted to exceed 1%. Conversely, the "regulated" should assume the upper 95% confidence limit value, to ensure that 95% of repeat measurements will predict concentrations lower than 1%. Clearly, for a sample examined without any matrix reduction, on the basis of a count of 400 non-empty points, compliance with a regulated value of 1% can only be demonstrated if no asbestos points are detected. If 1, 2, 3 or 4 asbestos points in 400 non-empty points are detected, there is a significant and progressively increasing probability that a repeat measurement will indicate that the concentration of asbestos is higher than 1%. Similarly, it would be unwise for a regulatory body to prosecute "the regulated" on the basis of fewer than 9 asbestos points in the 400-point count, because below this number there is a significant probability that a repeat measurement will show that the concentration of asbestos is lower than 1%. Therefore, using the 400-point count on a sample without matrix reduction, "the regulated" cannot prove compliance with a 1% asbestos regulation unless no asbestos points are detected, and the regulator could face challenges if prosecution is attempted below a measured asbestos concentration of approximately 2.3%. The EPA 400-point count, even from a theoretical standpoint, does not provide a legally-defensible basis for enforcement of a 1% asbestos regulation when the estimated asbestos concentration is below approximately 2.3%, and possibly a much higher concentration when obscuration of fibers by the matrix materials is a significant effect, or there are significant differences in the densities of constituent particles. Comparisons against a 0.1% asbestos regulation can be made if the vertical axis in Figure 1 is taken to range from 0 to 0.5%, and the number of asbestos points on the horizontal axis refers to a count of 4000 non-empty points.

In order to use compliance criteria similar to those which are conventionally applied in other fields of measurement, such as that used in determination of fiber concentrations by phase contrast microscopy, it is necessary to establish an analytical protocol which accounts for the statistical variability of the measurement method. Figure 2 shows the relationship between the calculated mean concentration of asbestos for which the 95% upper confidence limit corresponds to 1% and the number of asbestos points found in the point count. Measurements based on point counting, corresponding to the region below the curve, would be in compliance with a 1% asbestos regulation, at 95% confidence. For measurements indicating a mean asbestos concentration lower than 1%, but falling above the curve in Figure 2, there is more than a 5% probability that a repeat measurement will indicate that the asbestos concentration is higher than 1%. For example, if the mean asbestos concentration in a sample is determined to be 0.7% after 23 asbestos points have been counted, Figure 2 shows that compliance with a 1% asbestos regulation will become unambiguous at the 95% confidence level. However, this would have required a point count of 3286 non-empty points. Counting of such a large number of points would be very time-consuming and expensive, and it is also likely that the effects of operator fatigue would compromise the validity of the analysis. Figure 3 shows the corresponding curve for determining if a calculated mean asbestos concentration is higher than 1% at 95% confidence. Measurements based on point counting, corresponding to the region above the curve, would be out of compliance with a 1% asbestos regulation. For measurements indicating a mean asbestos concentration higher

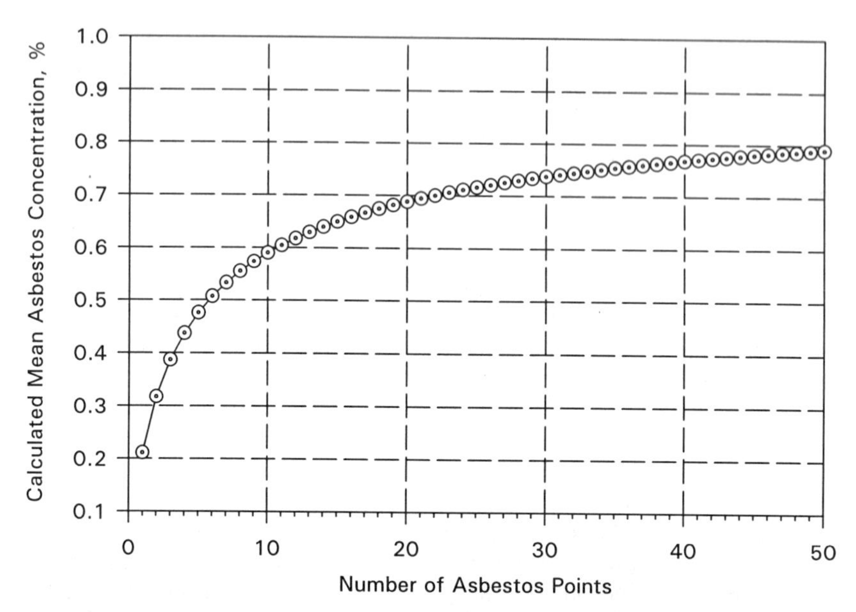

Figure 2--*Maximum calculated mean concentration of asbestos for which the 95% upper confidence limit of the concentration is lower than 1% for a given number of asbestos points*

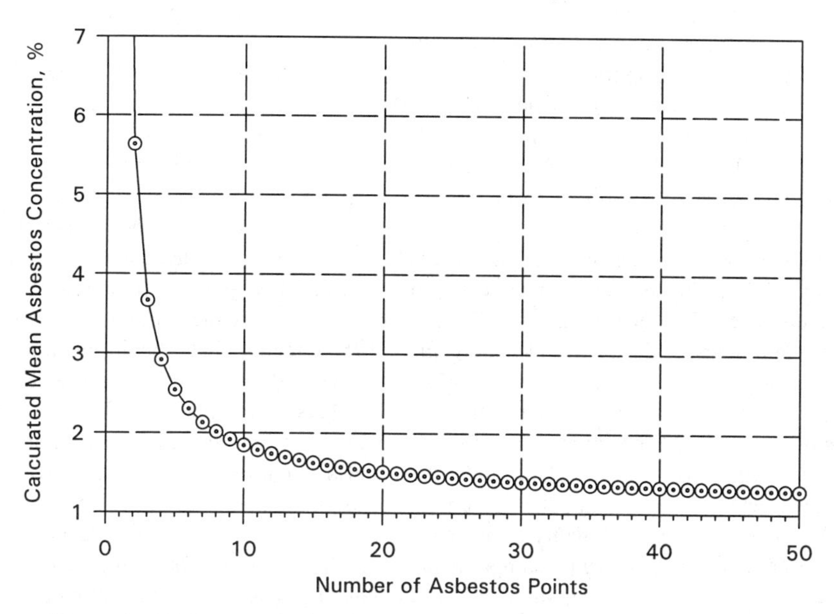

Figure 3--*Minimum calculated mean concentration of asbestos for which the 95% lower confidence limit of the concentration is higher than 1% for a given number of asbestos points*

than 1%, but falling below the curve in Figure 3, there is more than a 5% probability that a repeat measurement will indicate that the asbestos concentration is lower than 1%. Measurements which fall above the curve in Figure 2 but are below the curve in Figure 3 are ambiguous, and at 95% confidence cannot be determined to be above or below 1%. Similar arguments apply to a 0.1% asbestos regulation, if the values on the vertical axes of Figures 2 and 3 are divided by 10.

This paper describes a complete analytical method which has been validated for analysis of low concentrations of asbestos in a wide range of bulk sample matrices. The method, which is based on a combination of gravimetric matrix reduction and a new point counting protocol, can provide legally-defensible compliance data for the majority of bulk building material samples in which the asbestos concentration is close to 1% or 0.1% by weight.

Use of Gravimetric Matrix Reduction to Improve Accuracy and Precision

Gravimetric matrix reduction, followed by point counting of the residue from this procedure, reduces the effort required on point counting, and usually allows unambiguous compliance decisions to be made. Assuming ideal specimen preparation and a truly random dispersion of particles and fibers, the statistical precision of a point counting result depends only on the number of asbestos points detected. Gravimetric procedures which selectively remove non-asbestos matrix materials have the effect of reducing the number of non-empty points which must be examined in order to detect a specific number of asbestos points. Thus, in the example above of a mean asbestos concentration of 0.7%, if the residue remaining after removal of non-asbestos matrix components was 10% of the original material, 23 asbestos points would have been detected after only 329 non-empty points had been examined.

Matrix reduction procedures have the following effects which result in significant improvements in the accuracy and precision obtainable by a reasonable amount of point counting:

(a) ashing removes any organic components. These generally have low densities which compromise the accuracy of mass determination by point counting;

(b) acid dissolution removes acid-soluble constituents;

(c) sedimentation removes large, non-asbestos particles and fragments of sizes which otherwise compromise the preparation of satisfactory microscope slides, can obscure smaller particlesand fibers, can result in slides non-representative of the original material, and compromise the accuracy of mass determination by point counting;

(d) the overall matrix reduction procedure, depending on the nature of the sample, increases the sensitivity of the analysis so that measurements of low asbestos concentrations become statistically meaningful;

(e) the overall matrix reduction procedure results in homogenization of the fractions of the material which will contain any asbestos present;

(f) the overall matrix reduction procedure permits examination of a much larger amount of the bulk material than could be mounted on a reasonable number of microscope slides, and ensures that the sample examined microscopically is representative of the bulk material.

Summary of the Analytical Method

This method is appropriate for quantification of the asbestos in materials in which no asbestos or a low concentration of asbestos has been observed during routine PLM examination. The method is based on gravimetric matrix reduction, incorporating oxidation of organic materials, dissolution of acid-soluble components, and removal of large particle size materials by sedimentation or flotation, followed by analysis of the remaining material using a well-defined point counting procedure optimized for determination of mass concentration.

A known weight of the sample is heated in a muffle furnace to oxidize any organic materials. The temperature of the muffle furnace is controlled so that any asbestos present in the original material will still be identifiable after the treatment. The residual ash is then treated with dilute hydrochloric acid to dissolve soluble species such as gypsum, carbonates, cement and mineral wool. After the acid dissolution procedure, large and heavy fragments are allowed to settle, large and light fragments are allowed to float, and the remaining suspension is filtered on to a pre-weighed polycarbonate filter. The floating material, the sedimented material and the deposit on the polycarbonate filter are each examined for the presence of large asbestos fiber bundles. Any large asbestos fiber bundles detected are hand-picked from the sediment, from the floats, and from the filter, and these are combined and weighed. The deposit on the polycarbonate filter is also examined for large particles, and any detected are removed and added to the sedimented material. The final residue after all pre-treatment is examined by PLM and a defined point counting procedure is used to quantify the asbestos.

Computer modelling studies of the application of point counting to logarithmic-normal particle diameter and fiber diameter distributions have shown that the majority of the integrated volume of a particle or fiber species is contributed by a minority of particles or fibers with the largest diameters. In practice, after matrix reduction most of the volume, and therefore most of the weight, of particles in a logarithmic-normal size distribution can be accounted for by considering only those particles larger than approximately 10% of the largest particle detected. In the case of fibers with logarithmic-normal diameter and length distributions, the majority of the integrated volume is accounted for by fibers with diameters larger than approximately 20% of the maximum diameter fiber detected. The point counting protocol used in this method is based on these observations.

For comparison against a regulatory standard of 1% asbestos by weight, point counting is continued until either a minimum of 20 asbestos points or the **equivalent** of 1300 non-empty points on the untreated sample have been accumulated. For comparison against a regulatory standard of 0.1%, point counting is continued until either a minimum of 20 asbestos points or the **equivalent** of 13000 non-empty points on the untreated sample have been accumulated. For either of these regulatory standards, the **actual** number of non-empty points required depends on the degree of matrix reduction achieved during the chemical and sedimentation procedures, assuming that fewer than 20 asbestos points are encountered.

Analytical Method

Recording of Data

Use a form as illustrated in Figure 4 to record all gravimetric data and the results of point counting.

Select a Representative Sub-Sample

The amount of bulk material to be used for the analysis is optional. However, the initial weight should not exceed the capability of the acid and water to dissolve all of the soluble components. It is also important that the residue remaining after all of the separation steps can be weighed with sufficient accuracy (approximately 1% accuracy is desirable). Few bulk materials result in final residues lower than 2% of the original sub-sample weight. Therefore, an initial weight of 0.5 gram (g) will generally yield a residue weight exceeding 10 mg, which can be measured to 1% accuracy using a balance with 0.1 mg readability. For most analyses relating to a 1% asbestos criterion, a starting weight of 0.5 g is sufficient. If the results are to be compared with a 0.1% asbestos criterion, the starting weight should be higher if possible, and a weight of approximately 5 g is usually suitable. Use of a higher starting weight ensures that the effects of inhomogeneity in the distribution of asbestos in the original material are reduced, and these inhomogeneities generally become more serious at lower asbestos concentrations.

Take a representative sub-sample from the original sample. In order that the ashing and chemical treatment will be efficient, it may be necessary to lightly crush the representative sub-sample before weighing. Since separation of major proportions of the non-asbestos components by sedimentation is an essential feature of this method, it is important not to reduce the particle sizes any more than absolutely necessary. Sub-samples from samples received as pulverized material should be taken from the original sample by cone and quarter methods.

Ash the Sub-Sample

If it is determined that the organic content of the material is not significant, then the ashing step may be omitted. In this case, the sub-sample is weighed and taken directly to the acid dissolution step. However, if the sample contains some organic materials such as starch, omission of the ashing step may result in very slow filtration.

Label a glazed porcelain or fused silica crucible with a heat-resistant marker, and weigh it. Place the sub-sample of the bulk material, after crushing if necessary, into the

ASHING, SEDIMENTATION AND ACID EXTRACTION GRAVIMETRY

SAMPLE: ____________________	SAMPLE NO: 97M999-2
____________________	DATE: 1997-05-11
____________________	ANALYST: AL

INITIAL WEIGHTS			
Weight of Weighing Dish	18.74890		
Weight of Weighing Dish + Sample	23.30790		
Weight of Sample	4.55900		
ASHING			
Weight of Crucible	18.74890		
Weight of Crucible +Ash	23.03880		
Weight of Ash	4.28990		
Weight Loss During Ashing	0.26910		
Percent Organic Materials and Water	5.90261		
SEDIMENTATION/FLOTATION			
Weight of Sediment/Floats Container	3.77160		
Weight of Container + Sediment/Floats	6.84400		
Weight of Sediment/Floats	3.07240		
Percent Sediment/Floats	67.39197		
ACID TREATMENT			
Weight of Filter	0.01650		
Weight of Filter + Residue (After Hand-Picking)	0.11250		
Weight of Residue (After Hand-Picking)	0.09600		
Percent Acid-Soluble Materials	24.55363		
Percent Residue (After Hand-Picking)	2.10572		
Type of Asbestos	Chrysotile		
Weight of Container for Hand-Picked Asbestos	0.06240		
Weight of Container + Hand-Picked Asbestos	0.06450		
Weight of Hand-Picked Asbestos	0.00210	0.00000	0.00000
Percent Hand-Picked Asbestos	0.04606	0.00000	0.00000
POINT COUNT			
Total Points Counted	268	268	268
Asbestos Points Counted	41		
Point Count Lower 95% Confidence Limit	29.4210		
Point Count Upper 95% Confidence Limit	55.6220		
Asbestos in Sample (Weight Percent)	0.3682	0.0000	0.0000
Lower 95% Weight Concentration Limit	0.2772	0.0000	0.0000
Upper 95% Weight Concentration Limit	0.4831	0.0000	0.0000

COMMENTS

REPORT
ASBESTOS
OTHER COMPONENTS

Figure 4--*Example of form for data recording and calculation of results* crucible, and weigh it again to obtain the weight of the sub-sample. Cover the crucible with a crucible lid and place it in a muffle furnace at a temperature of 480±10°C for a minimum period of 10 hours.

Weigh the Sub-Sample After Ashing

Remove the crucible from the muffle furnace and allow it to cool to room temperature. Remove the crucible lid and weigh the crucible with its contents.

Dissolve the Acid-Soluble and Water-Soluble Components

For a sub-sample of approximately 0.5 g initial weight, transfer the contents of the crucible (or the entire sub-sample if no ashing treatment is required) to a 250 mL Erlenmeyer (conical) flask. Add 100 mL of dilute (2N) hydrochloric acid (HCl) and a polytetrafluoroethylene-coated magnetic stirring bar. Place on a magnetic stirrer for a period of 15 minutes. For larger sub-samples, it may be necessary to increase the volume of dilute acid and the size of the flask used, depending on the composition of the sample.

Weigh Polycarbonate Filter and Label Petri-dish

Weigh a 47 mm diameter, 0.4 μm pore size polycarbonate filter, and place it into a labelled plastic 50 mm diameter petri-dish.

Assemble Vacuum Filtration Systems

Assemble a 47 mm diameter vacuum filtration system and install the pre-weighed, 47 mm diameter, 0.4 μm pore size polycarbonate filter.

Assemble a 25 mm diameter vacuum filtration system and install a 0.45 μm nominal porosity mixed cellulose ester (MCE) filter.

Separate the Sedimented, Floating and Suspended Particles

This step is critical and requires some manual dexterity. The objective is to keep any asbestos fibers suspended in the liquid, while allowing the heavy components such as sand or aggregate to settle out, or light components such as perlite or vermiculite to float.

1. Remove the flask from the magnetic stirrer. A clean 1000 mL beaker is used to allow the acid suspension to be diluted and made up to a known volume.

2. If no floating particles are present, proceed to Step 4. If floating particles are present, add distilled water until the floating fraction is brought to the rim of the Erlenmeyer flask. Using a clean spatula, remove as much of the floating fraction as possible from the surface of the liquid, adding more water if needed, and place the floats in a pre-weighed plastic 50 mm petri-dish. Place

the petri-dish on a slide warmer at a temperature of approximately 55°C and allow the material to dry.

3. If Step 2 is used, decant approximately 50% of the supernatant liquid into the clean 1000 mL beaker.

4. Swirl the liquid in the flask with a circular motion to re-suspend any settled particles. Hesitate until the larger particles have settled. If no particles settle, pour all of the suspension into the 1000 mL beaker and proceed to Step 7.

5. If particles have settled, decant most of the supernatant liquid into the 1000 mL beaker. Add approximately 150 mL of distilled water to the Erlenmeyer flask and repeat the sedimentation and decanting of supernatant liquid. Repeat again with an additional 150 mL of distilled water.

6. Using a wash-bottle with distilled water, wash out the sedimented material from the Erlenmeyer flask into a glass 100 mm petri-dish. Decant the water into the 1000 mL beaker, and then place the petri-dish on to a slide warmer and allow the sediment to dry. Drying of the sediment may be accelerated by rinsing it with a small volume of ethanol or methanol, and decanting the ethanol or methanol into the 1000 mL beaker.

7. Add distilled water to the beaker to bring the volume to an appropriate value. This value is dependent on the nature of the sample, but should be sufficient to allow pipetting of aliquots for preparation of point counting filters without the need for serial dilution.

Prepare MCE Filters to be Used for Point Counting

Thoroughly disperse the particulate in the suspension in the 1000 mL beaker by stirring and blowing air through a pipette into the suspension. Withdraw an aliquot for filtration through the 25 mm diameter MCE filter. The particulate loading on the filter must be suitable for point counting. In general, aliquots of between 0.5 mL and 5 mL have been found to be suitable. Before filtration, dilute the aliquot to a volume of more than 5 mL with distilled water to ensure a uniform particulate deposit on the filter. Prepare a minimum of 4 filters from 4 separate aliquots. Place the filters into a labelled petri-dish, so that they are flat on the bottom of the dish and not touching each other, and allow them to dry on a slide warmer.

Filter the Suspension

Filter the balance of the suspension through the pre-weighed 0.4 μm pore size polycarbonate filter. Remove the polycarbonate filter from the filtration system and place it into its labelled petri-dish. Place the petri-dish on to the slide warmer and allow the filter to dry.

Separate Large Fiber Bundles for Weighing

Before weighing the sedimented and the floated materials and the polycarbonate filter with the deposit, examine them using a stereo-microscope. Using forceps, remove any large asbestos fiber bundles visible in the sedimented and the floated materials and place the fiber bundles into a pre-weighed container (a small, formed piece of aluminum foil is satisfactory). Examine the deposit on the polycarbonate filter for any large particles or asbestos fiber bundles, which would make sub-sampling of the filter deposit unrepresentative. Remove any large particles or fibers of non-asbestos material and add them to the sedimented material, and remove any large asbestos fiber bundles and add them to the pre-weighed container with any asbestos fiber bundles removed from the sedimented or floated materials. Transfer the sedimented material to a pre-weighed labelled plastic 50 mm petri-dish.

Weigh the Separated Fractions

Weigh the floated material (if applicable), the sedimented material (if applicable), the hand-picked asbestos fibers (if applicable) and the filter with the final residue deposit.

Identify the Asbestos Species Present

Before the point counting is carried out, it is necessary to determine whether any fibers present are asbestos. Use PLM methods [*3,4*] to identify suspect fibers. This can be done using either fibers selected from the untreated sample, if fibers can be seen in the original sample under the stereo-microscope, or fibers taken from the final residue deposit on the polycarbonate filter. It should be noted that the refractive indices of chrysotile fibers taken from the final residue will have been reduced by the acid treatment, but these fibers will still exhibit low birefringence, a positive sign of elongation and characteristic wavy morphology. The refractive indices of amosite or crocidolite may have been increased by the muffle furnace treatment, and partial oxidation of these fibers may have occurred, with some changes in color. In some cases, it will not be possible to identify suspect asbestos fibers by PLM, such as fibers of mixed crystallographic structure found in talc from some sources, and it will be necessary to confirm the identification of such fibers by TEM examination.

Alternative Rapid TEM Procedure if No Asbestos Fibers are Detected

If no suspected asbestos fibers are observed during the gravimetric reduction procedures, and if no suspected asbestos fibers can be seen in the final residue on the polycarbonate filter under the stereo-microscope, it is possible that either asbestos was not present in the original sample, or any asbestos fibers present are very small. In this case, the most expedient procedure to demonstrate a "none detected" or "trace asbestos" result is to examine a sub-sample of the residue on the polycarbonate filter by transmission electron microscopy (TEM), using a drop-mount specimen preparation technique. In this preparation technique, a small amount of the residue on the polycarbonate filter is dispersed in ethanol, and approximately 3 μL of the suspension is placed on a carbon-coated TEM specimen grid and evaporated to dryness. Tests have shown that as little as 0.1% chrysotile **in the final residue** can be detected by this TEM method, and this often allows an estimate of the maximum amount of asbestos which could be in samples in which either no asbestos is detected or trace levels of asbestos are detected.

Prepare Microscope Slides for Point Counting

Microscope slides for point counting can be prepared either by clearing the MCE filters, or from the residue collected on the polycarbonate filter. In practice, it has been found more satisfactory to use the MCE filters for the point counting, because it is sometimes difficult to redisperse material removed from the polycarbonate filter, particularly if the residue on the filter contains substantial quantities of chrysotile.

At least 4 slides are used for point counting. No more than 250 non-empty points are counted on each slide. The total number of non-empty points to be counted for comparison of the result with a 1% asbestos regulatory standard is calculated from the formula 13 x P, where P is the weight of residue on the filter expressed as a percentage of the weight of the original sub-sample. Similarly, for comparison with a regulatory standard of 0.1%, the number of non-empty points to be counted is 130 x P. Point counting may be terminated after accumulation of 20 asbestos points, but the non-empty points counted must be distributed approximately equally on each of the slides used. Regardless of the above, a minimum of 100 non-empty points is counted, even if more than 20 asbestos points are encountered. If the calculation indicates that more non-empty points are required after counting the maximum number of points on the slides prepared from the MCE filters, additional slides must be prepared from the residue on the polycarbonate filter.

Prepare microscope slides according to **either** (a) or (b).

(a) Place approximately 100 μL of a mixture of 35% dimethylformamide, 15% glacial acetic acid and 50% distilled water on a clean 75 mm x 25 mm microscope slide. Gently lower the edge of one of the MCE filters so that it touches the liquid, and lay it down on to the liquid so that there are no bubbles underneath it. Absorb any excess liquid from the slide using the edge of a paper towel. Repeat the procedure with the other MCE filters. Place the slides on a slide warmer at a temperature of between 65°C and 70°C for a period of 10 minutes. Remove them from the slide warmer. Place approximately 100 μL of Triacetin on each of the filters, and apply 22 mm x 22 mm cover slips. The slides are ready for point counting. When using this preparation method, the refractive index of the immersion medium is approximately 1.45. It is important to confirm that **all particle species present as significant proportions of the residue weight** are visible under the illumination conditions used for point counting. If it is found that any major particle or fiber species in the residue has refractive indices such that they are not visible, an alternative method of preparing slides must be used.

(b) Grinding of residues from the gravimetric reduction procedure should not be necessary, because liberation of the fibers from the matrix material during the ashing, acid dissolution and subsequent filtration should have resulted in a relatively uniform and homogeneous deposit on the polycarbonate filter, and any large particles should have already been removed in the flotation, sedimentation and hand-picking of large particles and large fiber bundles. The refractive index liquid to be used for point counting must be selected such that **all particle species present as significant proportions of the residue weight** are visible under the illumination conditions used for point counting. For the majority of samples containing chrysotile, amosite or crocidolite, a liquid of refractive index 1.605 has been found to be suitable, but occasionally it is found that a major particle species present cannot be reliably seen in this liquid. If this is the case, select another refractive index liquid. In many cases, the filtered deposit can be separated from the polycarbonate filter by curling back the filter using forceps. A representative sub-sample of the deposit can then be broken off using forceps and placed on to a clean 75 mm x 25 mm microscope slide. In other cases, the deposit may not be self-cohesive, and it will then be necessary to cut a representative area from the filter and then, using the edge of a scalpel blade, to scrape the deposit off the filter on to the microscope slide. After placing the sub-sample of the residue on to the microscope slide, apply one drop of the refractive index liquid to it. Place a second microscope slide parallel to the first one, but rotated at 90° to it, on to the area occupied by the refractive index liquid on the first slide. Press the slides together lightly, and rub them backwards and forwards parallel to each other in a rapid shearing motion. The liquid can then be moved back to the middle of the first slide using the edge of the second slide, and the process repeated several times. This procedure usually produces a uniform dispersion, and a 22 mm x 22 mm cover slip should then be placed on the dispersed deposit on the first slide. If the particle density is too high, the process can be repeated after the addition of more refractive index liquid, and some of the suspension can be removed and used to prepare additional slides. With a little practice, the amount of residue necessary to produce slides of acceptable particle density can be judged reliably. In some samples, the particles and fibers tend to re-aggregate after dispersal, but this does not appear to affect the point counting procedure. The number of slides to be prepared depends on the degree of matrix reduction achieved and the concentration of asbestos to be measured. Use a minimum of 4 slides.

ficroscope Adjustment

Use of the correct illumination conditions and checking of particle visibility are ·itical to the validity of the point count.

(a) Set up the PLM with crossed polars and a 530 nm plate in position.

(b) Select an objective to provide a total magnification of approximately 100.

(c) Adjust the illumination system to provide the maximum illumination, and close the sub-stage aperture as much as possible to provide maximum contrast.

(d) Use an eyepiece containing a cross-hair with scale divisions on at least one of the directions. Do **not** use a Chalkley reticle for the point count.

(e) Examine one of the prepared slides to determine if all of the particle and fiber species have sufficient contrast under the illumination conditions. Pay particular attention to isotropic particle and fiber species. If all particle and fiber species can readily be seen, proceed with the point count. If any one of the particle or fiber species exhibits insufficient contrast, select an alternative refractive index liquid and prepare new slides from the final residue deposit on the polycarbonate filter.

Perform the Point Count

The following point counting criteria ensure that the result is based on statistically-valid numbers of particles and fibers, and that the sizes of the particles and fibers included in the measurements are those which contribute the majority of the weight.

(a) Scan all of the slides to estimate the approximate diameters of the largest particle and the largest fiber.

(b) Carry out the point count.

1. For particles, record only those points which occur on particles exceeding 10% of the diameter of the largest particle detected in the initial scan.

2. For all fibers (asbestos and non-asbestos), record only those points which occur on fibers whose diameters exceed 20% of the diameter of the largest fiber detected in the initial scan. An exception to this rule must be made if a species of fiber is present which has a narrow range of diameters (for example, glass fibers) somewhat lower than the minimum fiber diameter criterion, and this fiber species constitutes a significant proportion of the residue. In such a case, it has been found satisfactory to adopt a minimum fiber diameter criterion (for all fibers) equal to the minimum diameter of the species with the narrow range of diameters.

3. When a point falls on a region where a particle and fiber overlap, record one point for each, provided that each is of sufficient diameter to record in accordance with 1 and 2.

4. When a point falls on an overlapped region of two fibers, each of sufficient diameter, record as one point for each.

5. When a point falls on an overlapped region of two particles, each of sufficient diameter, record as one point for each.

6. When a point falls on one segment of a split fiber, record it only if the diameter of the segment under the point meets the minimum 20% diameter criterion specified in 2.

7. Count no more than 250 non-empty points on each slide.

8. Count approximately an equal number of non-empty points on each slide used.

9. Continue the point count until at least 20 asbestos points have been recorded, or until the appropriate stopping point of either 13 x P non-empty points, for comparison of the result with a 1% asbestos regulatory standard, or 130 x P non-empty points, for comparison with a 0.1% asbestos regulatory standard, have been recorded, where P is the weight of the final residue on the polycarbonate filter expressed as a percentage of the original sub-sample weight. Regardless of the above, use a minimum of four slides and count a minimum of 100 non-empty points, distributed approximately equally on each slide used.

Calculate the Results

The percentage of asbestos is calculated from the formula:

$$\%\ Asbestos = 100\,(H + RA/N)/W$$

where:

H = weight of hand-picked asbestos (grams),
R = weight of residue on polycarbonate filter (grams),
A = number of asbestos points counted,
N = number of non-empty points counted,
W = weight of original sub-sample (grams).

For automatic calculation, all of the gravimetric and point counting data may be entered into either a spread sheet or a table containing embedded formulae in a word processing program.

Procedure for Interpretation of Results

For regulatory purposes, it is usually considered satisfactory if a compliance level is selected such that 95% of repeat measurements are predicted to comply with the regulation. The point counting procedures specified in this gravimetric/point counting method are designed to allow comparison of the analytical results with an asbestos regulatory standard of 1% or 0.1%, such that the same compliance decision will be predicted for more than 95% of repeat measurements. The minimum proximity of a result to the regulatory standard for which compliance can be demonstrated is determined by the number of **asbestos** points counted, and is unrelated to the total number of non-empty points counted. For this reason, a stopping rule of 20 asbestos points is specified in the method, this being a reasonable limit for routine analyses, above which further improvement of the precision becomes progressively slower.

After completion of the analytical work, determine where the point represented by the calculated percent asbestos and number of asbestos points counted occurs on either Figure 2 or Figure 3 (Figures 2 and 3 may also be used for comparison with a standard of 0.1% asbestos by simply dividing the vertical axis by 10). If the result is outside of the statistically-ambiguous region, the analysis is complete and the data may be reported. If the result is inside the statistically-ambiguous region, determine whether a reasonable amount of additional point counting, to accumulate additional asbestos points, would be likely to move the data point outside of the ambiguous region, and perform additional point counting if this is the case. If the result is likely to remain in the ambiguous region, report the calculated concentration and indicate the statistical limitations of the result.

Method Validation and Inter-Laboratory Data

A number of reference samples were prepared in the laboratory using known low concentrations of asbestos in a variety of matrices selected to simulate actual building materials. The samples were non-trivial, in that the non-asbestos components could not be completely removed by the gravimetric procedure, thus necessitating the use of point counting. The asbestos concentrations in these samples ranged from approximately 0.1% up to 6%. The majority of the samples contained chrysotile, because this is the most common asbestos variety encountered in materials containing asbestos at low concentrations. Other samples contained either amosite, crocidolite or tremolite. Figure 5 shows the analytical results obtained by Chatfield Technical Consulting Limited, plotted against the formula weight concentrations. A linear regression shows a correlation coefficient of 0.979, with a slope of 0.956. There is, however, a small positive bias which decreases as the concentration increases. This bias is approximately 20% in the 0.1% - 1.0% range, decreasing to almost zero at 10%, and the cause is not understood at this time. The same samples were also analyzed, using this method, by the Optical Microscope Laboratory of the Environmental Protection Department of the Government of Hong Kong (EPD), and these results are given in Figure 6. The EPD analyses show a closely similar relationship with the known weight concentrations of asbestos, with a correlation coefficient of 0.968 and a slope of 0.907. A comparison of the inter-laboratory data is given in Figure 7. Each data point in Figure 7 represents

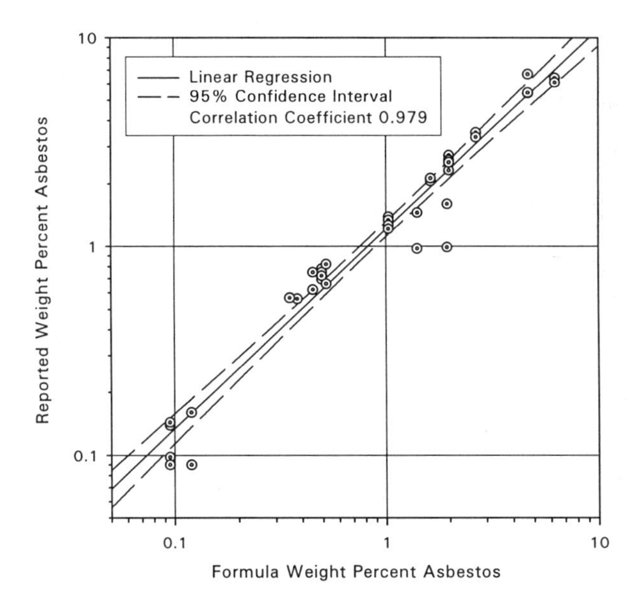

Figure 5--*Analyses of reference samples by Chatfield Technical Consulting Limited*

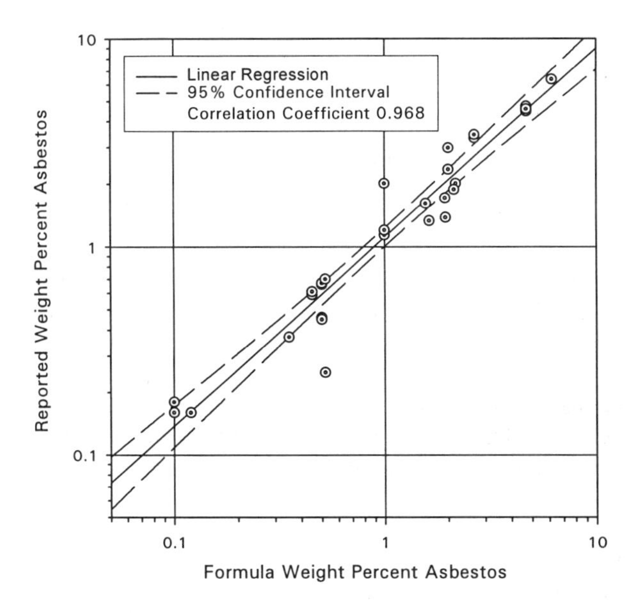

Figure 6--*Analyses of reference samples by the Hong Kong Environmental Protection Department*

either a single measurement or a mean of up to 4 results, wherever repeat measurements were made by either laboratory. The correlation coefficient of this inter-laboratory study is 0.971, with a slope of 0.929.

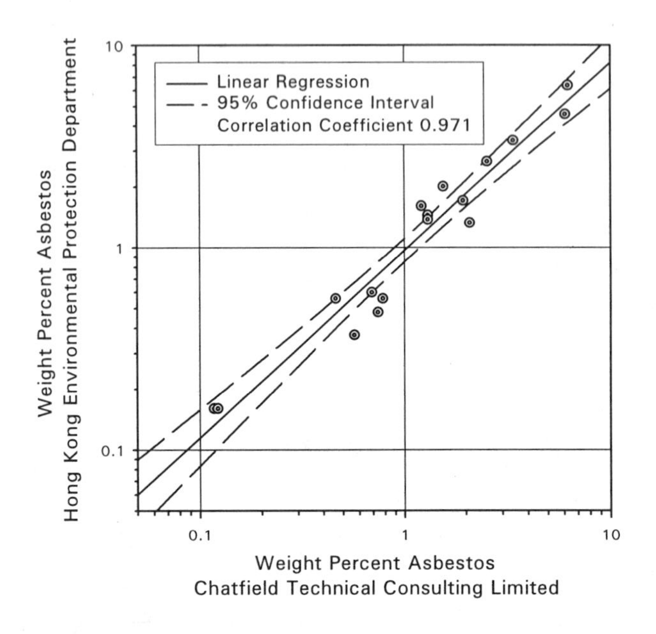

Figure 7--*Results of inter-laboratory analyses of reference samples*

Limitations of Accuracy and Precision

Inhomogeneity of the Original Sample

If the nature of the sample is such that, after the gravimetric matrix reduction, the hand-picked fraction consists of a small number of very large asbestos fiber bundles, then it is necessary to question:

(a) whether the sample and sub-sample were of sufficient size to be considered representative of the original material, since a statistically-unreliable number of these large bundles may constitute most of the weight of asbestos present; and,

(b) whether the weight of these few large fiber bundles is significant in terms of the final result and comparison with a control limit.

If, after the gravimetric matrix reduction, any very large asbestos fiber bundles are visible in any fraction, hand-picking of these large bundles and separate weighing of them is necessary. A representative portion of the final residue must also be analyzed by point counting. The concentration of asbestos can then be calculated from the measurements, with the proviso that the statistical relevance of the hand-picked material must be considered. If the weight concentration represented by a small number of hand-picked fiber bundles is significant in terms of the regulatory limit in use, the only recourse is to process a larger sub-sample of the original material, so that a statistically-valid number of the large fiber bundles is included. It would be possible to grind the residue from such a sample to homogenize the material for more accurate point counting, but this would not alter the fact that the original material was possibly inhomogeneous on the scale of the size of the original sample or sub-sample, and the **apparent** precision of the results would be very misleading. In some samples, it can be shown that the weight contribution made by the large fiber bundles represents a concentration significantly below the level of regulatory interest. In other cases, the contribution by the large fiber bundles to the measured concentration may always impose a statistical variability that causes the result to be ambiguous in terms of a regulatory control limit.

Asbestos Fibers with Diameters Below the Limit of Optical Resolution

Chrysotile originating from the New Idria region of California consists of large bundles and clusters of short fibrils, and only approximately 1% of these fibrils are longer than 10 µm. This variety of chrysotile disperses almost completely in water into isolated single fibrils, which are not within the range of optical visibility. Optical point counting cannot be used to determine chrysotile concentrations in materials which incorporate this variety of chrysotile. However, for analysis of such materials, the gravimetric procedures specified in this analytical method can be combined with a transmission electron microscopy (TEM) fiber counting protocol optimized for determination of mass [*5*].

Acknowledgements

The author wishes to express his appreciation to Mr. Benson Yeung, of the Hong Kong Environmental Protection Department (EPD), for permission to publish the analytical data produced by the EPD Optical Microscope Laboratory, and to Mrs. A. Liebert of Chatfield Technical Consulting Limited, for many hours of point counting during development of the protocol and in generation of the analytical data.

References

[*1*] U. S. Environmental Protection Agency, "National Emission Standards for Hazardous Air Pollutants", Federal Register, Vol. 55, No. 224, Tuesday, November 20, 1990, p. 48414 - 48433.

[*2*] Perkins, R. L., "Point-Counting Technique for Friable Asbestos-Containing Materials", *The Microscope*, Vol. 38, 1990, p. 29 - 39.

[3] U. S. Environmental Protection Agency, "Interim Method for the Determination of Asbestos in Bulk Insulation Samples", Report EPA-600/M4-82-020, December 1982.

[4] U. S. Environmental Protection Agency, "Method for the Determination of Asbestos in Bulk Building Materials", Report EPA/600/R-93/116, National Technical Information Service, Springfield, Virginia 22161, July 1993.

[5] International Organization for Standardization, "Ambient air - Determination of asbestos fibres - Indirect-transfer transmission electron microscopy method", ISO 13794, 1999.

Peter Frasca[1], John H. Newton[2], and Robert J. De Malo[3]

"Internal Standard Addition" Method for Identifying Asbestos Containing Materials In Bulk Samples Containing Low Levels of Asbestos.

Reference: Frasca, P., Newton, J. H., and De Malo, R. J., **"Internal Standard Addition Method for Identifying Asbestos Containing Materials In Bulk Samples Containing Low Levels of Asbestos"**, *Advances in Environmental Measurement Methods for Asbestos, ASTM STP 1342*, M. E. Beard and H. L. Rook, Eds., American Society for Testing and Materials, West Conshohocken, PA, 2000.

Abstract

This study introduces a new method for identifying asbestos containing materials (ACM) containing a low percentage of asbestos ($\leq$ 10 %). Currently used Polarized Light Microscopy (PLM) and Transmission Electron Microscopy (TEM) methodologies rely on area ratio estimation to determine asbestos concentrations and have limitations at low asbestos levels.

The "Internal Standard Addition" (ISA) method is designed to reduce this deficiency. Specifically, a small portion of sample is weighed and to it is added an amount of asbestos different in type to that present in the sample and equal in quantity to a "density adjusted 1% by weight". The asbestos containing material (ACM) determination is made with PLM visual comparison of the two types of asbestos and if necessary, by a modified point counting technique. Compared to the current methods, this new technique yielded a significant increase in precision and accuracy and reduction of false positives and false negatives in the range of 0.25 % - 5 % asbestos.

Key Words: Asbestos, Internal Standard, PLM, TEM, Low Asbestos Levels

[1]President, EMSL Analytical, Inc., 107 Haddon Avenue, Westmont, NJ 08108.
[2]Laboratory Manager, Materials Science Division, EMSL Analytical, Inc., 107 Haddon Avenue, Westmont, NJ 08108.
[3]Industrial Hygienist and Director of Business Development, EMSL Analytical, Inc., 107 Haddon Avenue, Westmont, NJ 08108.

Introduction

The quantitation of low levels of asbestos (≤10 %) by PLM and TEM methods as currently used presents a problem to the asbestos laboratory industry, which (except for California) revolves around the Environmental Protection Agency's (EPA's) 1% asbestos content rule. Problems in quantifying asbestos by PLM were first reported in 1989, in a different study, by Frasca et al.[*1*]. The limitations of the PLM analyst to quantitate using the stratified point count method at low levels of asbestos was documented by Webber et al.[*2*], by Perkins and Beard [*3*], Harvey et al. [*4*] and Perkins et al. [*5*] in their studies of the analysis of prepared standards by both PLM visual area estimation "PLM (VAE)" and PLM point count "PLM (PC)" methods. The subsequent introduction of gravimetric reduction for non-friable samples, containing a matrix of organic binders and/or calcium carbonates, significantly reduced those difficulties resulting from the obscuration of asbestos fibers by matrix, and also improved the quantitation of asbestos in the remaining residue. The TEM gravimetric method "TEM (GR)" was first introduced by Chatfield [*6*] and later incorporated by New York State as its 198.4 method [*7*]. In an effort to reduce the number of samples that had to be analyzed by TEM, the PLM gravimetric method was first introduced by New York State as its method 198.1 [*8*]. Gravimetric methodology represents a considerable advancement in the field of asbestos determination.

Despite the introduction of these significant improvements, quantitation of low levels of asbestos still remains a problem. This fact was suggested by a previous study by these authors in which low levels of asbestos (ranging from 0 to 20%) in prepared standards were analyzed by PLM (VAE) [*9*], PLM (PC) [*10*] and TEM (GR) methods.

In this paper, we discuss the application of the "Internal Standard Addition" (ISA) method to the analysis of bulk samples containing low levels of asbestos. The addition of a standard to a sample or "spiking" is not a new concept in analytical procedures but was rarely, if ever, applied to the microscopical analysis of asbestos prior to its use in a 1990 study by Frasca et al.[*11*]. The analytical method used in that study utilized TEM and was rather laborious, requiring hours of analysis. For the current research we sought a faster and more practical method using PLM i.e., PLM (ISA)..

The advantage of the PLM over TEM lies in its ability to analyze much more material in an equal amount of time and thus obtain better statistical data. Its disadvantage compared to the TEM is of course its inability to resolve asbestos fibers below 0.25 μm in diameter. Therefore, as long as the predominant asbestos mass content in the sample is due to asbestos fibers visible with the PLM, the PLM has an inherent advantage over the TEM in the analysis of bulk samples.

Method

Using the PLM (ISA) method, samples containing low levels of asbestos can be identified as ACM (>1%) or Non-ACM (≤1%) by adding a finite, known amount of asbestos, different in type (the internal standard) to that present in the sample and by subsequently comparing the relative amount of the two types of asbestos to determine which one of the two is greater. For the EPA's 1% asbestos content rule, the internal standard is the 1% asbestos by weight (density adjusted) spike while for the California 0.1% asbestos content rule, the internal standard is the 0.1% asbestos spike. Thus the internal standard can be adjusted to satisfy whatever percent is called for in a particular regulation. In situations which may warrant extra care, multiple samples, each containing spikes of different percentages, can be prepared so as to determine an upper and lower value for the laboratory sample.

If the analyst, by simple visual analysis, is not able to determine with confidence which of the two types of asbestos is present in larger quantity, than he or she will proceed to a modified point counting technique in which only the two types of asbestos fibers are point counted separately. For this study, sample standards containing 0.1%, 0.25%, 0.5%, 0.75%, 1.0%, 1.25%, 1.5%, 1.75%, 2.0% and 4.0% by weight chrysotile asbestos were prepared and spiked with a 1% by weight (density adjusted) crocidolite asbestos and given to sixteen (16) analysts for analysis. The sample preparations and methods are described in detail below.

Preparation of Standards

Standards were prepared by mixing chrysotile asbestos with a matrix consisting mostly of fine clay and a small amount of quartz. The matrix was prepared from soil, which was ashed at 480°C, treated with HCL^- and sieved through a 63 μm mesh. It was shown to be asbestos free by both PLM and TEM examination. The ten (10) standards ranging from 0.1% to 4% chrysotile asbestos were prepared by mixing appropriate amounts of NIST traceable chrysotile asbestos fibers (which could be resolved with PLM) with the fine matrix particulates.

The following sample preparation section describes the steps taken for a laboratory sample, which may contain layers (unlike our standards) and which may contain asbestos of a type different from chrysotile asbestos, which is present in our standards. For the purpose of the PLM (ISA) study, the study samples were the ten (10) prepared standards containing chrysotile asbestos and the spike was crocidolite asbestos. Study samples were prepared by applying the steps listed below in the sample preparation section.

Sample Preparation

1. If the laboratory sample is layered, separate all sample layers as per EPA requirements [*12*] and perform PLM analysis to determine the type of asbestos present.
2. Reduce the sample mechanically as much as possible by grinding with mortar and pestle or by tearing and cutting with tweezers and scalpel.
3. Homogenize the sample with a steel spatula and place a minimum of 500 mg into a preweighed crucible.
4. Record the sample weight.

5. Add one percent (1%) by weight of an asbestos type not found during the initial PLM analysis of the sample. An adjustment in the weight of the spike must be made to account for differences in the specific gravities of the two asbestos types. For example; 10 gm of chrysotile with a specific gravity of 2.6 is the equivalent of 11.5 gm of crocidolite with a specific gravity of 3.0 by volume. This correction is performed in order to equalize the volume of the two types of asbestos since the analysis is being performed by visual means and relies on volume not weight. The following types of asbestos are recommended as spikes:

sample found to contain:	*spike with:*
Chrysotile	Crocidolite
Crocidolite	Chrysotile
Amosite, Tremolite or Actinolite	Anthophyllite
Anthophyllite	Amosite

6. Homogenize the sample with a glass stir rod or steel spatula taking care not to lose any material.
7. Place the sample into a muffle furnace set at 480°C.
8. Ash the sample until no measurable weight loss can be detected by re-weighing the sample.
9. Remove the sample and allow to cool.
10. Slowly drip concentrated HCl^- into the crucible until the sample is completely covered and allowed to sit until bubbling stops. *Stirring with a glass rod may facilitate the dispersion of the sample. Caution should be used at this point to avoid prolonged exposure to the HCl^- as magnesium leaching may result. Two minutes is usually sufficient to remove the majority of the acid soluble material without severe damage to the asbestos.*
11. Pour the sample into a beaker containing 100 to 200 ml of particle free, de-ionized water.
12. Rinse the crucible to remove any remaining particulates and place rinsate into the beaker with the sample.
13. Agitate the sample to break up clumped particles and pour into an industrial blender. *Care should be taken at this point. The purpose of this step is to separate any asbestos clumps and homogenize the sample without producing a significant number of fibers below the resolvable limit of the PLM.*
14. Blend the sample to homogenize the particles and reduce any large clumps. *A "drop mount" preparation[6] can be quickly prepared and checked by TEM to verify that the asbestos fibers of both types which are below the resolvable limit of the PLM(fine fibrils) do not contribute significantly to the asbestos mass content. If the fine fibrils for either type of asbestos are dominant in mass over the larger fibers visible with the PLM, then a TEM (ISA) method must be used. This work is still in progress and it will not be presented here.*
15. Vacuum filter the entire sample through a 0.45µm MCE filter using a vertical walled filter unit.
16. Dry the sample.

PLM Visual Analysis

17. Examine the sample by PLM for asbestos concentrations by blindly preparing several preparations onto a clean glass microscope slide. *Do not choose portions using the stereoscope or a bias may result.*
18. If the concentration of the sample asbestos is, by PLM visual estimate, clearly greater than or less than the concentration of the asbestos in the standard, then the analysis may be halted and the results reported as ≤1% or >1%, respectively.
19. If the concentrations are not easily distinguishable, proceed with the modified point counting method described below.

PLM Point Counting

19.1 Prepare a minimum of 10 preparations of the sample for point counting analysis taking the subsamples from various areas of the filter.

19.2 Point count only the asbestos fibers that fall on the superimposed reference points. *Ignore everything else, i.e. the non-asbestos particulates.*

19.3 Document the number of counts separately for the sample asbestos and for the spike asbestos and establish a ratio between them.

19.4 Count a minimum of one thousand (1000) non-empty points over the ten (10) preparations.

19.5 If the difference in counted points between the sample asbestos and the spike asbestos is greater than one percent, halt the analysis and record the results.

19.6 If the difference in counted points between the sample asbestos to spike asbestos is less than one percent, prepare five (5) additional preparations and count 500 additional non-empty points.

19.7 Continue following steps 19.5 and 19.6 until either a difference in counted non-empty points of at least one percent is achieved, or a total of 5,000 non-empty points have been reached.

19.8 Determine the ratio of counted points of the sample asbestos to those of the spike and multiply times 1% and report the result as ≤1% or >1%.

Results

The results obtained from the PLM (ISA) study are presented below and are also compared to results obtained from the previous study by the authors involving the use of PLM (VAE), PLM (PC) and TEM (GR). Compared to sixteen (16) analysts participating in the PLM (ISA) study, (generating from fifteen (15) to nineteen (19) data points for each standard) nine (9) analysts contributed to the TEM (GR) data, eighteen (18) analysts generated the PLM (VAE) data, and twenty-two (22) analysts contributed twenty-eight (28) data points. The error rate (false positives or false negatives) resulting from the

PLM (ISA) analysis of standards ranging from 0.1% to 4% is shown in Figure 1 to be much less than that obtained from the PLM (VAE), PLM (PC) and TEM (GR) studies of prepared standards ranging from 0-20% asbestos content. False positives are depicted by points falling on or to the left of the 1% vertical line and false negatives are shown to the right of the line. No "none detected" results were reported for the PLM (ISA) analyses and occurred infrequently in the other three methods. "None detected" was reported for only one case in the sixteen (16) PLM (VAE) analyses of the 0.5% standard, in one of the twenty-eight (28) PLM (PC) analyses for the same standard and in one of the eight (8) TEM (GR) analyses for the 1% standard. The PLM (ISA) data displayed in this graph and the subsequent graphs were generated by the modified point count technique.

Utilizing PLM (ISA) visual observation, the analysts were able to determine, without error, and with confidence, that the sample contained more or less asbestos than the 1% asbestos added as the internal standard only for the asbestos standards outside of the range of 0.25 % - 1.75 % asbestos. Within this narrow range the modified point count technique had to be used and although an error rate persisted it was less than that incurred from the other methods, as seen in Figure 1. In contrast, for the other three methods, false positives occurred well below 0.5% and false negatives were reported to occur up to 2% for TEM (GR) and beyond 5% for both PLM methods.

Error Rate vs. Asbestos Weight %

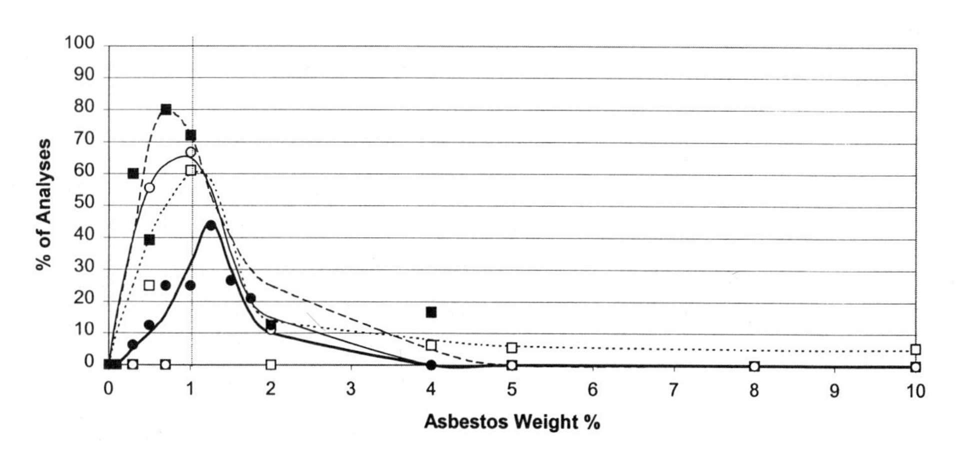

- - □ - - PLM (VAE) -- --■-- -- PLM (PC) ----o---- TEM (GR) ----●---- PLM (ISA)

Figure 1 – *Error Rates Consisting of False Positives and False Negatives for PLM (ISA), PLM (VAE), PLM (PC) and TEM (GR) Analysis as a Function of Asbestos Weight %*

On a prepared standard containing 1% by weight of chrysotile asbestos fibers, large enough to be visible with PLM, analysts reported a higher precision (Figure 2) and accuracy (Figure 3) for the PLM (ISA) method than for the PLM (VAE), PLM (PC) and

TEM (GR) methods. PLM (VAE) yielded reported values from "none detected" to 10% on the 1% known standard. PLM (PC) also yielded values from trace to approximately 10%, while TEM (GR), i.e. TEM analysis of gravimetric residues, generated results ranging from "none detected" to 15%. In contrast to these methods, PLM (ISA) yielded values for the sixteen (16) analysts ranging from 0.2% to 2.2%. The mean value for the 1% asbestos standard was 2.5% by PLM (VAE), 3.2% for PLM (PC), 3.8% for TEM (GR) and 0.95% of PLM (ISA).

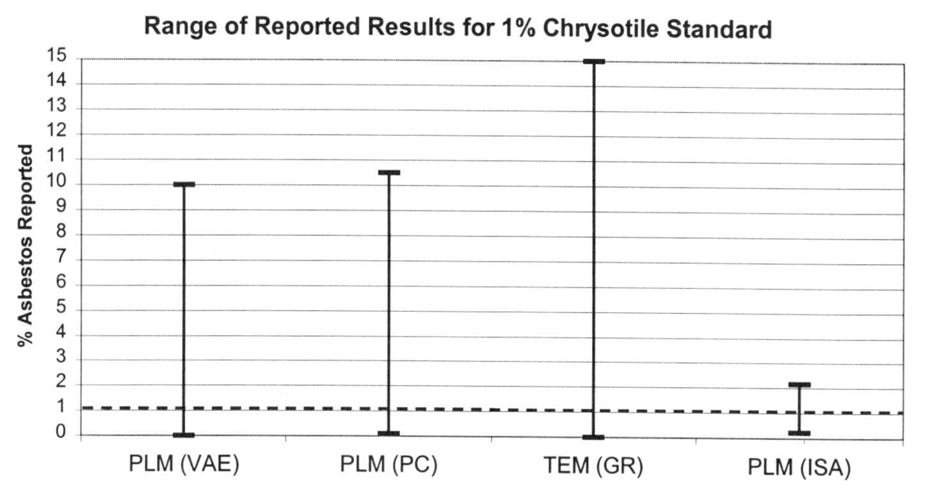

Figure 2 – *Upper and Lower Values Reported by Analysts for a 1% Chrysotile Asbestos Standard*

Note that for the PLM (ISA), most of the sixteen (16) analysts were reporting in the proximity of 0.5% to 1.0%. For PLM (VAE) and TEM (GR) most analysts reported values around 2% and 5%, respectively. The PLM (PC) results were spread out over the 0-15% range with a broad peak occurring over 2-10%. For all methods, except the PLM (ISA) method, the overestimation problem is evident. The PLM (ISA) method shows a slight underestimation as noted from the slight shift (≈0.25%) of the peak from the ideal 1% to 0.75%.

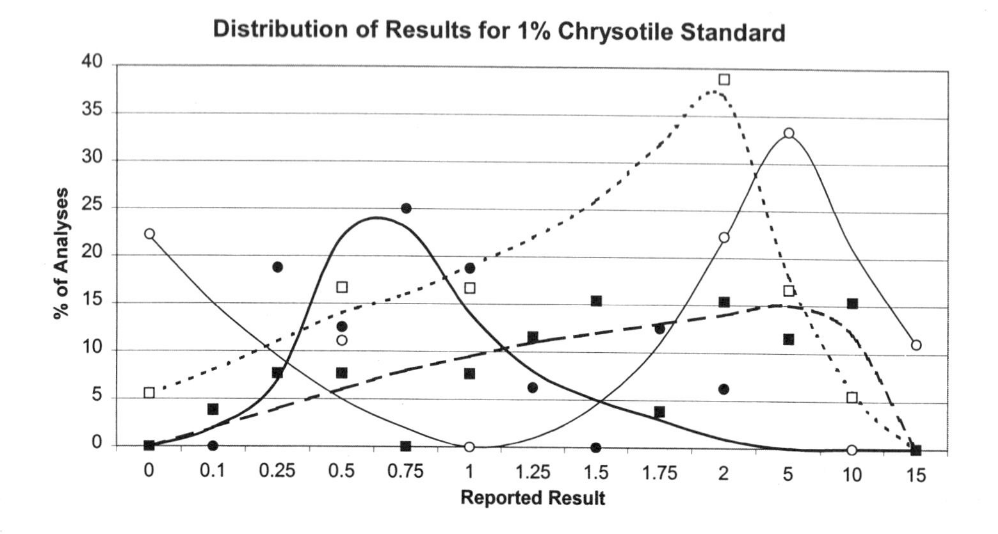

Figure 3 – *Range of Results Given by Analysts for a 1% by Weight Chrysotile Asbestos Standard (note: the "X" axis is non-linear)*

Discussion

For minimizing time and expense, PLM analysis is still the best screening method available for estimating the amount of asbestos in a bulk sample. However, this methodology should not be pushed beyond its inherent limitations, mainly its inability to resolve asbestos fibers with diameters less than the practical resolution limit of the PLM i.e., ≤0.25 μm and its inability to produce accurate and precise quantitation at low levels of asbestos, i.e., 0.25% to 5%. Both of these limitations often lead to false positives and false negatives, as far as the 1% rule is concerned, even in the hands of experienced PLM analysts.

To remedy this problem, if the initial PLM (VAE) analysis fails to reveal any asbestos at all, a qualitative "drop mount" TEM analysis should be performed, using chemical matrix reduction if necessary, to verify or refute the absence of asbestos. If the TEM analyst does not detect any asbestos fibers then the result can be considered conclusive. If asbestos fibers are noted and these are large enough to be visible with the PLM and are dominant in mass over any asbestos fibers below the resolution limit of the PLM, then the PLM (ISA) method should be applied. Only if the asbestos fibers not resolvable by PLM dominate in mass over those that are resolvable, can the PLM (ISA) method not be applied. In this case only, a TEM (ISA) method must be implemented to handle this situation. Fortunately, bulk samples containing a low percent of asbestos and consisting predominantly (in terms of mass) of asbestos fibers not resolvable by PLM, occur very infrequently. If two types of asbestos are present in the sample, matters certainly become more complicated. One plausible approach would be to analyze the sample twice, i.e., separately for each type of asbestos. If in the first analysis the sample asbestos exceeds that of the spike, the analysis can terminate.

If the initial PLM (VAE) analysis reveals a low level of asbestos, i.e., 0.1 to 5%, the data from our previous study, some of which has been presented in this paper as a comparison to the PLM (ISA) data, show that neither PLM (PC) nor TEM (GR) can provide reliable results and remedy this PLM (VAE) deficiency at such low asbestos concentrations. In particular, the TEM (GR) method and the New York State 198.1 PLM gravimetric method are very successful in those samples in which the asbestos content of the sample and the degree of matrix reduction is such, that the ratio of asbestos in the residue to total residue is high enough for the analyst to make a confident determination of its percent content. The types of samples for which the gravimetric approach was intended [*6,7,8*], are for resilient vinyl floor tiles, mastics and asphaltic materials, such as most roofing materials. However, some samples, even though they appear to fall under the general category of non-friable organically bound (NOB) matrices, (floor tiles and asphaltic materials included) are not suitable for a reliable gravimetric analysis when they present the TEM and PLM analysts (using the New York State 198.1 method) with a residue consisting of an unfavorable ratio (<10%) of asbestos to total residue. Of course, misapplication of the gravimetric method to unsuitable samples in which matrix reduction is not significant, will not improve the outcome of the analysis. It is in these situations that this method, either through analytical limitation or misapplication, falls short and the PLM (ISA) method is recommended.

Overestimation, which has been shown in our previous study in PLM (VAE), PLM (PC) and TEM (GR) analysis as well as by previous authors for PLM (VAE) and PLM (PC)[*3,4,5*], leads to false positives for samples truly containing < 1% asbestos content. The tendency to overestimate is especially true for TEM (GR) as evidenced by the incidence of 55% false positives for a 0.5% asbestos standard. Thus a building, or parts of it, may be abated due to a laboratory result of 2% asbestos by the currently used methods when it could be shown by a more accurate method to contain < 1% asbestos.

The use of known weight percent standards as a visual calibration aid to PLM analysts is a recommended practice in asbestos testing laboratories however, it is very difficult for a PLM analyst to remember what a 0.5% or a 1%, 2%, 3% look like and it is not practical to constantly refer to these standards. However, by inserting the standard within the sample to be analyzed, the analyst has a constant reference for comparison.

The advantage of the ISA method over the currently used PLM and TEM(GR) methods is based on the notion that a visual or quantitative (point count) comparison of the asbestos fibers from the sample to a finite, known amount of asbestos fibers added to the sample, leads to a more precise and accurate quantitation than a visual or quantitative (point count) comparison of asbestos fibers to nonasbestos fibers and/or matrix material. An advantage of the ISA approach is that it bypasses the problem of density variation among the samples' constituents and reduces the area/volume problem, which leads to overestimation. The authors can not presently explain the slight bias toward underestimation for the PLM (ISA) method, however this phenomenon is being studied further.

Conclusion

The currently used PLM and TEM (GR) methods fall short in their ability to generate reliable results at asbestos concentrations of ≤5%. Results vary greatly among analysts and accuracy is generally poor. The client is often left questioning which analyst or laboratory is correct. The "Internal Standard Addition" (ISA) method yields more precise and accurate results as well as significantly lower false positives and false negatives than the current methods in the range of 0.25% - 5% asbestos. Thus it represents a significant advancement in the quantitation of low levels of asbestos in bulk samples. Based on the results obtained by the authors, it is their opinion that results presently reported by laboratories using the current methods in the trace to 5% range, be examined more carefully with this new method.

References

[1] Frasca, P., Baltz, A., Megill, J., Scarano, J., Faulseit, B., Wells, L., Shelmire, D., Mastovich, M. and Marcus, M., "Asbestos Misdiagnosis of Bulk Samples by PLM". *National Asbestos Council Journal,* 1989, Vol. 7, No. 1, pp. 21-24. J.

[2] Webber, J.S., Janulis, R.J. and Laurie, J. "Quantitating Asbestos Content in Friable Bulk Samples: Development of a Stratified Point-Count Method.",*American Industrial Hygiene Association,* 1990,J. 51(8),pp.447-452.

[3] Perkins, R.L. and Beard, M.E., "Estimating Asbestos Content of Bulk Samples". *National Asbestos Council Journal,* 1991, J. 9(1),pp.27-31.

[4] Harvey, B.W., Perkins, R.L., Nickerson, J.G., Newland, A.J., and Beard, M.E., "Formulating Bulk Asbestos Standards", *Asbestos Issues*, 1991, 4(4)pp..22-29.

[5] Perkins, R.L., Harvey, B.W., and Beard, M.E., "The One Percent Dilemma",*Environmental Information Association Journal,* 1994, pp.5-10.

[6] Chatfield, E., Chatfield Technical Consulting Limited, "Standard Operating Procedure; SOP-1988-02: Analysis of Vinyl Floor Tile", Presented to ASTM Committee D22.05.07.007; November 1988".

[7] New York State Environmental Laboratory Approval Program, "Method 198.4; Transmission Electron Microscope Method for Identifying and Quantitating Asbestos in Non-Friable Organically Bound Bulk Samples",1995.

[8] New York State Environmental Laboratory Approval Program "Method 198.1; Polarized-Light Microscope Method for Identifying and Quantitating Asbestos in Non-Friable Organically Bound Bulk Samples", 1995.

[9] U.S. Environmental Protection Agency: "Method for the Determination of Asbestos in Bulk Building Materials", EPA/600/R-93/116, July 1993.

[10] U.S. Environmental Protection Agency: "Interim Method for the Determination of Asbestos in Bulk Insulation Samples", EPA/600/M4-82-020, 1982.

[11] Frasca, P., Mahoney, R.K., Faulseit, B.K., Megill, J. and Baltz, A., "A New Method for Determining Low Levels of Asbestos (Near 1%) in Asbestos Containing Material", Poster Presentation, *National Asbestos Council Conference*, Phoenix, AR., 1990.

[12] U.S. Environmental Protection Agency: Asbestos NESHAP Clarification Regarding Analysis of Multi-Layered Systems. *Code of Federal Regulations*, 40 CFR, Part 61, Vol. 59, No. 3. USEPA, Washington, DC (1994).

Ian M. Stewart[1]

ASTM'S BULK METHOD: WHERE ARE WE, WHERE ARE WE HEADED, WHERE SHOULD WE BE HEADED?

Stewart, I. M., "ASTM's Bulk Method: Where Are We, Where Are We Headed, Where Should We Be Headed?" *Advances in Environmental Measurement Methods for*

REFERENCE: Stewart, I. M., **"ASTM's Bulk Method: Where Are We, Where Are We Headed, Where Should We Be Headed?"** *Advances in Environmental Measurement Methods for Asbestos, ASTM STP 1342,* M. E. Beard, H. L. Rook, Eds., American Society for Testing and Materials, 2000.

ABSTRACT: The analysis of bulk samples for asbestos was originally considered best conducted by polarized light microscopy (PLM). EPA published a methodology for this analysis which also permitted the use of x-ray diffraction (XRD). An awareness of the limitations of both PLM and XRD led many laboratories to supplement their PLM analyses with other methods including scanning and transmission electron microscopy and gravimetry combined with any of the previously mentioned techniques. To address these methods, EPA contracted with the Research Triangle Institute to prepare a methodology for the analysis of bulk materials. ASTM is now converting this methodology into an ASTM Standard Method. Carlton Hommel has placed the original RTI document into ASTM format but the resulting document is somewhat lengthy and cumbersome. Current efforts are aimed at splitting this document into more manageable pieces as separate Standard Methods whose applicability will be addressed in an overall Guidance Document. There remains the question of the usefulness of an ASTM bulk method in light of both government and industry's current acceptance of the RTI protocol and their failure to adopt the ASTM PLM method previously published in The Gray Pages.

KEYWORDS: asbestos, bulk analysis, polarized light microscopy (PLM)

---

[1] Vice President, Analytical Services, RJ Lee Group, Inc., 350 Hochberg Road, Monroeville, PA 15146

Introduction

ASTM Committee D22.07 is currently in the early stages of developing a Standard Method for the Analysis of Bulk Samples for Asbestos. One rationale given for the development of an ASTM method, when an existing government sponsored and approved method already exists, is the recent government initiative that requires the use, for regulatory purposes, of such a voluntary standard method when one is available. This paper will review some of the historical background to the development of bulk asbestos methods and poses the question whether the efforts of D22.07 might better be directed to pursuing other analytical procedures which currently are not standardized.

Historical

In 1970, Julian and McCrone [1] published their methodology for the identification of asbestos using polarized light microscopy combined with dispersion staining. This became, in essence, the reference method for the determination of asbestos in building materials and remained so for several years. In 1979, the EPA published its first guidance to schools, "The Orange Book" [2], which advised schools to request the analysis of friable materials for asbestos by polarized light microscopy and x-ray diffraction "as necessary to supplement the PLM method." Acknowledging that no standard protocol for bulk sample analysis was currently available, the Orange Book provided, in Appendix H of Part 1 [3], guidelines for such an analysis using PLM and XRD. At the same time, EPA indicated that it had set in motion the development of a standard method for the analysis of bulk samples through a contract with Research Triangle Institute.

In 1980, McCrone [4] published his Asbestos Particle Atlas which had been developed as a result of a photomicrography project sponsored by EPA's Office of Toxic Substances. This Atlas became an instructional tool for those entering the field of asbestos bulk sample analysis as well as a reference work for the experienced. The need for documentation of the techniques used was demonstrated by the success of this Atlas; nearly 5000 copies were sold between the time of its introduction and its revision in 1987 [5].

In 1982, the method developed for EPA by Research Triangle Institute (RTI) was published as EPA 600/M4-82-020 [6]. The RTI document covered the preliminary examination of the material by stereo microscopy, the identification of asbestos by PLM and quantitation of the asbestos using point counting. The document also described an x-ray diffraction procedure for the identification and quantitation of those mineral species which, in their asbestiform habit, are regulated as asbestos. EPA 600/M4-82-020 was adopted by EPA in their Asbestos in Schools Rule [7] as the method by which the determination was to be made that a building material contained asbestos and remained the method of choice when the Asbestos Hazard Emergency Response Act (AHERA) [8] was promulgated. EPA 600/M4-82-020 is still the only approved method in current regulatory legislation, including the latest revisions to NESHAP [9].

Experiences with EPA 600/M4-82-020 pointed up several shortcomings in the method, particularly for those materials in which the asbestos content was very low and/or the size of the asbestos present very small. Accordingly, RTI was again assigned the task by EPA to develop a more comprehensive procedure which would address some of these problems. This new procedure now includes TEM and gravimetric methods, that were already in common use, to determine the weight percentage of asbestos present in samples with low asbestos concentrations, short or extremely fine fibers and/or difficult matrices. The draft of this method was circulated for review to all participating laboratories in the NVLAP[2] program, and their comments were considered in finalizing the method. In 1991 this method was published by EPA as EPA /600/R-93/116 [10].

In the meantime, ASTM Committee D22.07 had not been idle. In the mid to late 1980s work began on an ASTM method based, substantially, on the work of Rohl, Langer and Wylie [3], and stewarded through the ASTM process by Richard Hatfield. This process did not run the full ASTM Standard course but was published as a Proposed Test Method for Asbestos-Containing Materials by Polarized Light Microscopy (P236) in ASTM's Gray Pages in 1993. ASTM proposed methods have a "life" of two years, thus in late 1994 the status of P236 was reviewed and the decision was made to again attempt to develop an ASTM Standard Method for the Determination of Asbestos in Bulk Materials. In this instance, the basis for the method is the most recent RTI/EPA method. Carlton Hommel undertook the formidable task of translating this document into ASTM format and produced a first draft which was passed to the committee early in 1995 and is now under the stewardship of the present author.

As received from Hommel, the document, without references, exceeds 50 pages in length and Carlton is to be commended on the stalwart effort he has put in. However, a standard of over 50 pages in length is cumbersome, unwieldy and not likely to be read in sufficient depth to be of value. It has been proposed, therefore, that the document be split into sub units each of which would provide a standard method to perform particular aspects of the procedure, e.g. a standard method for optical examination, a standard method for gravimetric analysis, a standard method for TEM analysis, etc., with a Guideline document to tie them altogether. It is now over a year since the present author sought direction from Committee D22.07 as to the advisability of splitting the method into sub-units and offered the first drafts of two of these sub-units for comment. The response has been underwhelming.

The Future

I believe ASTM Committee D22.07 must look hard at the necessity for an ASTM Standard Method for the Analysis of Asbestos in Bulk Materials. As stated above, the impetus for the development of the method is cited as the government requirement that,

---

[2] National Voluntary Laboratory Accreditation Program, a program administered by the National Institute for Standards and Technology in which laboratories analyzing asbestos bulk samples for schools are required to participate by the Asbestos Hazard Emergency Response Act

whenever available, published voluntary standard methods be used in the regulatory process. It should be noted, however, that this mandate is predicated on the existence of such methods and is not a mandate to outside agencies to develop new methods, particularly if an existing method already exists. In addition, we must take cognizance of the historical fact that there has not been a wild rush by governmental agencies, either at state or federal levels, to embrace ASTM standards that were already in existence. In the field of environmental and industrial hygiene chemistry, for example, both EPA and NIOSH publish their own methods, many of which are substantially similar to ASTM methods. While the new directive may change this, there has been little encouragement so far to believe that this will happen: the most recent iteration of OSHA's asbestos regulations, for example, continues to ignore ASTM's long standing PCM method - D4240 Standard Test Method for Airborne Asbestos Concentration in Workplace Atmosphere - which is currently under revision by D22.04. Additionally, when considering a bulk asbestos method, one can legitimately claim that, at least numerically, the RTI method has passed a more widespread peer review process than any method emanating from D22.07 is likely to see.

Bearing in mind the several other tasks underway in Committee D22.07 and the presence of an existing, accepted methodology for the analysis of bulk materials, the question I put to this Boulder Conference, therefore, is whether the potential for acceptance and use of an ASTM bulk method justifies the dilution of D22.07's efforts in other areas.

References

[1] Julian, Y and McCrone, W. C. "Identification of Asbestos Fibers by Microscopical Dispersion Staining", Presented at Inter-Micro '69, July 1969, *The Microscope,* vol. 18, pp. 1-10, 1970.

[2] *Asbestos-Containing Materials in School Buildings: A Guidance Document, Parts 1 and 2*, U.S.E.P.A., Washington, DC, March, 1979

[3] Rohl, A. N., Langer, A. M. and Wylie, A. G., *"Mineral Characterization of Mineral Spray Finishes"*, ibid., Appendix H to Part 1, pp. 59-64.

[4] McCrone, W. C., *The Asbestos Particle Atlas*, Ann Arbor Science Publishers, Inc., Ann Arbor, MI, 1980.

[5] McCrone, W. C., *Asbestos Identification*, McCrone Research Institute, Chicago, IL, 1987.

[6] *Interim Method for the Determination of Asbestos in Bulk Insulation Samples*, U.S.E.P.A. Report EPA600/M4-82-020, 1982.

[7] *Friable Asbestos Containing-Materials in Schools, Identification and Notification Rule*, Federal Register, vol. 47, p. 23360 et seq., May 27, 1982

[8] *Asbestos-Containing Materials in Schools; Final Rule and Notice*, 40 CFR Part 763, Federal Register, Vol. 52, p. 41826 et seq., October 30, 1987

[9] *National Emission Standards for Hazardous Air Pollutants; Asbestos NESHAP Revision; Final Rule.* Federal Register, Volume 55, p. 48406 et seq., November 20, 1990.

[10] *Method for the Determination of Asbestos in Bulk Building Materials (Test Method)*, U.S.E.P.A. Report EPA/600/R-93/116, 1991.

Measurement Methods for Asbestos in Ambient, Indoor, and Workplace Air

Garry J. Burdett and Graham Revell

STRATEGY, DEVELOPMENT AND LABORATORY CALIBRATION OF A PERSONAL PASSIVE SAMPLER FOR MONITORING THE ASBESTOS EXPOSURE OF MAINTENANCE WORKERS

REFERENCE: Burdett, G. J. and Revell, G. **"Strategy, Development and Calibration of a Personal Passive Sampler for Monitoring the Asbestos Exposure of Maintenance Workers,"** *Advances in Environmental Measurement Methods for Asbestos, ASTM STP 1342,* M. E. Beard, H. L. Rook, Eds. American Society for Testing and Materials, 2000.

ABSTRACT: Maintenance workers have been recognised as having an increased risk to asbestos related disease due to the frequency with which they may encounter asbestos containing materials and the amount of disturbance they are likely to cause. While significant attempts have been made to reduce their risk (e.g. the OSHA 'Presumptive' approach and advertising campaigns in the UK), the current asbestos exposures of maintenance workers are not well quantified.

Conventional pump and membrane filter sampling of maintenance workers presents a logistic challenge, as it requires prior knowledge that asbestos is to be disturbed and detailed co-ordination between the building owner, sampling personnel and maintenance workers. This is thought to produce a bias, as only good work practices can be monitored and asbestos which is being unknowingly disturbed at other times will not be sampled. A more effective sampling strategy would be to use a longer duration sampler to continuously monitor a group of workers over a week or more, so any asbestos release will be sampled and an average exposure obtained.

A passive dust sampler developed at the Health and Safety Laboratory (HSL), and operating on the principle of electrostatic attraction, has the ability to fulfil such a role. The performance of the passive sampler has been compared under laboratory conditions to the conventional sampling method, to see whether it can achieve sufficient sensitivity and meet proposed performance requirements, so that it can be used in a general survey of maintenance workers.

KEYWORDS: passive, sampler, asbestos, fibres, measurement, comparability, membrane filter, PCM, SEM, TEM.

Health and Safety Laboratory, Broad Lane, Sheffield, S3 7HQ, U.K.
© British Crown Copyright 1997.

Introduction

Background

In recent years there has been considerable concern over the mesothelioma rates in groups of maintenance workers and building tradesmen [1-3], who in the past may have unknowingly come into frequent contact with asbestos containing materials (ACM's) resulting in uncontrolled releases and exposures to airborne asbestos fibres. Initial UK predictions [1] are that the male mesothelioma numbers seemed set to rise to an annual peak of 3,300 deaths by about 2020; by date of birth, the cohort most at risk was that of men born in the late 1940's; and for men born in the late 1950's the risks appeared to be about half the peak risk. However, some features of the data suggested there may have been some diagnostic trend contributing to the growth in case numbers so far and on what was described as "an extreme but arguable case" [1] the peak of annual male deaths might be reduced to 1,300, reached around the year 2010. An average of the two estimates, approximately 2000 male mesotheliomas, is perhaps the best current estimate and if the relationships from cohort studies of asbestos manufacturing apply, the total asbestos related mortality is expected to be about three times the number of mesotheliomas, some 6000 premature deaths per year. Data from last occupation on death certificates showed that construction and related trades (e.g. plumbers, gas fitters, carpenters and electricians) accounted for 24% of the mesotheliomas.

There is limited historical exposure data for maintenance workers (much of which was summarised in reference [2]), as most of the exposure database is from occupational hygiene surveys in the asbestos manufacturing industry. The two largest studies are summarised below. A study by CONSAD [4] using data collected in the early 1980's estimated the number of US exposed maintenance workers and their exposures for different maintenance activities. This gave estimated exposures for 740,000 maintenance workers of between 0.75 f/ml for repairing dry walling to 0.11 f/ml for repairing plumbing, assuming no respiratory protection was worn. A later study from data supplied to HEI [5] reported levels from an operations and maintenance programme at a hospital gave a mean of 0.11 f/ml for the 107 tasks monitored. These results in some respects reflect the different regulatory requirements at the time of sampling. The Occupational Health and Safety Administration (OSHA) permissible exposure limit [3] was set at 2 f/ml when the CONSAD work was carried out and 0.2 f/ml during the hospital sampling. Therefore it is likely that current asbestos exposures to maintenance workers will lead to a much reduced asbestos related mortality than the predicted rise based on the 1940 cohort, which may be based on historical high exposures, which are no longer prevalent. It is, however, too important an issue to be complacent about and further work to measure maintenance worker exposures has been funded by the UK Health and Safety Executive (HSE).

Maintenance Worker Surveys

As the regulatory asbestos sampling methods and strategies were devised for monitoring exposure in the manufacturing industry [3, 6], there are some difficulties in applying these methods to a survey of maintenance workers. Personal samples for determining worker exposure to asbestos and other fibres are collected by drawing air through a membrane filter with a battery driven pump. The membrane filter - pump sampling strategy can be easily used for monitoring 4 or 8 hour time weighted average (TWA) exposures in a manufacturing industry. However, the method requires considerable planning and logistic support when used to sample maintenance workers who are constantly moving between workplaces and are subject to intermittent and unforeseen exposures. At the very minimum to conduct a conventional monitoring survey it is required that:

a) The area where the work is to be carried out must have been surveyed for asbestos containing materials and a register of such materials and their locations established.
b) A permit to work system is in use, and the building supervisor has planned that work was going to be carried out on a certain date.
c) The building supervisor was able and willing to contact the sampling personnel in advance, to give details of the time and place of work.

The above requirements are usually only met by good employers with well-controlled operations. Also as the sampling personnel attend the site in person to attach the sampling pump and cassette and to monitor the worker activity and pump performance, the workers are well-aware that they are under scrutiny. This means that most studies are likely to have a low bias and may underestimate the average and peak exposure levels of the majority of maintenance workers. A personal passive sampler which can be left with the worker to sample over a period of 1 week to 1 month offers an alternative sampling strategy, where maintenance workers wear the badge but may not / or need not know whether they are disturbing asbestos. This allows much more flexible, random and non - invasive strategies to be used, which will allow information to be collected on the frequency which different groups of workers are disturbing asbestos and whether significant exposures are produced.

Description of the HSL Passive Sampler.

The development, theory and construction of the HSL passive sampler has been described by Brown et al. [7,8]. In its current form [9] shown in figure 1, it consists of a 25 mm diameter polypropylene electret, which is charged to 1000 V by scanning several times under a wire held at this potential. An infinitely extended charged sheet has no external field but one is created by placing a conductor close to the surface. Hence the electret is placed in a stainless steel surround or cage which also protects it from damage or interference. A 40 mm diameter circular metal plate is placed about 1 cm in front of the electret in preference to a mesh, to give a more even field over the whole electret. This plate also prevents direct impaction onto the electret by large particles. The passive sampler when worn on the upper torso is essentially two vertical parallel discs separated by 1 cm, between which (due to the characteristic flow of air currents and the bodies convective heating) respirable sized particles will move. The collection of particles depends on their electrical mobility (and polarity) assuming a minimum velocity through the sampler is achieved.

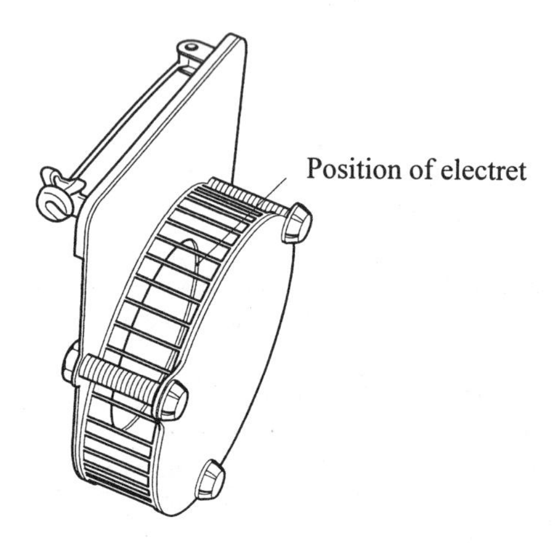

Figure 1: Diagram of the passive sampler

Initial Passive Sampling Strategy

It was originally anticipated that as field sampling studies would last one or more weeks, it would be possible to analyse the fibres as collected on the electret surface using phase contrast microscopy analysis (PCM) to screen the fibre concentration and scanning electron microscopy (SEM) with energy dispersive x-ray analysis to determine the possible asbestos fibre concentration. Sample preparation methods to do this were developed and tested. However, to carry out laboratory comparisons between the passive sampler and the membrane filter MDHS 39/4 reference method [6] a much shorter sampling period is necessary and after two initial comparative runs, it was found that there were insufficient fibres on the electret to make analysis feasible. Therefore it was necessary to make a substantial increase in the analytical sensitivity of the method before the performance of the passive sampler could be compared to the MDHS 39/4 reference method. This would also be a very positive step forward in extending the range and uses of the passive sampler for fibre sampling.

Any increase in the analytical sensitivity of the passive sampler when carrying out analysis directly on the electret can only be achieved by increasing the area of the electret analysed, the time of sampling or the collection rate. Only the first is possible but is rather expensive and impractical if we want to achieve an order of magnitude improvement. It is far better to concentrate the electret deposit using an indirect sample preparation method, where the deposit is washed off the electret and re-filtered onto a membrane filter with a much smaller deposit area. Use of an indirect method has several positive attributes, in that it allows the sample density to be optimised, which will reduce the analytical time and improve the analytical sensitivity. The resuspension in water also removes soluble materials and if filtered carefully should give a random distribution of particles, which will increase the analytical precision. There is also the opportunity to introduce additional stages, such as to remove organic fibres and particles by ashing (but this was not done in the following experiments).

The main disadvantage of a re-suspension step is that fibre agglomerates and matrices can be broken up and in particular chrysotile fibres and bundles may divide longitudinally, increasing the fibre number count. However by adopting careful methods of re-suspension any increases to the >5 μm long fibre number count can be minimised [10].

Aims and Objectives

For the passive sampler to be considered as a suitable sampler for conducting surveys of asbestos exposure, it must be shown to give valid results in laboratory comparisons with the current membrane filter - pump methods [6] for sampling asbestos. Therefore the initial aim of the work was to compare the performance of the passive sampler using an indirect sample preparation method, with the current membrane filter - direct preparation method MDHS 39/4, using a variety of fibre types. This work is referred to as the initial experiments.

The second aim was to determine the analytical sensitivity that can be achieved with the indirect preparation method and to investigate any refinements that may be necessary. This work is referred to as the sensitivity experiments.

The third aim was to test the performance of possible fibre discrimination methods which would be needed for field studies. In particular the use of analytical scanning electron microscopy (SEM) and analytical transmission electron microscopy (TEM).

Fibre Generation and Sample Loading.

Experimental Equipment

All the samples were exposed to asbestos in a dust box facility with a working section 2 m high with a 1 m x 1 m cross-section. It is fitted with a HEPA filter inlet filter at the top and is kept under negative pressure by an extract system at the bottom. The flow rate within the box is generally adjusted to give a downward flow of a few cm/sec to prevent the build up of fine fibres and to maintain a good negative pressure for safety reasons. The floor of the box is fitted with a rotating mesh turntable, on which are 8 posts suitable for holding both a cowled membrane filter cassette and a passive sampler approximately 10 cm apart. During each sampling run, the turntable rotates at 0.2 revs per minute and is fitted with a micro- switch, which will reverse the direction after each complete rotation.

The fibres were generated outside the dust box using a fluidised bed aerosol generator. The aerosol produced was fed into the top of the box and dispersed in an air stream of + ve and - ve ions to part neutralise the aerosol. Further mixing takes place before the aerosol passes through aluminium honeycomb flow straighteners into the top of the 2 m high working section. The rate of production of fibres with time was usually monitored with a fibre/particle counter (Harley Scientific FC2 Fibrecheck) located outside the dust box but connected via a sampling port to a fixed point inside the dust box.

Mixed esters of cellulose membrane filters (0.8 μm nominal pore size), were used to sample the dust box atmosphere. Each filter was held in a three piece conducting plastic cowl conforming to the specifications in MDHS 39/4. Each cowl was placed face downwards and held in a vertical position by the spring clip on the post. The cowl extract was connected via flexible tubing to connections at one side of the box; the other side of the connectors were connected to a flow compensated personal sampling pump (Rotheroe and Mitchell L2SF). In some experiments, two additional cowls were also held in a fixed position. Each filter head and pump was individually calibrated against a bubble flow meter and set at either 1 L/minute or 2 L/minute (to within ± 5%).

The passive samplers were held on the same posts as the cowls but towards the centre of the box and were suspended vertically as they would be when pinned to the upper torso. The electrets were kept in semi-airtight boxes before use and transferred into sealable plastic bags after use.

Sampling and Filter Loading Protocol

Two protocols were used for exposing the samplers to fibrous aerosols:

A) For the initial set of experiments eight membrane filters in cowled samplers and eight electrets in passive samplers were simultaneously exposed, for either a 30 minute or a 2 hour period. At this point half the samplers would be removed and replaced with new samplers, before a second fibre type or a different concentration of the same fibre was generated for a similar time. This protocol would therefore produce three different fibre loadings from the one experiment. Fibre types used included amosite, chrysotile, crocidolite and RCF fibres.

B) To give a greater range of concentrations and to test the sensitivity and reliability of a single sample, one run was carried out with amosite, where the generation of amosite was held constant and pairs of cowled and passive samplers were exposed for increasing times at approximately 20 minute intervals (e.g. 20, 40 60........180 minutes).

Analytical Method

Membrane Filter - Direct Sample Preparation

The membrane filters exposed in the cowled samplers were removed and cut in half with a type 22 scalpel blade. A half filter was mounted for PCM analysis using the acetone-triacetin mount described in MDHS 39/4. The collapsed filters were then analysed by X500 magnification to count the number of >5 μm long, < 3 μm wide particles with aspect ratios >3:1. The normal stopping points were a count of 200 fibre ends, or a minimum of 20 and a maximum of 200 Walton - Becket graticule counted.

Passive sampler - Indirect Sample Preparation method for the initial experiments.

The 25 mm diameter polypropylene electret was removed from a protective metal cage of the passive sampler and cut into half with a scalpel. A half-electret was placed in a pre-cleaned 15 ml glass screw-top vial and a measured quantity of ultra-pure water (reverse osmosis and 0.2 μm filtered) added to the vial. The vial was then sealed and the contents shaken for 30 seconds and placed in an ultrasonic bath for 30 seconds and then shaken for a further 30 seconds. 1 ml and 2.5 ml aliquots were taken from the suspension using a pre-cleaned disposable 2.5 ml syringe. To filter, the Leur tip was pushed against a 0.22 μm pore size membrane filter held on a Millipore glass filter system with a 5 μm pore size backing filter. A vacuum is applied to the filter and hand pressure is used to push the contents of the syringe through the filter, over a period of approximately 1 minute. The filter was then left in a warm air cabinet to dry overnight. Normally on each filter a 2.5 ml laboratory blank , two 2.5 ml aliquots and one 1 ml aliquot were filtered. In the sensitivity protocol 12 ml of water was used so that a 1 ml and 2.5 ml aliquot could be prepared on each of three filters, which were then prepared for the three methods of analysis (PCM, SEM and TEM).

Sample preparation blanks: Laboratory blanks were prepared for each filter by placing the 9 ml of water in one of the batches of cleaned bottles and shaking and ultrasonic treating it as above. After rinsing the syringe with pure water from another container a 2.5 ml aliquot was filtered onto a membrane filter. The same syringe was then used to draw a 1 ml, then a 2.5 ml aliquot. With each batch of electrets one or two unused electrets were prepared using the same procedures as for the exposed electrets and are referred to as electret blanks.

Passive samples - PCM analysis: When dry, the filter is cut into half or quarters and the syringe water blank, 1 ml and 2.5 ml deposit is mounted using the acetone-triacetin method, as described in MDHS 39/4 and analysed by phase contrast microscopy. With careful step scanning and taking adjacent fields of view 200 Walton-Becket graticules can be observed in the 2.4 mm^2 deposit, although normally only 100 were examined.

Passive samples - SEM analysis: For SEM analysis the DM 450 filters were mounted using a collapsing and etch method [11] and coated with gold. 100 fields of view at X4000 magnification were scanned in a Philips XL 20 SEM, and all >5 µm fibres were counted. On some of the samples, an Oxford Instruments ISIS energy dispersive x-ray dispersive analyser (EDXA) was used to determine the fibre chemistry and discriminate between asbestos and non-asbestos fibres.

Passive samples - TEM analysis: The TEM deposits were mounted onto electron microscope (EM) grids using a collapse, etch, carbon - coat and dissolution method [11]. The grids have approximate 0.009 mm^2 grid openings and between 20 to 50 grid opening were examined at 11,800 magnification in a Philips CM12 TEM. The counting, sizing and identification of fibres was carried out in accordance with the International Standards Organisation (ISO) methods for asbestos in ambient air [12,13]. Each particle (fibre) with an aspect ratio of >3:1 and length >5 µm was sized and analysed/identified using an EDAX DX4 energy dispersive analyser. Either 50 fibres or 50 grid openings were counted. Each fibre was classified into the following groups: all fibres > 5µm long, all fibres > 5µm long not attached to >3 µm particles, asbestos fibres > 5µm long and asbestos fibres > 5µm long not attached to >3 µm particles

Calculations

Fibre concentration on directly prepared membrane filter samples.

The fibre density C_F per unit area of exposed membrane filter (f/mm^2) is given by:

$$C_F = N / n\ a \qquad (1)$$

where: N = the number of fibres counted
n = the number of graticule areas examined
a (mm^2) = Area of the Walton - Becket graticule (0.00785 mm^2)

The airborne concentration C_A is given by:

$$C_A = 1000\ N\ D^2 / V\ n\ d^2 \qquad (2)$$

where: D (mm) = diameter of the filter deposit (22 mm)
d (µm) diameter of the Walton - Becket graticule (100 µm)
V = Volume of air sampled (Litres).

The upper and lower confidence interval for each membrane filter count was calculated using the experimentally determined values for intra-laboratory counts in MDHS 39/4.

Passive sampler

Passive sampler - PCM: The density of PCM fibres in f/mm^2 on the indirectly prepared membrane filter was calculated using equation (1) above.

The concentration on the electret C_E is given by :

$$C_E = C_F \, d_s^2 \, V_w / v \, F \, D_e^2 \qquad (3)$$

where: D_e (mm) = diameter of the electret (25 mm)
d_s (μm) = diameter of syringe deposit (1.75 mm)
F = fraction of electret used (normally 0.5)
V_w (ml) = Volume of water in vial (usually 9ml)
v (ml) = volume of aliquot filtered (usually 1 or 2.5 ml)

Two normalisation's to the initial raw count data were made. The diameter of the syringe deposit was measured when analysed and for most samples was close to an average of 1.75 mm but on a few occasions was found to be more than 10% smaller than the average. The fibre density of these deposits were recalculated based on the average diameter. Most electret samples were filtered at two concentrations 2.5 ml and 1 ml. In most cases the 2.5 ml deposit has been counted but in some instances due to the formation of air bubbles, it was not possible to count the 2.5 ml deposit and the 1 ml deposit was counted. For the purposes of comparison all the electret concentrations were normalised to 2.5 ml.

Passive sampler - SEM: The calculations for the passive sampler are as in equations (1) and (3) above, except that:
n = the number of SEM fields of view examined.
a (mm^2) = Area of the SEM field of view

Passive sampler - TEM: Again equation 3 can be used by substituting:
n = the number of grid openings examined.
a (mm^2) = Average area of the grid opening.

Results

The results from the tests are summarised in terms of the fibre densities on the direct membrane filter from the MDHS 39/4 analysis and indirect membrane filter prepared from the passive sampler electret.

Initial PCM tests (protocol A)

Figure 2 shows the results from the initial tests, where the average values for three replicate samples were usually calculated, along with the associated 95% confidence interval. All results have been standardised based on the membrane filter sampling at 2 L/minute and a half electret being washed off into 9 ml of water from which a 2.5 ml aliquot is filtered onto a 2.4 mm^2 filter area. In general the graph showed a good correlation ($R^2 = 0.8$) between the membrane filter and electret method of sampling, given the wide 95% confidence interval associated with fibre counting. The indirect filter deposits prepared from the electrets still only had one third (slope of the graph = 0.33) of the fibre density of the directly sampled membrane filters collected at a flow rate of 2 L/minute. The scatter of results from individual electrets is given in Figure 3 and when compared to the calculated

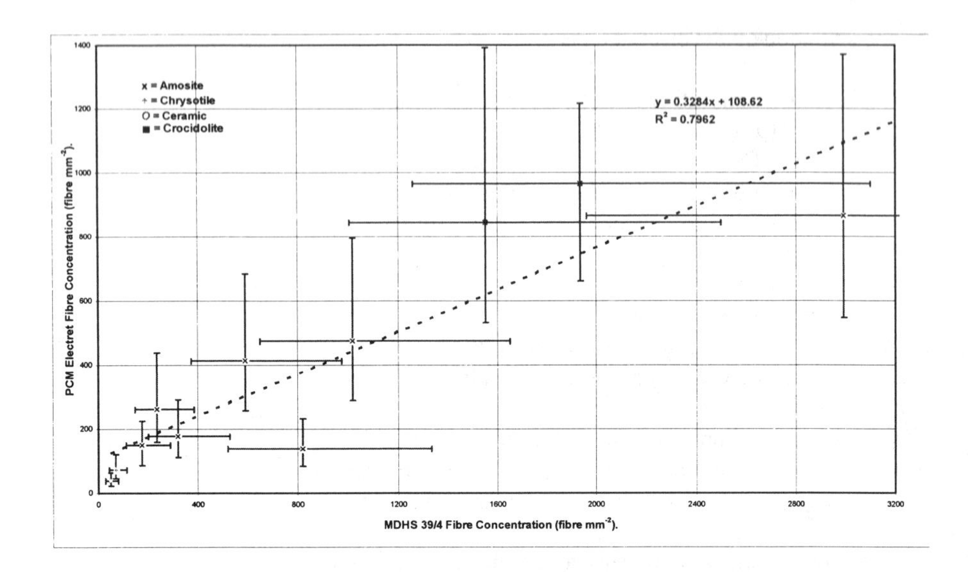

Figure 2: Electret fibre density v MDHS 39/4 fibre density: Initial tests

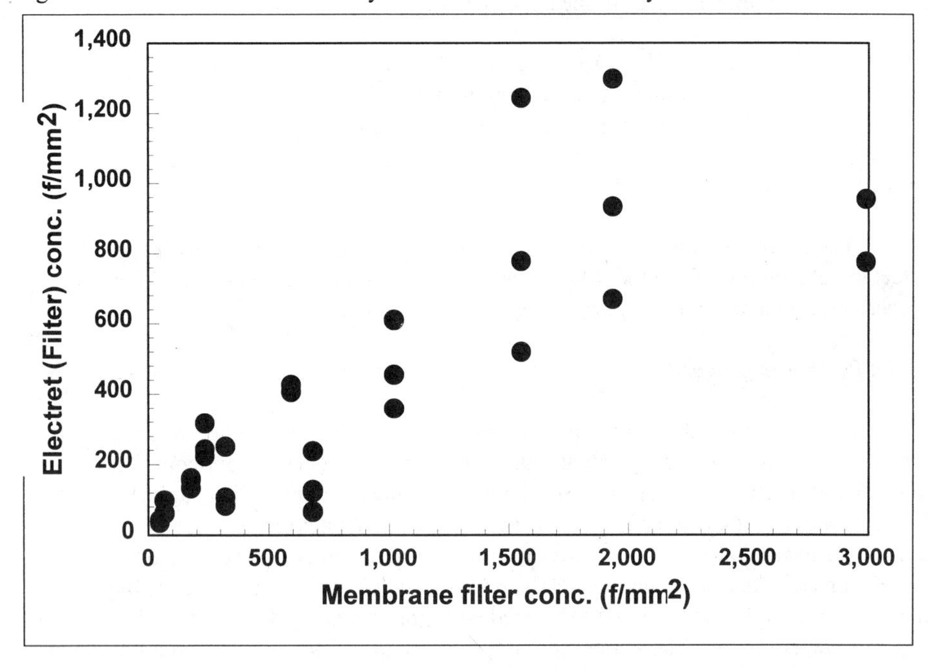

Figure 3: Values of electret indirect filter results v average membrane filter concentration. (Standardised to 2.5 ml and 2 L/min)

95% confidence intervals in Figure 2, can be seen to give variations within the expected counting precision.

It was noted that the regression line did not go through the origin with an intercept of 108.6, suggesting that there was a background of fibres. The reason for this background of fibres was investigated.

Fibre contamination of electrets

It has been found [6] that for directly prepared membrane filters a background count of up to 7 f/mm^2 can be obtained when analysed by PCM using MDHS 39/4. However, the indirect preparation method filtered 2.5 ml through 2.4 mm^2 filter area which is equivalent to filtering about 400ml of water through a 25 mm diameter filter. Therefore, considerable care is necessary to minimise any fibre contamination from the water, syringes and containers.

As described in the method, for every filter prepared a water blank using the same syringe and water and one of the pre-washed bottles from the batch was also prepared on the same filter. In the initial experiments there was no evidence in the water blanks for any significant contamination, with most blanks below 10 f/mm^2 and rarely exceeding 15 f/mm^2 . However, from the PCM results in the initial experiments it was apparent that there were additional fibres entering the system from the electret. The source of these fibres was carefully tracked and it was apparent that the electrets being prepared were randomly contaminated with one or more fibre types. This was attributed to either insufficient cleaning of the electret in the initial stage or failure to keep the electret in clean environments from the time it was charged up and sampling.

Large paper fibres were particularly common on some of the electrets but as they were >3 μm wide they were not considered a problem. However, in later electrets which had been prepared with greater care to clean them, there appeared to be a source of organic fibres which split into many fine fibres, just visible by X500 PCM. These were particularly found on the filters used for the sensitivity test (and on one or two earlier filters). Unfortunately these fibres seem particularly able to split into finer fibres with ultrasonic treatment, and as the ultrasonic treatment time had been increased from 0.5 to 10 minutes for the amosite sensitivity test, there was widespread contamination of fine fibres on the filters.

Although these contamination's were extremely disruptive, it gave a chance to practice the use of discrimination counting to determine between asbestos and non-asbestos fibres. This would be needed to be used for any field survey, where non-asbestos fibres will also be collected. For the PCM analysis of the sensitivity samples only straight rod - like fibres were counted on the indirect preparations from the electrets. SEM and TEM analysis with EDXA identification of the individual fibre chemistry was also used to discriminate fibres on these samples.

Sampling rate of the electret

The sampling rate of the electret is based on the electrical mobility of the particles being sampled and the polarity and intensity of charge on the surface of the electret. This can be estimated theoretically but can be calculated directly from the fibre count results. For the initial experiments a regression line was obtained with a gradient of 0.33 between the area concentration on a membrane filter sampling at 2 L/minute and an half electret (area 491 mm^2 /2) washed into 9 ml of water from which 2.5 ml was syringe filtered onto an area of 2.40 mm^2. This gave an area concentration of 490.87 x 0.5 / 2.4 = 102.26 and a liquid dilution of 2.5/9 = 0.277 with a total electret surface to filter deposit concentration of 102.26 x 0.2777 = 28.4. When divided by the gradient of the regression line the ratio of membrane filter: electret collection efficiency = 28.40/0.32 = 88.7. Therefore as the membrane filters were sampled at 2 L/minute the average sampling rate of the electret in the dust box was equivalent to 22.5 ml/minute.

Sensitivity Test (protocol B)

PCM results: The protocol for the sensitivity test required that a near constant fibre production rate was produced over several hours, so that the exposed filters would receive loading proportional to the time exposed. This even production of fibres was achieved only after a number of design changes to the fluidised bed generator and the use of a real time fibre monitor to feedback and adjust the fluidised bed output.

The PCM results from the sensitivity test for the direct membrane filter MDHS 39/4 method v time of sampling gave an R^2 = 0.74 while the electret method (after indirect sample preparation) v time gave an R^2 = 0.56. A comparison of the results from each pair of samples obtained at each time point is given in Figure 4. It can be seen that the results from the two samplers were within 95% confidence limits with a R^2 = 0.86.

SEM and TEM results: Further aliquots filtered from the electret wash off were analysed by SEM and TEM for fibres > 5 μm long and the fibre densities are plotted against the PCM counts in Figure 5. The SEM count is for all fibres > 5 μm long but when selected counts were discriminated using EDXA, it suggested that there was a constant background contamination level of non-asbestos fibres from the electret, equal to some 60 fibres/mm^2 on the prepared indirect filter deposits. The TEM analysis discriminated the fibre types on all the samples analysed and the asbestos PCM equivalent (PCME) fibres are plotted in Fig 5.

All three methods show much the same trend, the SEM results are the highest as the non-discrimination count is plotted. The TEM asbestos PCME fibre concentrations are close to the discriminated PCM counts giving added confidence that the PCM discrimination for amosite was successful. The variation between the PCM and TEM counts are within normal 95% confidence limits suggesting that the 0.2 μm lower detection limit used to define optically visible fibres is about right.

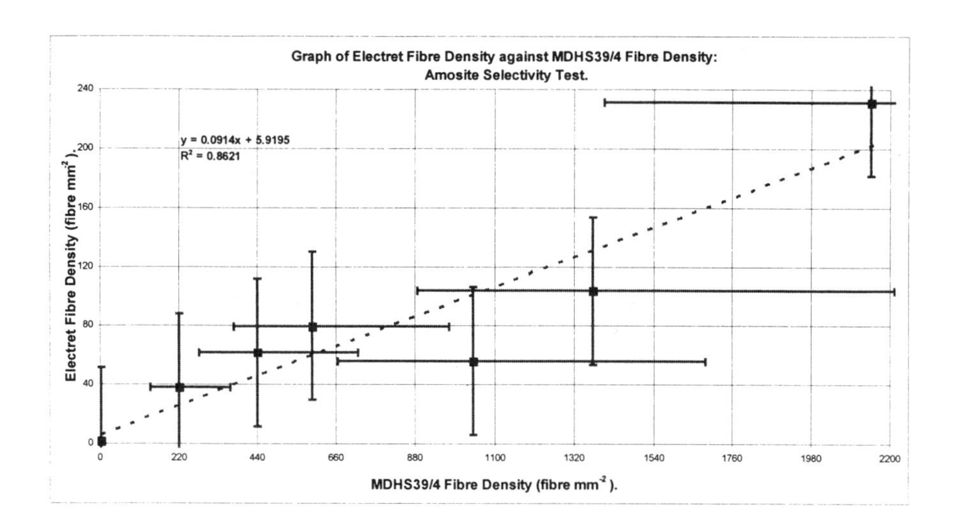

Figure 4: Electret fibre density v MDHS 39/4 fibre density - Amosite sensitivity test.

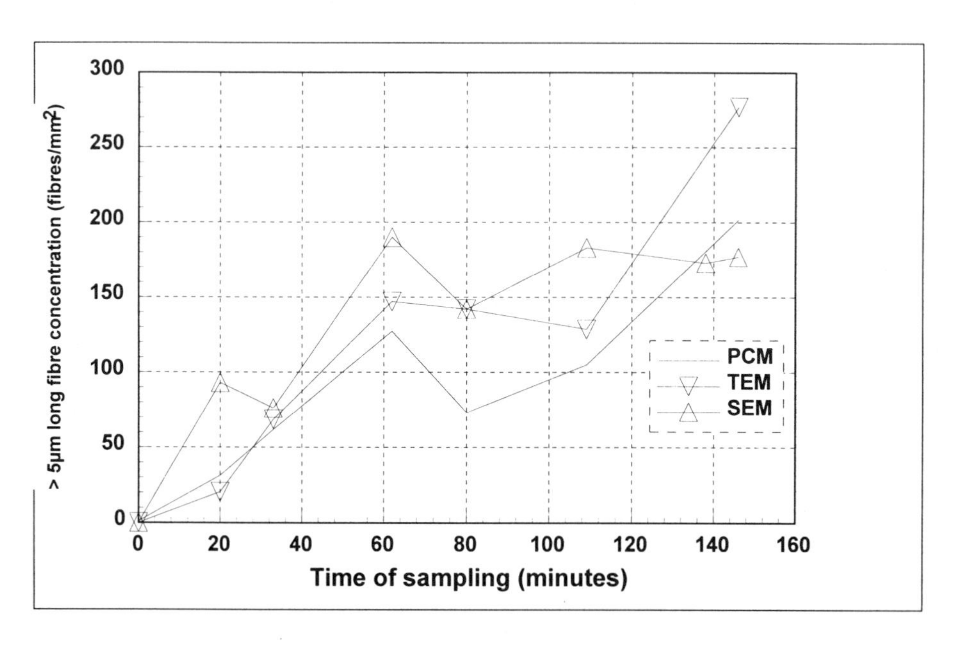

Figure 5 :Comparison of PCM, SEM and TEM >5 µm long fibre concentrations v Time of sampling for electrets

Discussion

Sampling capability

The data from the sensitivity experiment, which uses an indirect sample preparation method to concentrate the deposit, demonstrated that the passive sampler can be used to detect and monitor short duration exposures above the 10 minute control limit for amosite (0.6 f/ml). For longer sampling surveys sample volumes up to about 1 m^3 can be seen as an upper loading for most occupational sampling surveys, which is equivalent to 740 hours sampling at 22.5 ml/minute (i.e. one month of continuous sampling). Experience with electrets left out in rooms shows that sufficient charge is retained for at least two weeks and normally 4 weeks for efficient sampling to take place.

Semi-quantitative measurements of the electrical mobility of amosite fibres in the dust box showed that there was a reasonable charge remaining on the particles produced by the fluidised bed generator, even though attempts to neutralise the charge had been taken. This may mean that the sampling rate calculated from the laboratory experiments is different from what may be achieved in the field. But this is dependent on the electrical mobility of the dust aerosol monitored. At the present time it is suggested that the dust box value of 22.5 ml / minute is used, but changes in the electrical mobility can be expected.

Analytical Sensitivity

Based on the calculated sampling rate of 22.5 ml / minute, that an electret to filter deposit concentration of 28.4 is achieved and it is feasible to examine about half the syringe deposit by step-scanning the Walton-Becket graticule: about 0.4 ml of air will be analysed for every minute of sampling. This means that the approximate analytical sensitivities in table 1 should be achieved.

Table 1: Calculated analytical sensitivity.

Time of sampling (minutes)	Air volume sampled (l)	Air volume analysed (ml)	Analytical sensitivit (F/ml)
10	0.23	4	0.25
240 (4 hour)	5.4	96	0.01
480 (8 hour)	10.8	192	0.005
2400 (working week)	54	960	0.001

For PCM counting the electrets have been found to be contaminated with other non-asbestos fibres and even if this contamination can be overcome, the laboratory blank samples alone may be expected to produce counts of up to 10 f/mm^2 or 12 fibres in 150 Walton - Becket graticules. This means that 10 minute and four hour samples will have a background concentration of 3.0 and 0.12 f/ml respectively. Therefore discrimination will be necessary, as any PCM screening of the electrets will be of limited use because of the frequency which non-asbestos fibres are likely to be found.

Discrimination methods and probable limit of detection

Only in well-controlled cases (as in this study) can PCM discrimination be used and in field studies, at the very minimum, the chemistry of each countable fibre must be obtained using EDXA. An ISO method which uses the SEM to discriminate fibres based on the EDXA spectrum is in preparation [14] but is limited to fibres widths greater than 0.2 - 0.3 μm and should only be used with known sources of fibres. ISO methods for discriminating and positively identifying asbestos fibres by analytical TEM using EDXA and selective area electron diffraction have been published [12, 13] and can be applied to all fibres from both known and unknown sources.

The discrimination counts applied to the amosite samples contaminated by organic fibres showed that both methods can be used, although there will always be uncertainty with the SEM counts. On balance it will be quicker to analyse and discriminate the fibres by analytical TEM, which will also give a much more definitive identification of the asbestos fibre types present, if required.

The biggest advantage of applying discrimination to field samples is that the background of > 5 μm long asbestos fibres is usually well-below the analytical sensitivity of the method, so any asbestos fibre found is likely to have some significance. Obviously cross-contamination from other samples or with asbestos must be closely controlled and monitored. Therefore it is expected that the published ISO methods [12,13] will be used to analyse field samples. The ISO method reports low fibre counts based on the upper Poisson 95% confidence limit (e.g. 0 fibre is reported as <2.99 fibres etc.). This means that the limit of detection will be 0.75 and 0.03 f/ml for the 10 minute and 4 hour sample respectively and will be expected to be of the order of 0.01 f/ml for sampling periods of greater than 12 hours.

Reliability of the passive sampler results

The precision of a fibre count depends on the number of fibres counted. As the passive sampler has a much lower flow rate than the membrane filter - pump method, it will always have poorer precision, even though the sample may be concentrated using an indirect preparation method. In the laboratory experiments, a range of airborne concentrations were generated to compare the precision of both replicate (Figure 2) and single passive samples (Figure 4) with the current MDHS 39/4 membrane filter - reference method. Figure 3 showed that replicate passive samples were within the 95% confidence envelope (62.5 - 163 %) for PCM counts and correlation coefficients of 0.8 were found, when compared to the current MDHS 39/4 method (Figure 2). When a comparison was made based on single pairs of passive and membrane filter samples, again the values were within the MDHS 39/4 95% confidence limits and gave a R^2 = 0.86. Therefore it would seem reasonable to conclude from the laboratory investigations that the passive sampler can be used for surveys in much the same way as conventional membrane filter sampling.

It must be recognised that the sampler has only been tested in the laboratory and a number of practical difficulties may exist in the working environment. For instance water vapour and rain could have a large effect on the charge retained on the electret and the sampling efficiency, it may also disrupt any deposit collected by the sampler. The sampling rate of the passive sampler is dependent on the electrical mobility of the fibres and this can be expected to vary between sites. The range of electrical mobility at maintenance worker sites is unknown at present, but previous work on mass measurements [9] have reported quite a wide spread in various industries, with a tendency for the passive sampler to overestimate. However, it is likely that with the longer sampling periods envisaged in any field study and the relatively limited range of maintenance environments the effective sampling rate will be more constant.

Conclusions

The passive sampler performed reasonably well in laboratory tests when compared with the conventional MDHS 39/4 membrane filter - pump sampling method.

The use of an indirect sample preparation method, which concentrated the fibres collected by the passive sampler showed that the analytical sensitivity of the method could be increased to make it suitable for most short term and long term field sampling uses.

The charged electret inside the passive sampler, as supplied for use, was found to be randomly contaminated with two or more types of non-asbestos fibres, which made the PCM analysis difficult to interpret. These difficulties presented an ideal test for the proposed methods of fibre discrimination, which will be essential in field surveys, where a number of fibre types are likely to be encountered. Both SEM and TEM with fibre discrimination made on the basis of fibre chemistry obtained from the energy dispersive x ray analysis of individual fibres, showed that they could successfully be used.

The sampling rate of the passive sampler is dependent on the electrical mobility of the aerosol. For the initial experiments a sampling rate of 22.5 ml/minute was calculated but it is not known how well this will correlate to fibres released by maintenance work. Further studies of the passive sampler in the field and different laboratory charging conditions are presently underway.

Acknowledgement

The authors would like to thank Mike Hemingway and Richard Brown for the supply of passive samplers and Joe Brammer who counted many of the samples.

References

[1] Peto, J., Hodgson, J.T., Mathews, F.E. and Jones, J.R., "Continuing increase in mesothelioma mortality in Britain." *The Lancet,* 345, March 4, 1995, 535 - 539.

[2] Health Effects Institute - Asbestos Research, "Asbestos in public and commercial buildings: A literature review and synthesis of current knowledge." 1991.

[3] OSHA "Preamble and final rules for asbestos exposure in general industry and construction (revised) and shipyard employment (new)". 59 FR 40964, Aug, 10th, 1994.

[4] CONSAD Research Corporation. "Economic analysis of the proposed revisions to the OSHA asbestos standard for construction and general Industry. OSHA J-9-8-0033, U.S. Dept. of Labor, Washington, DC. 1990.

[5] Health Effects Institute - Asbestos Research, "Asbestos in public and commercial buildings: Supplementary analyses of selected data previously considered by the literature review panel." 1992.

[6] MDHS 39/4 "Asbestos Fibres in Air: Light Microscope Methods for use with the Control of Asbestos at Work Regulations: Methods for the Determination of Hazardous Substances. Health and Safety Executive, Bootle, Merseyside, L20 30Z. (ISBN 1 7176 0356 3). (revised 1995)

[7] Brown, R.C., Wake, D., Thorpe A., Hemmingway, M.A. and Roff, M.W. "Theory and measurement of the capture of charged particles by an electret". Journal of Aerosol Science. vol. 25, no. 1, pp. 49 - 163, 1993.

[8] Brown, R.C., Wake, D., Thorpe A., Hemmingway, M.A. and Roff, M.W., "Preliminary assessment of a device for passive sampling of airborne particulates". Annals of Occupational Hygiene, vol. 38, no. 3, pp. 303 -318. 1994.

[9] Brown, R.C., Hemmingway, M.A., Wake, D. and Thompson, J., "Field trials of an electret-based passive dust sampler in metal processing industries". Annals of Occupational Hygiene, vol. 39, no. 5, pp. 603 - 622, 1995.

[10] Chatfield E.J. "Measurement and interpretation of asbestos fibre concentrations in ambient air". in: Fifth International Colloqium on Dust Measuring Technique and Strategy. Held in Johannesburg, South Africa, October 29 -31, pp.269 - 296, 1984.

[11] Burdett, G. J. (1982) Methods of sample preparation for SEM, TEM and optical analysis of asbestos on membrane filters. in Fourth International Colloqium on dust measuring technique and strategy.Held in Edinburgh, September 20th-23rd. Asbestos International Association.

[12] ISO 13794:98 "Ambient air: - Determination of asbestos fibres - Indirect - transfer transmission electron microscopy method".International Standards Organisation, Geneva.1998.

[13] ISO 10312:95 "Ambient atmospheres: -Determination of asbestos fibres - Direct -transfer transmission electron microscopy method". International Standards Organisation, Geneva.1995.

[14] ISO DIS 14966 "Ambient atmospheres: -Measurement of inorganic fibrous particles- Scanning electron microscopy method". ISO TC146/SC3. International Standards Organisation, Geneva.1999

Paul A. Baron,[1] Gregory J. Deye,[1] Joseph E. Fernback,[1] and William G. Jones[2]

DIRECT-READING MEASUREMENT OF FIBER LENGTH/DIAMETER DISTRIBUTIONS

REFERENCE: Paul A. Baron, Gregory J. Deye, Joseph E. Fernback, and William G. Jones, **"Direct Reading Measurement of Fiber Length/Diameter Distributions,"** *Advances in Environmental Measurement Methods for Asbestos, ASTM STP 1342,* M. E. Beard, H. L. Rook, Eds., American Society for Testing and Materials, West Conshohocken, PA, 2000.

ABSTRACT: A new technique for classifying fibers according to length has opened up opportunities for improving fiber measurement technology. The classification principle is dielectrophoresis, which moves conductive fibers at a velocity proportional to the square of their length. Even non-conductive fibers can be made sufficiently conductive by increasing the humidity in the classifier. A device based on this principle has been constructed. It can classify fibers from a broad distribution into length distributions with CVs in the range of 0.2-0.3 and with mean lengths selectable from about 4 µm to >50 µm. To complement this device, diameter classification readily can be achieved by gravitational or inertial techniques in combination with length classification. An inertial classifier was used to produce relatively monodisperse diameter fibers in the aerodynamic diameter range of 3 - 7 µm. This device was used to calibrate the response of the Aerodynamic Particle Sizer (APS) so that real-time diameter distributions were obtained for each length distribution from the dielectrophoretic length classifier. By varying the voltage on the length classifier, a length/diameter distribution can be produced in a matter of minutes

KEYWORDS: fiber, fiber measurement, fiber classification, fiber size distribution, dielectrophoresis

---

INTRODUCTION

Fiber measurement has traditionally been carried out using sample collection and subsequent analysis by microscopy. This approach has been taken by necessity in the absence of real time measurement techniques. Sample collection is prone to biases due to inertial, gravitational and electrostatic effects. Sample preparation is a source of bias that typically causes a reduction of measured fibers. Light microscopy has limited sensitivity to small diameter fibers; scanning electron microscopy has improved sensitivity and allows elemental analysis of individual fibers, though with increased cost. Transmission electron microscopy allows detection of all fiber sizes, permits measurement of fiber elemental composition and crystal structure, though with a further increase in cost. All forms of microscopy are limited in precision due to small number of fibers analyzed. Under optimum conditions, a sample can be collected, prepared, and analyzed by light microscopy

[1] Scientist, US Department of Health and Human Services, US Public Health Service, Centers for Disease Control and Prevention, National Institute for Occupational Safety and Health, Cincinnati OH 45226,
[2] Scientist, US Department of Health and Human Services, US Public Health Service, Centers for Disease Control and Prevention, National Institute for Occupational Safety and Health, Morgantown WV 26505

in less than an hour, while typical SEM and TEM analysis times are typically longer than a day, especially when a good estimate of the size distribution is desired. These various drawbacks limit the type of measurement and research that can be carried out on fiber behavior and toxicity.

A direct reading instrument, the fibrous aerosol monitor (FAM) was developed in the late 1970s[1] and is still commercially available (Model FM-7400, MIE, Inc. Bedford MA). It aligns individual airborne fibers in an electrostatic field and measures their light scattering patterns at 90°. Theoretical calculations have indicated that this approach can be used to extract size information regarding the fibers.[2] Scanning electron microscope measurements on 0.85 μm diameter glass fibers indicated that the FAM gave comparable length distributions [3]. Further measurements are needed to confirm this capability for a range of fiber diameters and types.

There are clearly gaps in the technology of accurate, real-time measurement of fiber length and diameter. Measurement of compact particles was in a similar state in the early 1970s. Development of the differential mobility analyzer, inertial classifiers, and the aerodynamic particle sizer allowed direct reading measurement of submicrometer and larger particles, respectively [4,5]. These measurement instruments produced an explosion in research and understanding of airborne particles in a range of environments, including atmospheric, workplace and clean room situations. These instruments also allowed development and testing of simpler devices for measurement and classification of particles.

Size classification of fibers by length and diameter is difficult. Early attempts to size segregate fibers from a liquid suspension met with very moderate success. The primary impediment is the ability to separate fibers according to length. Diameter separation appears likely once length separation can be effected, because several studies have shown that gravitational and inertial behavior depends primarily on fiber diameter and only weakly on length. Recently, a technique was developed using fiber charging to separate monodisperse diameter fibers by length [6]. This appeared to be successful, but no follow-up work has been performed. We have taken the suggestion of Lipowicz and Yeh [7] that fibers can be classified by length using dielectrophoresis and developed a differential fiber length classifier. The principles, implementation and application of this technique are described below.

MEASUREMENT PRINCIPLES

Conductive fibers placed in an electric field can be readily polarized and aligned parallel to the field. The degree of conductivity required for this polarization is relatively small, since it only requires the motion of a few charges over the length of the fiber. This polarization was observed in the fibrous aerosol monitor for asbestos fibers, even though asbestos is normally considered a good insulator. Glass fibers on the other hand did not always polarize so easily. Relative humidity levels on the order of 50% were required to align fibers in early versions of the FAM. Later versions used an increased voltage to align glass fibers at lower humidity levels. The higher humidity caused water condensation on the surface of the fibers that allowed sufficient current to flow, resulting in fiber alignment.

Conductive fibers placed in a gradient electric field will first align parallel to the field, as indicated above, and then move toward the region of higher field intensity. This technique was described by Lipowicz and Yeh and shown to work for aluminum fibers in liquid suspension [7]. The velocity of fibers in a gradient electric field was described by the equation

$$v = \frac{K_m \varepsilon_0}{36\eta} DL \left\{ g(\beta) \left(\frac{\alpha}{\alpha - 1} - f(\beta) \right) \right\}^{-1} \nabla(E^2) \quad (1)$$

where D and L are the fiber length and diameter, β is the aspect ratio, f and g are complex functions of β, E is the electric field, and the rest of the parameters are physical constants. The net result of this equation for conductive fibers in a specific classifier is that the velocity is proportional to fiber length squared and to voltage squared.

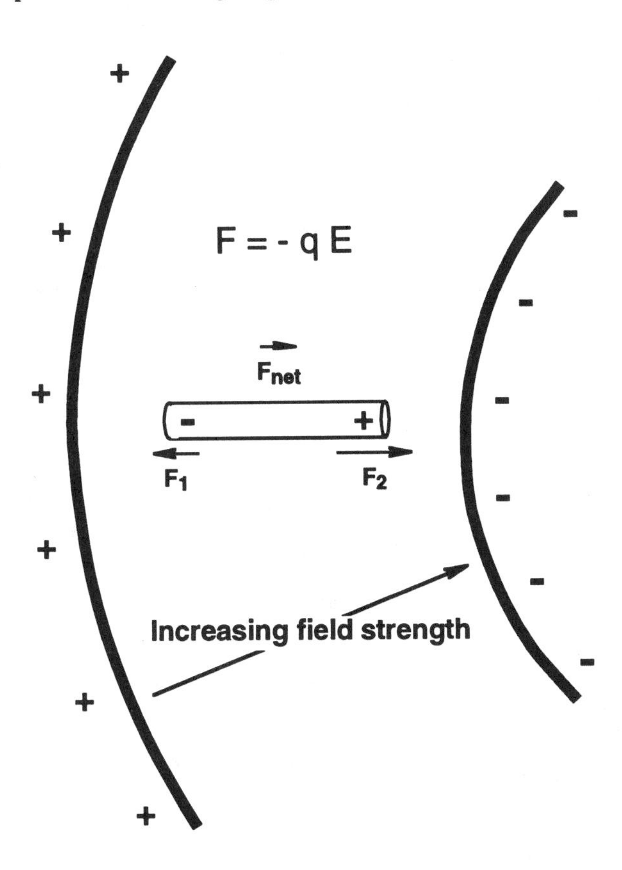

FIG. 1--*An overview of the fiber length classifier configuration with a fiber placed in the annular space between two concentric cylinders*

The dielectrophoresis technique was applied to airborne chrysotile fibers and shown to work for fibers in the length range of about 5 to 50 μm [8]. The classifier section consisted of two concentric tubes with a space of about 3 mm through which the fibers passed (Figure 1). Longer fibers deposited at the upper end of the inner tube, while shorter fibers deposited at progressively longer distances down the tube. Fibers deposited at several distances on the inner tube were removed and analyzed by TEM.

The voltage applied between the tubes was originally a sine wave, but was then changed to a square wave (50 - 100 Hz), with the peak voltages ranging from 0 to about ±7000 V. The square wave produced twice the rms electric field as the sine wave with the same peak voltage. The peak voltage was limited by corona or discharge breakdown in the classifier. The alternating voltage was used to reduce the likelihood that charged particle motion (electrophoresis) would dominate the fiber behavior. Han et el. showed that fibers with low charge levels will oscillate about the trajectory they would have taken if no charge were present [9]. Since it was relatively straightforward to reduce the charge level to Boltzmann equilibrium using a ^{85}Kr source, the use of an alternating electric field ensured proper classification of fibers. Balancing the ac voltage to give zero average voltage was also found to be important to obtain optimum resolution.

Several additional factors were found to be important in separating fibers by this technique. Fibers had to be sufficiently conductive. As noted with the FAM, glass fibers were not sufficiently conductive at relative humidity below 50% to allow optimum separation. At high input aerosol concentrations, a second neutralizer was sometimes used. In addition, the high voltages required to classify fibers were close to the breakdown voltage of air. Therefore, all surfaces that could exacerbate corona discharges were rounded and coated with a thin dielectric layer. In addition, surfaces on which fibers could deposit were coated with an oil and grease mixture to prevent fiber resuspension.

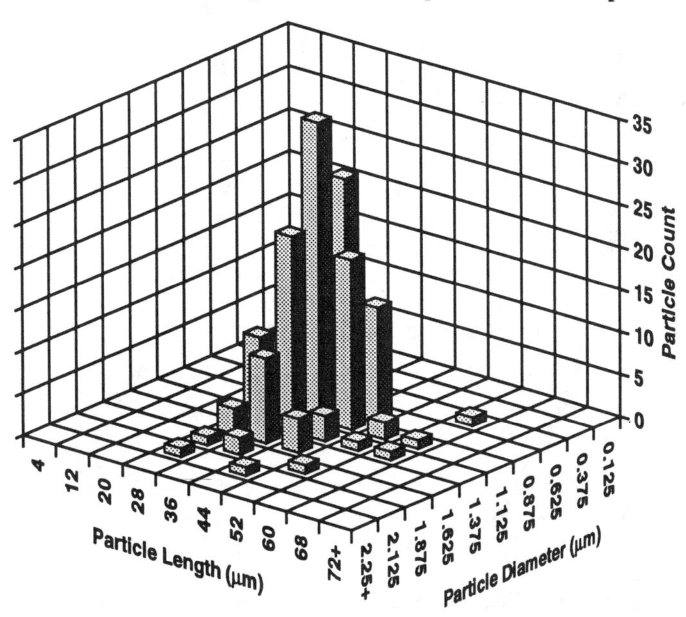

FIG 2--*A size distribution of glass fibers produced with the fiber length classifier and measured using SEM.*

The classifier developed by Baron et al. [8] was modified to operate in a differential mode so that a minor flow was extracted at the bottom of the classifier near the inner tube. This minor flow contained fibers that had been attracted to the inner tube, but had not yet deposited. This flow contained narrow length-distribution fibers, as opposed to the earlier

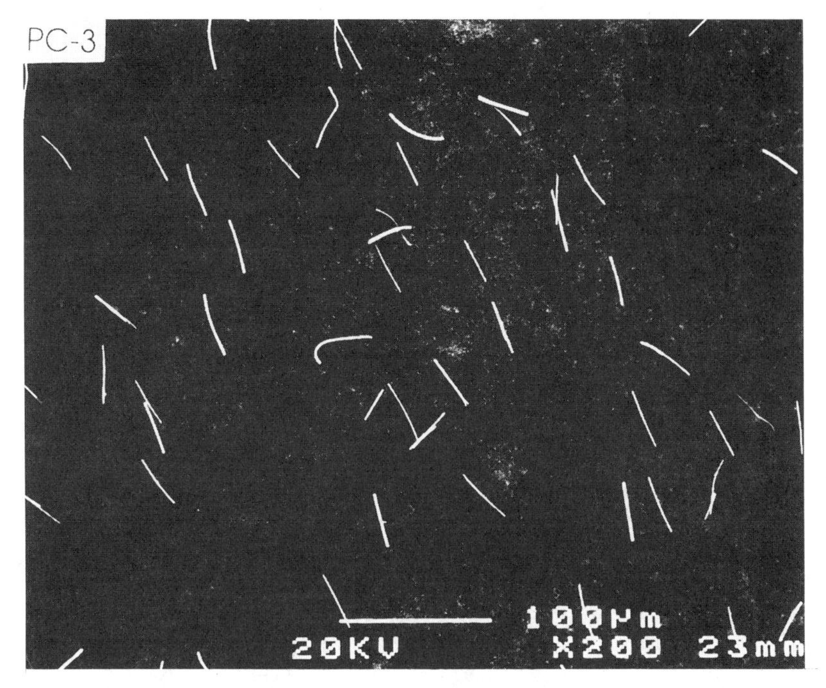

FIG. 2b--*A picture of glass fibers produced with the fiber length classifier*

classifier in which the exiting flow contained all fibers shorter than the length depositing on the inner tube. In the differential mode, the major flow contained fibers shortest fibers, the longest fibers were deposited on the inner tube, and the intermediate length fibers were in the minor flow.

The fibers in the minor flow were measured as a function of flow rate and voltage. Equation 1 appeared to give an adequate description of the classifier behavior. However, there are still some problems with the classifier since the spread of the classified fiber lengths is about twice that predicted. It is suspected that the region in which the minor and major flows are split may contain turbulence which reduces the resolution of the classifier. An example size distribution is shown in Figure 2a and a picture of the fibers in this distribution is shown in Figure 2b.

CLASSIFIER APPLICATIONS

A fibrous aerosol classifier can be used in two ways. Once the behavior of the classifier has been calibrated, the device can be used to measure fiber lengths. Alternatively, the output of the classifier can be used for other experiments.

The fiber length classifier has been used with several particle counters to indicate size distributions. An optical particle counter was used to indicate the concentration at each of several classifier voltage levels to indicate the fiber length distribution. However, it is often more useful to obtain both length and diameter for a distribution. Several possibilities are available. Inertial sizing has been shown to separate fibers primarily according to diameter. Therefore, with length-classified fibers, an inertial classifier can be used to produce narrow diameter distributions. If this inertial classifier is tunable, then the output of the two classifiers in series can be used to determine a size distribution. We have used a tunable virtual impactor to produce monodisperse diameter fibers. This device was developed by V.Toporkov (Novosibirsk, Russia) and marketed by GIV (Breuberg Germany). This classifier was capable of separating particles in the aerodynamic diameter range of 3 - 7 μm and producing a narrow diameter distribution with a standard deviation of about 10%. One of the problems encountered with the use of both classifiers simultaneously is that the number of fibers penetrating each classifier is on the order of 1% of the aerosol entering the classifier. Thus the concentration at the inlet of the first classifier must be high in order to detect a significant number of fibers at the exit of the second classifier. It was found that high concentrations caused mixing of the aerosol particles in

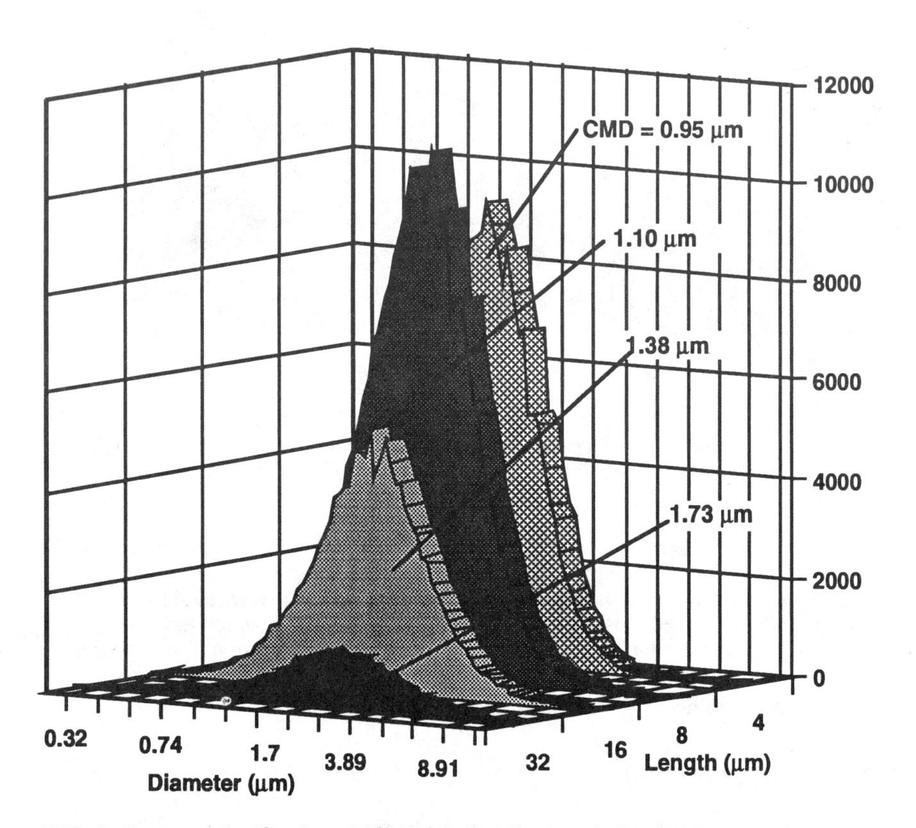

FIG. 3--*An example fiber length/diameter distribution of Owens-Corning AAA10 microfiber measured using a combination of the fiber length classifier and the Aerosizer. The diameter measured with the Aerosizer has not been calibrated and is assumed to be approximately proportional to physical diameter.*

the virtual impactor, producing a distribution with a large fraction of small particles in addition to the expected mode. Reduction of the challenge concentration resulted in poor counting statistics of particles exiting the system.

Another way to make length/diameter distributions is to use an aerodynamic sizing instrument to measure the output of the length classifier. Two commercial instruments (Aerodynamic Particle Sizer [APS3300], TSI, Inc. St., Paul MN; Aerosizer, Amherst Process Instruments, Hadley MA) measure the velocity of particles accelerated through a nozzle. This velocity is interpreted in terms of particle aerodynamic diameter. The behavior of fibers in such high acceleration flow fields has not been examined in detail, but it appears possible that these instruments can be calibrated to indicate fiber diameter for known fiber lengths. A calibration curve of this type is being developed for the APS3300. Figure 3 shows a size distribution measured with the Aerosizer. The diameters indicated by the Aerosizer have not been calibrated in this figure, so that the physical diameters cannot be directly estimated. However, it should be noted that approximately 10^6 particles were measured over a time of several minutes, indicating good precision in the description of the distribution. Measurement of such a large number of fibers is simply not feasible by microscopy. Further efforts at calibration of the APS3300 are under way. Figure 4 shows a comparison of APS measured size distribution versus SEM-measured diameter distribution for 10 µm long fibers. The physical diameter observed using SEM had to be multiplied by 2.6 to approximately match the aerodynamic diameter observed in the real time instrument. A comparison of this type must be made for a range of lengths and diameters in order to use the APS3300 for reliable diameter distribution measurements.

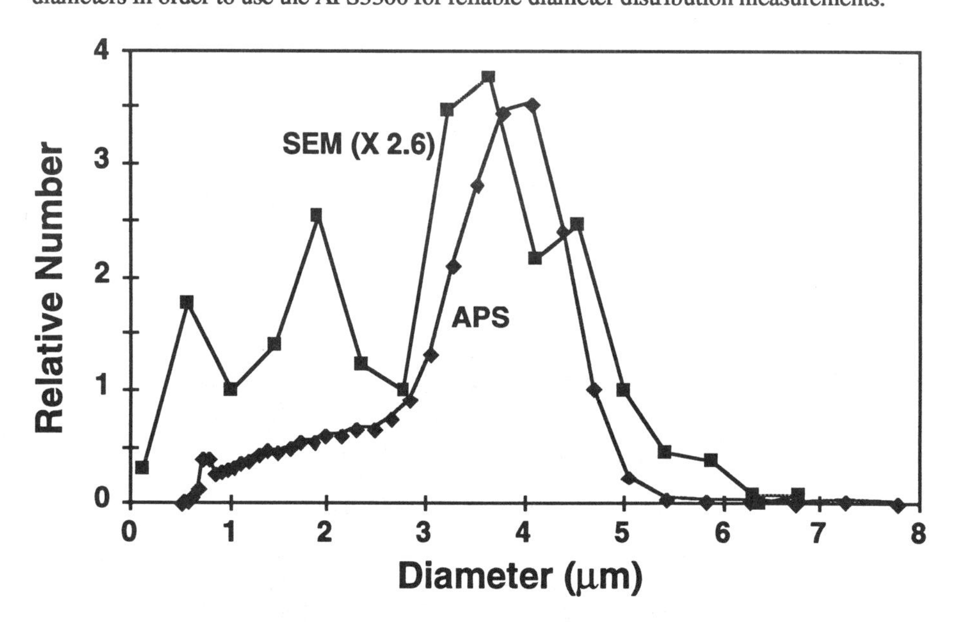

FIG. 4--*Glass fibers (length-classified 10 µm long) were classified with a virtual impactor system and collected for SEM analysis. The diameter of the SEM-measured fibers were adjusted using a factor of 2.6 to match the APS-measured diameter distribution.*

The fiber length classifier can be used for production of small quantities of selected lengths. An evaluation of the physical separation mechanism indicates that it is not feasible to scale the classifier to large dimensions that larger quantities can be generated. However, a successful effort was made to generate milligram quantities of fibers of <5, 5, 8, 22, 35, 50 µm length. These fibers were used in scaled-down *in vitro* assays to determine the effect of fiber length on macrophage response [10]. It was found that there was a clear difference in response between fibers 8 μm and shorter versus the fibers 22 μm and longer. From SEM images of fibers in this study, it was clear that the shorter fibers could be engulfed by the macrophages, while the longer ones penetrated the macrophage walls, causing measurable damage.

Other experiments are also planned. The classifier can be used to quantitatively assess penetration of fibers through other systems and devices. The development of a sampler is planned for the collection of only fibers that might reach the lung region of the respiratory system, defined as thoracic fibers by the International Standards Organization and the American Conference of Governmental Industrial Hygienists [11]. The fiber classifier will aid in the assessment of this sampler. This work is being carried out in collaboration with the Health and Safety Executive in the UK. The length- and diameter-dependent response of the FAM also will be investigated.

CONCLUSION

The fiber length classifier shows great promise in opening up areas of research and understanding of fiber behavior. The classifier has been demonstrated to separate fibers primarily according to length over the range of 5 μm to greater than 50 μm. The instrument can be used for fiber size distribution measurement as well as for production of fibers for other experiments.

DISCLAIMER

The mention of product or company name does not constitute endorsement by the Centers for Disease Control and Prevention.

REFERENCES

[1] Lilienfeld, P., Elterman, P., and Baron, P., "Development of a Prototype Fibrous Aerosol Monitor," American Industrial Hyiene Assocociation Journal, Vol. 40, No. 4, 1979, pp. 270-282.

[2] Lilienfeld, P., "Light Scattering from Oscillating Fibers at Normal Incidence," Journal of Aerosol Science, Vol. 18, No. 4, 1987, pp. 389.

[3] Marijnissen, J., Lilienfeld, P., and Zhou, Y., "A Laser Monitor for the Fiber Deposition in a Lung Model," Journal of Aerosol Science, Vol. 27, No. Suppl. 1, 1996, pp. S523-S524.

[4] Yeh, H.-C., "Electrical Techniques," in Aerosol Measurement, eds. K. Willeke and P. A. Baron (New York: Van Nostrand Reinhold, 1993), pp. 410-426.

[5] Baron, P. A., Mazumder, M. K., and Cheng, Y. S., "Direct Reading Techniques Using Optical Particle Detection," in Aerosol Measurement: Principles, Techniques and Applications, eds. K. Willeke, and P. A. Baron (New York: Van Nostrand Reinhold, 1993), pp. 381-409.

[6] Chen, B. T., Yeh, H. C., and Hobbs, C. H., "Size Classification of Carbon Fiber Aerosols," Aerosol Science and Technology, Vol. 19, No. 2, 1993, pp. 109-120

[7] Lipowicz, P. J. and Yeh, H. C., "Fiber Dielectrophoresis," Aerosol Science and Technology, Vol. 11, No. 3, 1989, pp. 206-212.

[8] Baron, P. A., Deye, G. J., and Fernback, J., "Length Separation of Fibers," Aerosol Science and Technology, Vol. 21, No. 2, 1994, pp. 179-192.

[9] Han, R. J., Moss, O. R., and Wong, B. A., "Airborne Fiber Separation by Electrophoresis and Dielectrophoresis: Theory and Design Considerations," Aerosol Science and Technology, Vol. 21, No. 3, 1994, pp. 241-258.

[10] Blake, T., Jones, W., Schwegler-Berry, D., Baron, P., and Castranova, V., "Effect of Fiber Length on Glass Microfiber Cytotoxicity," American Journal of Respiratory and Critical Care Medicine, Vol. 155, 1997, p. A959.

[11] Baron, P. A., "Application of the Thoracic Sampling Definition to Fiber Measurement," American Industrial Hyiene Assocociation Journal, Vol. 57, No. 9, 1996, pp. 820-824.

Eric J. Chatfield[1]

INTERNATIONAL ORGANIZATION FOR STANDARDIZATION METHODS FOR DETERMINATION OF ASBESTOS IN AIR

REFERENCE: Chatfield, E. J., "**International Organization for Standardization Methods for Determination of Asbestos in Air,**" *Advances in Environmental Measurement Methods for Asbestos, ASTM STP 1342,* M. E. Beard and H. L. Rook, Eds., American Society for Testing and Materials, West Conshohocken, PA, 2000.

ABSTRACT: Two International Standards for measurement of asbestos in air by transmission electron microscopy (TEM) have been approved. A key feature of the ISO TEM methods is the ability, from a single examination, to derive fiber concentrations for comparison with all known legislated control limits, and to provide a coded description of the size and nature of each asbestos structure to assist in the interpretation of health effects. The direct-transfer TEM method for determination of asbestos in air (ISO 10312) was published as an International Standard in May 1995. Voting on an indirect-transfer TEM method (ISO 13794) has been completed, and the final draft has been submitted for publication. A method for determination of the concentration of inorganic fibers in air, based on scanning electron microscopy (SEM), has recently been submitted for voting as a Draft International Standard, currently designated as ISO/DIS 14966. A new working group, ISO/TC 146/SC 6/WG4, was formed in late 1996, to develop International Standards for sampling strategy for determination of asbestos and other mineral fibers in building atmospheres.

KEYWORDS: Asbestos, fiber, air, water, ambient, atmosphere, transmission electron microscopy, scanning electron microscopy, international standard, building, indoor air

Introduction

The first meeting of a Joint Working Group ISO/TC 146/SC 3/WG1 - ISO/TC 147/SC 2/WG18 was held in Luxembourg in 1978, to develop International Standards for determination of asbestos in ambient air and water. At that time, major differences of opinion existed as to the utility of optical microscopy, scanning electron microscopy (SEM) and transmission electron microscopy (TEM) for determination of asbestos. For water sample analysis, these differences of opinion were eventually resolved by conducting an international inter-laboratory study in which filters were sent

---

[1]President, Chatfield Technical Consulting Limited, 2071 Dickson Road, Mississauga, Ontario, Canada L5B 1Y8

to the participating laboratories for analysis. The filters were prepared using reference dispersions of chrysotile and amosite, and also samples of drinking water from Sherbrooke, Québec (naturally-occurring chrysotile) and Beaver Bay, Minnesota (naturally-occurring amphibole). The results of this study clearly demonstrated to the proponents of optical microscopy that, in water samples, almost all of any asbestos fibers present were below the limits of optical resolution. The filters prepared from the drinking water samples also convinced most of the proponents of the SEM that this instrument was not appropriate for determination of asbestos in this type of sample.

Following the largely negative results of animal feeding studies, interest in analysis of water for asbestos eventually declined and the Working Group then concentrated on development of TEM and SEM methods for determination of asbestos in ambient air. However, one important legacy of the inter-laboratory study was the recognition that, in determination of fiber concentrations, in either water or air, by any microscopical counting method, it is important to specify a minimum fiber length in the counting criteria. In the ISO inter-laboratory analyses, a minimum aspect ratio of 3:1 was specified for definition of a fiber; as for all such analytical methods of the time, no other restrictions were applied. In retrospect, it now seems obvious that in using an instrument capable of resolving better than 0.2 nm, specification of a countable fiber as any particle with a minimum aspect ratio of 3:1, no matter how small, would provide considerable opportunity for individual interpretation. The data in Table 1 clearly show that if a minimum fiber length is not specified in the fiber counting criteria, different analysts will select their own, resulting in considerable variability in the reported fiber concentrations. The proportion of fibers shorter than 0.5 μm in the reported data varied from 0 to 45%. A minimum fiber length criterion of 0.5 μm has since been incorporated in all methods for determination of asbestos by TEM.

Table 1--*ISO inter-laboratory study (1981) - percentage of reported fibers shorter than 0.5 μm*

Laboratory Number	Chrysotile Sample 1	Chrysotile Sample 2	Sherbrooke Drinking Water	UICC Amosite	Beaver Bay Drinking Water
1	11	6	31	11	21
8	38	44	21	11	16
10	18	3	2	0	0
12	31	27	45	23	25
13	7	10	15	3	15
15	14	2	25	5	6

The other important result from the inter-laboratory study, relevant to the development of measurement methods for both water and air, was confirmation that the basic TEM analytical method was reproducible. Table 2 shows the results from the six participating laboratories which had used the polycarbonate preparation procedure and TEM instrumentation with selected area electron diffraction (SAED) facilities. Data from three other laboratories were rejected. One of these laboratories used an obsolete TEM without selected area electron diffraction (SAED) facilities, and one prepared TEM specimens from cellulose ester filters using a method known to incur unreproducible fiber losses. The preparation method used by the third laboratory was reported to have yielded TEM specimens exhibiting significant numbers of empty replicas from which particles had been lost, and data from all five samples indicated that the deposits of fibers on these TEM specimens were uneven.

Table 2--*ISO inter-laboratory study (1981) - reported concentrations of fibers on filters after rejection of fibers shorter than 0.5 μm, fibers/mm²*

Laboratory Number	Chrysotile Sample 1	Chrysotile Sample 2	Sherbrooke Drinking Water	UICC Amosite	Beaver Bay Drinking Water
1	1211	8558	1299	297	260
8	1189	8111	1477	673	291
10	1127	4081	1576	478	303
12	1148	7086	2669	399	339
13	1161	5093	1694	443	497
15	1520	6790	1534	424	419
Mean	1226	6620	1708	452	352
Coefficient of Variation	0.12	0.26	0.29	0.27	0.25

Development of ISO 10312 - Ambient Air - Determination of Asbestos Fibres-Direct-Transfer Transmission Electron Microscopy Method

Historical Development

A procedure for preparation of TEM specimens from polycarbonate filters [*1*] and a protocol for identification of asbestos fibers in the TEM had been incorporated in the U.S. EPA Analytical Method for Determination of Asbestos Fibers in Water [*2*]. These procedures were adopted in ISO 10312 for air samples. Concurrently, an improvement of the filter collapsing method for preparation of TEM specimens from cellulose ester (CE) filters published by Ortiz and Isom [*3*] had been under development in the United Kingdom [*4,5*]. In these methods, the problem of fiber losses due to engulfment of embedded fibers during solvent collapsing of CE filters was addressed by the addition of a plasma etching step to the procedure. The Burdett and Rood procedure [*5*] was adopted by the ISO working group as the preparation method for samples collected on CE filters.

For CE filters, methods for collapsing the filter other than the Burdett and Rood procedure were considered. The filter collapsing method using acetone vapor in a petri-dish as originally described by Ortiz and Isom [*3*] was not incorporated as an optional method for two reasons. It was found that filters collapsed in this way often tended to adhere strongly to the glass slide, and stretching of the carbon-coated, collapsed filter during separation from the slide frequently resulted in broken carbon replica in the final TEM specimens. Also, the requirement to restrain the filter from curling during the collapsing step incurs the destruction of a significant proportion of the filter medium which otherwise could be used for repeat preparations when required. Collapsing of filters by use of acetone vaporizing devices [*6*], such as those used for clearing filters for phase contrast microscopy (PCM), was considered unsatisfactory because of particle movement, evidence for which can be observed around the edges of the filters after this procedure has been used. It was also considered that the Burdett and Rood filter collapsing method, in which a liquid medium is absorbed into the filter structure by capillary action from below, could result in partially embedded fibers being moved closer to the filter surface.

Other methods for direct-transfer of particles and fibers from CE filters to TEM specimens, such as the early methods in which portions of a CE filter were placed on carbon-coated TEM specimen grids and subjected to either Jaffe or condensation washer dissolution [*7*] were rejected. These methods had been shown to produce very high and variable losses of fibers, to the point that repeat analyses of identical filters were unreproducible [*8*].

Specification of the plasma etching procedure was difficult, because plasma etching units are variable in performance, and many do not have instrumentation which allows accurate monitoring of the process. A performance-based procedure was adopted, in which the time to completely ash an uncollapsed CE filter of the same type is measured, and the power is adjusted to a value such that oxidation of the filter is complete in a period of approximately 15 minutes.

In view of the inability of the medical and biological community to arrive at a consensus regarding the relevant variables to be measured, it was decided that, so far as possible, the method should permit a record to be made giving a complete description of each asbestos structure detected during the TEM examination. The structure counting criteria eventually adopted permit the recorded data to be interpreted in terms of any known legislated control limits of any country.

In May 1995, the first of the ISO methods for determination of asbestos in air was published [*9*].

ISO 10312 - Method Description

Air samples may be collected on either 0.4 μm pore size polycarbonate (PC), capillary-pore filters, or cellulose ester (CE) filters of 0.45 μm porosity. The use of conductive filter cassettes with short cowls is specified. Since sampling characteristics are a function of face velocity, the range of face velocity is limited to 4 - 25 cm/s. For TEM specimen preparation, PC filters are carbon-coated, and a conventional Jaffe washer [*10*] is used for dissolution of the filter medium. Chloroform, 1-methyl-2-pyrrolidone, or a mixture of 1-methyl-2-pyrrolidone and 1-1-diaminoethane are permitted as solvents. TEM specimens are prepared from CE filters by the Burdett and Rood procedure [*5*], using either acetone or dimethylformamide as the solvent in the Jaffe washer.

Fiber identification is achieved by a combination of fiber morphology, SAED and energy dispersive x-ray analysis (EDXA). It is recognized in the method that some applications require more definitive fiber identification than others. For example, routine monitoring of a well-characterized emission source does not require the unequivocal identification of fibers which is needed when legal challenges are possible. When an attempt is made to identify every fiber, it is found that, for instrumental reasons, it is usually not be possible to do so. The method takes account of instrumental limitations by requiring a declaration of the target degree of identification before the analysis is commenced, and, for each fiber, a statement of the successful identification actions that were achieved. The identification level achieved for each fiber is recorded as a two or three letter code. The identification codes are particularly useful in the interpretation of quality assurance analyses, where a prior examination of the specimen may have resulted in damage to fibers such that SAED patterns cannot be obtained from fibers in subsequent examinations of the same specimen, or where the same specimen is successively examined in instruments with different specimen stage geometries.

A key feature of ISO 10312 is the ability, from a single TEM examination, to derive fiber or structure concentrations for comparison with all known legislated control limits, and to provide a coded description of the size and nature of each asbestos structure to allow re-evaluation of the data in terms of new risk criteria which may become available. It was recognized that classification of fibrous structures into simple fiber, bundle, cluster and matrix categories [*11*] was not sufficiently descriptive to achieve this. If, for example, a control limit is specified in terms of asbestos structures [*12*], with no legislative importance assigned to either the nature or size of the asbestos structures, a simple count of isolated fibrous entities is sufficient. However, if the control limit is specified in terms of fibers longer than 5 μm, as in Ontario, Canada [*13*], all fibers longer than 5 μm must be included in the count, regardless of whether they are present as isolated fibers or as components of more complex fibrous structures. Accordingly, the morphological fiber counting criteria of ISO 10312 are very detailed, but conceptually very simple. A fibrous structure is first classified as a fiber, bundle, cluster or matrix. Clusters are further subdivided into disperse clusters and compact clusters. In a disperse cluster, it is possible to visualize both ends of at least one of the constituent fibers or bundles, whereas for a compact cluster it is not possible to associate any two ends as belonging to one fiber. The cluster is assigned a code "CDxy" or "CCxy", depending on whether it is a disperse or compact cluster, where x and y are each integers 0 - 9 or a

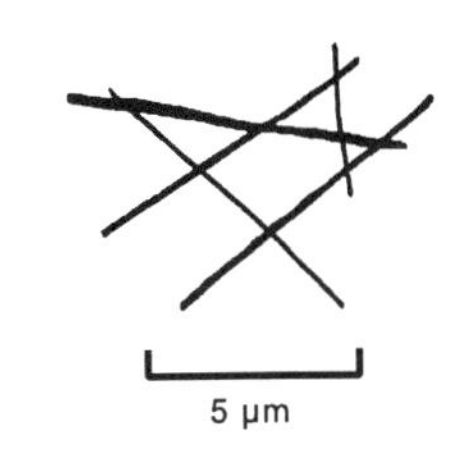

Counted as a disperse cluster consisting of 5 fibers, 4 of which are longer than 5 μm.

Recorded as CD54, followed by 5 fibers, each recorded as CF along with its dimensions.

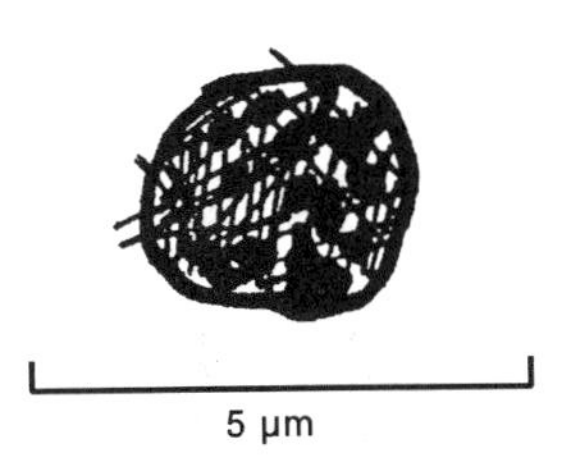

Counted as one compact matrix, containing more than 9 fibers, none of which is longer than 5 μm.

Recorded as MC+0

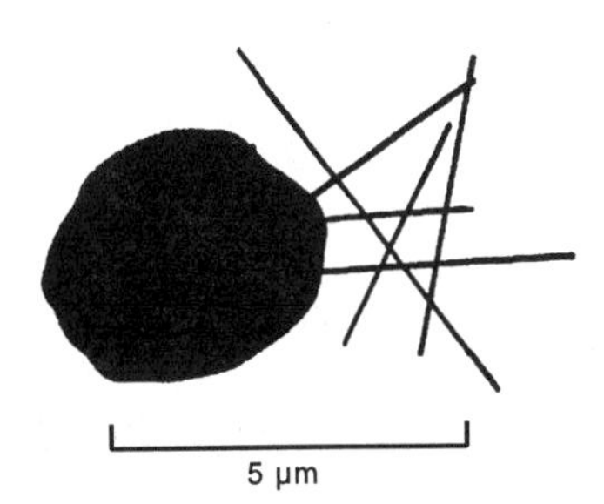

Counted as a disperse matrix, containing 6 fibers, one of which is longer than 5 μm.

Recorded as MD61, followed by 3 fibers each recorded as MF along with its dimensions, and one matrix residual (the lengths of 3 fibers cannot be defined and they remain as part of a matrix). The matrix residual is recorded as MR30.

Figure 1--*Examples of ISO fiber counting procedure and morphological coding system*

"+" sign. The value x is the total number of constituent fibers or bundles in the cluster, and the value y is the number of these longer than 5 μm. If the value of x or y exceeds 9, a "+" sign is used. The same procedure is used for matrices, which are also subdivided into disperse and compact varieties. The corresponding codes used for matrices are "MDxy" and "MCxy". Figure 1 illustrates several examples of the procedure. For disperse clusters and matrices, there is a requirement to tabulate each measurable fiber or bundle, along with its dimensions, up to a maximum of 5, in decreasing order of length. Constituent fibers or bundles in a cluster or matrix are designated by the codes CF, CB, MF or MB, to distinguish them from isolated fibers and bundles. When additional fibers or bundles are present in a cluster or a matrix, the balance of the structure is recorded as a cluster residual (CRxy) or a matrix residual (MRxy).

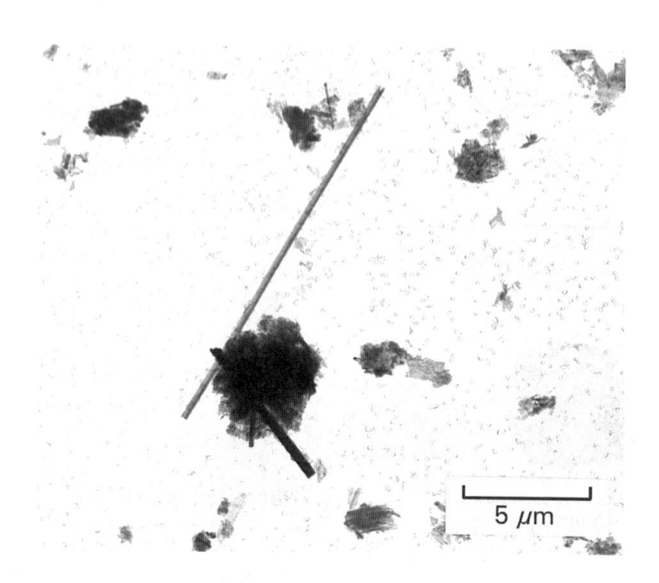

Figure 2--*TEM micrograph showing amphibole matrix type MD32*

Figure 2 shows an example of a simple matrix found in an air sample taken close to an emission source. The identification code assigned is ADQ, representing identification of one of the amphibole fibers in the structure as amosite by both electron diffraction and quantitative EDXA. The morphological code for this structure is MD32, indicating that it is a disperse matrix containing 3 fibers, 2 of which are longer than 5 μm.

The ISO structure counting criteria were designed to provide concise information concerning the nature of each airborne asbestos structure as it relates to risks due to inhalation, and to permit interpretation of the measurement in terms of all known legislation. The coding system is an expedient way to provide the relevant information in a concise form. The overall dimensions of the structure are recorded to provide a means of assessing the respirability of the structure. The number of constituent fibers and their dimensions are also recorded to allow calculation of the contribution to the inhalation exposure made by the structure. Although drawings or photographs, as used in verified fiber counting, can be more descriptive, they do not provide a basis for computer generation of fiber size and concentration data. Moreover, making a drawing or recording a micrograph for every fibrous structure in the analysis of samples is time-consuming and expensive.

In the Asbestos Hazards Emergency Response Act (AHERA) structure counting criteria [*12*], the result includes all asbestos structures regardless of size, and a fiber 0.5 μm in length carries the same weight in the calculation of airborne asbestos concentration as a large cluster or bundle of fibers longer than 5 μm. Because of this lack of fiber size discrimination, the result of a structure count following the AHERA criteria is meaningless in terms of health effects, and the result is also not useful for determining compliance with regulations in which particular size ranges of fibers are specified. A measurement performed in accordance with ISO 10312 contains all of the information necessary for comparison against any known legislated standards.

Development of ISO 13794 - Ambient Air - Determination of Asbestos Fibres-Indirect-Transfer Transmission Electron Microscopy Method

Historical Development

Some countries, notably France, the U.K., Canada and the U.S.A., requested that a parallel indirect-transfer TEM method be standardized. The early drafts of ISO 10312 contained an Annex which specified the details of an indirect-transfer procedure. This Annex was deleted during development of the standard, because the working group determined that it was not possible to publish an International Standard which yielded two different results for the same sample.

ISO 13794 - Method Description

The indirect-transfer method ISO 13794 [*14*] is very similar to ISO 10312, with respect to the fiber identification and structure counting criteria. ISO 13794 includes an additional fiber counting protocol for use if it is required to determine asbestos mass concentration. In early measurements of asbestos mass concentration [*15, 16*], attempts were made to disperse chrysotile into single fibrils, and then the integrated fiber volume was calculated from the length and width measurements. However, it is not always possible to achieve complete dispersal of chrysotile, and amphibole fiber types always exhibit a wide range of fiber widths. If fibers of all sizes are counted from a distribution containing a wide range of fiber widths, the contribution to the integrated volume by one fiber or fiber bundle with a large diameter can outweigh the total contribution to the integrated volume made by all of the other fibers [*17*]. In such a case, the value of the reported mass concentration is essentially determined by one fiber. The majority of the mass in any distribution of fibers is contributed by those with the largest diameters, and a total fiber count made at a magnification suitable for detection of the smallest fibers does not produce a statistically-valid determination of the mass, unless the fibers are all within a very narrow range of diameters.

The mass counting protocol incorporated in ISO 13794 requires an initial scan of the TEM specimens to determine the maximum fiber or bundle diameter. The magnification to be used for the fiber count is then adjusted such that a width of 1 mm on the fluorescent screen corresponds to approximately 10% of the width of the largest fiber or bundle. Fibers and bundles are then counted and measured at this magnification, disregarding any fibers or bundles less than 1 mm in diameter. Fiber counting is continued until the integrated volume of all fibers and bundles is at least 10 times the volume of the original large fiber. In this way, the fiber count includes only those fibers and bundles which contribute significantly to the mass concentration, and a statistically-valid value for the mass concentration is obtained.

Although ultrasonic baths have often been used to disperse asbestos fibers during indirect-transfer preparation, analytical methods have not usually specified any particular power. Moreover, the energy absorption by the specimen depends on the container material and the temperature of the water in the bath. Therefore, the energy imparted to the specimen in dispersing fibers has been variable. Where asbestos is already present as a dispersion of single fibrils, as is usually the case with water samples, the power of the ultrasonic bath is relatively unimportant, because it is very difficult to break chrysotile fibrils into shorter lengths [*8*]. However, in air samples fiber bundles are present, and it is important to limit the power of the ultrasonic treatment so that the dimensions of fiber bundles are not significantly changed. Ideally, the ultrasonic treatment should separate the individual components of clusters and matrices, but not result in significant breakdown of fiber bundles. In order to standardize the level of ultrasonic power used in the analysis, a calibration procedure for the ultrasonic bath is specified in ISO 13794. Aqueous dispersal and ultrasonic treatment, carried out during indirect-transfer specimen preparation of air filters, have been shown to result in significant increases in the concentrations of short chrysotile fibers, but to have only a marginal effect on the concentrations of chrysotile fibers longer than 5 μm [*18*].

In ISO 13794, a portion of the filter is ashed in a plasma asher. If CE filters are ashed too rapidly, a point is reached at which an explosive reaction occurs and material is lost from the container. Under some other ashing conditions, the filter shrinks to become a fragment of material which does not disperse in water when it is exposed to treatment in an ultrasonic bath. The procedure for ashing of CE filters is specified in ISO 13794. It has been found that, for some batches of CE filters, the residual ash consists of fragments of a thin silica film, in which the fibers on the original surface of the filter are bonded, but the fibers are much closer to each other than they were on the original filter.

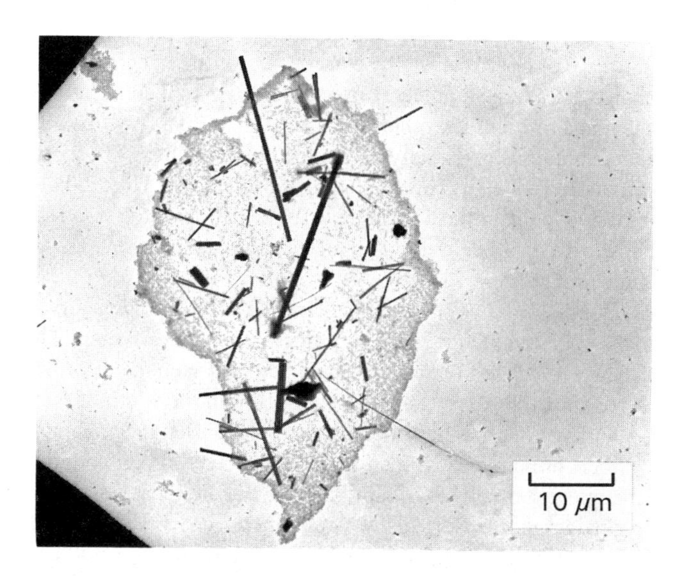

Figure 3--*TEM micrograph showing amosite fibers trapped in a silica film, after indirect-transfer preparation*

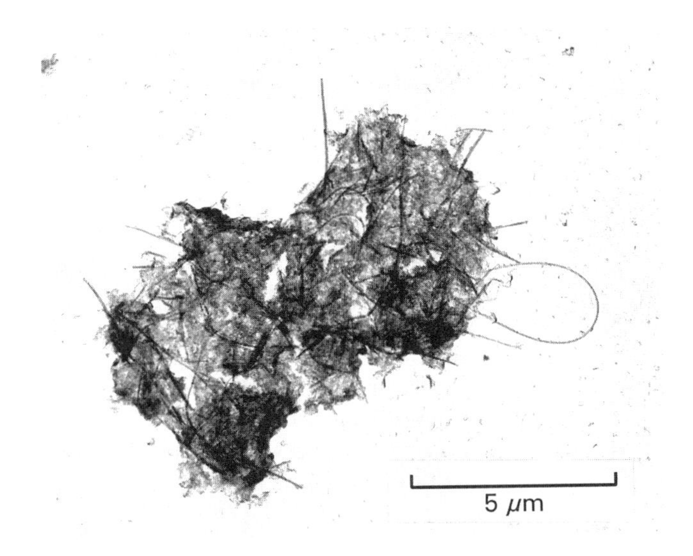

Figure 4--*TEM micrograph showing chrysotile fibers trapped in a silica film, after indirect-transfer preparation*

The thin film of silica is sufficiently strong that ultrasonic treatment in water does not release all of the fibers it contains, and large areas of intact silica film remain. The TEM micrograph in Figure 3 shows a specimen prepared by ashing a CE filter of this type which had been used to collect airborne amosite fibers. The fragment of silica film illustrated contains many closely-spaced amosite fibers. Direct-transfer TEM specimen preparations from the original filter showed that, on the original filter, the fibers were much more widely separated than they are in this fragment of silica film. Figure 4 shows a similar fragment of silica film which resulted after ashing of a filter which had been used to collect airborne chrysotile. The origin of the silica film is not certain, but it is thought to be a consequence of transfer of silicone materials to the surface of the unused CE filter from the parchment used to separate the filters in the package. The separator parchment is known to contain a silicone release agent. Filters which exhibit this effect are not satisfactory for use with ISO 13794, and accordingly a test for filter suitability, to be applied prior to selection of filters for air sampling, is incorporated in the method.

Development of ISO/DIS 14966 - Ambient Air - Determination of Inorganic Fibres - Scanning Electron Microscopy Method

Historical Development

In the early years of the ISO working group, there was disagreement concerning the relative merits of the TEM and SEM methods for determination of asbestos in air. However, most of the working group members considered the detection and identification capabilities of the TEM to be superior, and therefore the draft TEM method proceeded to standardization first. In 1980, a method based on the use of gold-coated PC filters, followed by SEM examination, was published in Germany [*19*]. This method eventually was adapted and published by the Asbestos International Association (AIA) [*20*]. In 1989, at the request of France and Germany, work was commenced on a draft of an ISO analytical method for determination of asbestos based on the SEM. The initial draft was prepared using the AIA method as a starting point. In 1991, a method based on the SEM was published by the German Association of Engineers (VDI) as VDI Guideline 3492 [*21*]. The working group decided that, since VDI Guideline 3492 was at a much later stage of development than the first ISO draft, it would be more expedient to start again, and to produce a new ISO draft based on the VDI document. Voting on the new draft yielded a large number of comments. In particular, there was consensus that the title of the document must be changed to reflect the fact that asbestos cannot be identified using the SEM. In the SEM, fibers can only be classified on the basis of their morphology and semi-quantitative observations of chemical composition. Accordingly, the title now refers to “inorganic fibres”, and within the document the term “identify” has been replaced by “classify”.

ISO/DIS 14966 - Method Description

Air samples are collected using gold-coated, 0.8 μm pore size polycarbonate filters. The filters are gold-coated **prior to** collection of the air samples. A face velocity of approximately 35 cm/s is used.

After sample collection, the filter is placed in a plasma asher for a period of approximately 30 minutes to oxidize any organic material which has been collected on the filter. If any organic fibers or other organic materials remain after this treatment, the plasma ashing treatment may be extended.

SEM examination is conducted at an accelerating voltage of approximately 20 kV and a screen magnification of between 2000 and 2500. A reference specimen of chrysotile fibers is used to adjust the SEM such that fibers with a diameter of 0.2 μm are just visible at the magnification used for fiber counting.

The fiber counting criteria are based on those used in the ISO phase contrast microscopy (PCM) method, ISO 8672 [*22*]. In the SEM method, a fiber is defined as any object longer than 5 μm, with a diameter less than 3 μm, and with a minimum aspect ratio of 3:1. There is an implicit minimum diameter of 0.2 μm, because the operating conditions of the SEM are adjusted such that fibers smaller in diameter than 0.2 μm are not visible. Each fibrous structure is classified morphologically as a fiber, bundle, cluster or matrix. Individual fiber components of clusters or matrices are counted separately if their dimensions conform to the fiber definition. A feature of the fiber counting criteria of both ISO 8672 and ISO/DIS 14966 which has not met with universal acceptance is the rejection of fibers in contact with particles exceeding 3 μm in diameter. The justification for this criterion is that particles exceeding 3 μm in diameter, of densities consistent with silicate mineral chemistry, are above the upper limit of respirability. This feature of the counting criteria is somewhat controversial, and may be subject to change in future revisions of both documents.

Fibers are classified into compositional groups on the basis of their EDXA spectra. The groups are: serpentine, amosite, crocidolite, tremolite/actinolite, anthophyllite or talc, calcium sulfate, and "other inorganic fibres". It is recognized that, depending on the low energy efficiency of the x-ray detector with respect to its ability to detect the sodium peak of crocidolite, it may not be possible to discriminate between crocidolite and amosite. It is also recognized that discrimination between anthophyllite and talc is not possible using the SEM. Calcium sulfate fibers are often present at significant concentrations in air samples. In this method, calcium sulfate fibers are recorded, but their dimensions are not measured and they are not included in the final result for other inorganic fibers, because on the basis of current knowledge, they do not represent any health hazard. In general, however, the numerical concentration of calcium sulfate fibers must be determined, since a high concentration of these fibers can negatively bias the results for other types of fiber, and in some circumstances a high concentration of calcium sulfate fibers may be cause to reject the sample.

The method provides an analytical sensitivity of 100 fibers/m^3 if an air volume of 1 m^3 per cm^2 of active filter area is collected, and an area of 1 mm^2 of filter area is examined in the SEM. This sensitivity is sufficient to allow measurements for comparison with the German ambient air standard of 500 fibers/m^3 (fibers longer than 5 μm and between 0.2 μm and 3.0 μm in diameter).

Development of Sampling Strategy Standards

The three ISO analytical methods for determination of asbestos or inorganic fibers do not address air sampling strategy in detail. There are a number of reasons why air sampling may be conducted to determine the concentration of airborne asbestos in building indoor air. These include determination of prevalent fiber concentrations, investigation of fiber emissions during routine maintenance activities, and determination of habitability in accordance with a clearance procedure following asbestos abatement. In response to the requirement for sampling strategy standards for asbestos in indoor air, a new working group, ISO/TC 146/SC 6/WG4, was formed in late 1996, to develop International Standards for sampling strategy for determination of asbestos and other mineral fibers in building atmospheres. This working group is reviewing existing documents, and is actively seeking contributions.

References

[1] Cook, P. M., Rubin, I. B., Maggiore, C. J., and Nicholson, W. J., "X-ray Diffraction and Electron Beam Analysis of Asbestiform Minerals in Lake Superior Waters", *Proceedings of the International Conference on Environmental Sensing and Assessment*, IEEE, New York, 2, 1975, 34-1-1.

[2] Chatfield, E. J. and Dillon, M. J., "Analytical Method for Determination of Asbestos Fibers in Water", Report EPA-600/4-83-043, Order No. PB83-260471, National Technical Information Service, Springfield, Virginia, 1983.

[3] Ortiz, L. W. and Isom, B. L., "Transfer Technique for Electron Microscopy of Membrane Filter Samples", *American Industrial Hygiene Association Journal*, 1974, 35:423.

[4] Middleton, A. P. and Jackson, E. A., "A Procedure for the Estimation of Asbestos Collected on Membrane Filters Using Transmission Electron Microscopy (TEM)", *Annals of Occupational Hygiene*, 1982, 25(4):381-391.

[5] Burdett, G. J. and Rood, A. P. "A Membrane Filter, Direct-Transfer Technique for the Analysis of Asbestos Fibres or Other Inorganic Particles by Transmission Electron Microscopy", *Environmental Science and Technology*, 1983, 17:643-648.

[6] Asbestos International Association, "Reference Method for the Determination of Airborne Asbestos Fibre Concentrations at Workplaces by Light Microscopy (Membrane Filter Method)", AIA Health and Safety Publication, Recommended Technical Method No. 1 (RTM1), Asbestos International Association, 68 Gloucester Place, London W1H 3HL, United Kingdom, 1979.

[7] Beaman, D. R., and File, D. M., "Quantitative Determination of Asbestos Fiber Concentrations", *Analytical Chemistry*, 1976, 48, 101.

[8] Chatfield, E. J., Glass, R. W., and Dillon, M. J., "Preparation of Water Samples for Asbestos Fiber Counting by Electron Microscopy", Report EPA-600/4-78-011, National Technical Information Service, Springfield, Virginia 22161, 1978.

[9] International Organization for Standardization, "Ambient air - Determination of asbestos fibres - Direct-transfer transmission electron microscopy method", ISO 10312, 1995.

[10] Jaffe, M. S., "Handling and Washing Fragile Replicas", *Journal of Applied Physics*, 1948, 19:1187.

[11] Yamate, G., Agarwal, S. C., and Gibbons, R. D., "Methodology for the Measurement of Airborne Asbestos by Electron Microscopy", Draft Report, Environmental Monitoring Systems Laboratory, Office of Research and Development, U.S. EPA, Research Triangle Park, North Carolina, 1984.

[12] Environmental Protection Agency, "Asbestos-Containing Materials in Schools; Final Rule and Notice", Federal Register, 40 CFR Part 763, Vol. 52, No. 210, Friday, October 30, 1987, 41826-41905.

[*13*] Ontario Ministry of the Environment, "List of Ambient Air Quality Criteria, Standards, Tentative Design Standards, Guidelines and Provisional Guidelines as of 29 March, 1989", Ontario Ministry of the Environment, Air Resources Branch.

[*14*] International Organization for Standardization, "Ambient air - Determination of asbestos fibres - Indirect-transfer transmission electron microscopy method", ISO 13794, 1999.

[*15*] Nicholson, W. J., Rohl, A. N. and Weisman, I., "Asbestos Contamination of the Air in Public Buildings", Report EPA-450/3-76-004, National Technical Information Service, Springfield, Virginia 22161, 1975.

[*16*] Sébastien, P., Billon-Galland, M. A., Dufour, G., and Bignon, J., "Measurement of Asbestos Air Pollution Inside Buildings Sprayed With Asbestos", Report EPA-560/13-80-026, National Technical Information Service, Springfield, Virginia 22161, 1980.

[*17*] Burdett, G. J., "The Measurement of Airborne Asbestos Releases From Damaged Amosite Insulation Subjected to Physical Attrition", *Asbestos Fibre Measurements in Building Atmospheres,* Ontario Research Foundation, Mississauga, Ontario, Canada, 1987, 209-239.

[*18*] Chatfield, E. J., "Limits of Precision and Accuracy in Analytical Techniques Based on Fibre Counting", *Asbestos Fibre Measurements in Building Atmospheres,* Ontario Research Foundation, 1987, 115-137.

[*19*] Battelle-Institut e.V, "Untersuchungen des Asbestanteils im Staub der Außenluft. Entwicklung und Erprobung kostengünstiger Meßverfahren für Asbest und ähnliche faserförmige Stoffe", Battelle-Institut e.V, Frankfurt/Main. Forschungsbericht 10402613 im Auftrag des Umweltbundesamtes, Juni 1980.

[*20*] Asbestos International Association, "Method for the Determination of Airborne Asbestos Fibres and Other Inorganic Fibres by Scanning Electron Microscopy", AIA Health and Safety Publication, Recommended Technical Method No. 2 (RTM2), Asbestos International Association, 68 Gloucester Place, London W1H 3HL, United Kingdom, 1984.

[*21*] Verein Deutscher Ingenieure, "Measurement of Inorganic Fibrous Particles in Ambient Air, Scanning Electron Microscopy Method", Guideline VDI 3492 Part 1, VDI Handbuch Reinhaltung der Luft, Band 4, August 1991.

[*22*] International Organization for Standardization, "Air quality - Determination of the number concentration of airborne inorganic fibres by phase contrast optical microscopy - Membrane filter method", ISO 8672, 1993.

James R. Millette[1], W. Randy Boltin[1], Patrick J. Clark [2] and Kim A. Brackett[3]

PROPOSED ASTM METHOD FOR THE DETERMINATION OF ASBESTOS IN AIR BY TEM AND INFORMATION ON INTERFERING FIBERS

REFERENCE: Millette, J. R., Boltin, W. R., Clark, P. J., and Brackett, K. A., **"Proposed ASTM Method for the Determination of Asbestos in Air by TEM and Information on Interfering Fibers,"** *Advances in Environmental Measurement Methods for Asbestos, ASTM STP 1342,* M. E. Beard and H. L. Rook, Eds., American Society for Testing and Materials, 2000.

ABSTRACT: The draft of the ASTM Test Method for air entitled: "Airborne Asbestos Concentration in Ambient and Indoor Atmospheres as Determined by Transmission Electron Microscopy Direct Transfer (TEM)" (ASTM Z7077Z) is an adaptation of the International Standard, ISO 10312. It is currently undergoing a revision in regard to the reporting of confidence intervals for analysis results. The method differs considerably from the AHERA air clearance analysis method in that the counting rules provide a fairly complete characterization of the sizes of structures and structural components of asbestos-containing particles in the air. With the information derived from a full analysis, it is possible to determine the total asbestos structure count, the total fiber count, the PCM equivalent fiber count, the AHERA equivalent structure count and a number of other count values based on specific criteria. Examples of non-asbestos fibers which might interfere with the analysis (antigorite, palygorskite, halloysite, hornblende, sepiolite and vermiculite scrolls) are also included.

KEYWORDS: Asbestos, Air, Interferences, TEM, Detection Limit

A number of methods for the analysis of asbestos in air (and in other media) have been developed over the years [*1*]. Occupational monitoring for asbestos in air is done with a light microscope with methods such as the National Institute for Occupational Safety and Health (NIOSH) 7400 method and the ASTM "Standard Test Method for Airborne Asbestos Concentration in Workplace Atmosphere", D4240 - 83, (reapproved 1994). [*2-3*]. Although the phase contrast light microscopy (PCM) techniques for airborne asbestos monitoring required by the U.S. Occupational Safety and Health

[1] Executive Director, MVA, Inc., 5500 Oakbrook Parkway, Suite 200, Norcross, GA 30093
[2] U.S. Environmental Protection Agency, 26 W. M.L. King, Dr., Cincinnati, OH, 45268
[3] TN, Inc., 26 W. M.L. King Dr. Cincinnati, OH 45268

Administration (OSHA) are relatively inexpensive and can provide timely on-site information useful in monitoring asbestos exposures, they are limited in specificity and sensitivity. PCM methods allow the analyst to count fibers, asbestos and non-asbestos, regardless of type. They are sensitive only to those fibers thicker than approximately 0.25 mm in diameter and the methods limit themselves to those over 5 mm in length. In order to distinguish between asbestos and non-asbestos fibers, procedures using electron microscopy have been developed [*4,5*]. The capability of transmission electron microscopy (TEM) to resolve the thinnest asbestos fibers and identify asbestos on the basis of morphology, crystal structure, and/or chemical composition, was used in the development of the EPA procedures for ambient air monitoring for asbestos. However, while a provisional method [*6*] and a draft method [*7*] were released, no final standard EPA method was promulgated for asbestos in air. In 1987 when it became necessary to require a rapid TEM asbestos method that would be used to clear school buildings after asbestos abatement actions under the Asbestos Hazard Emergency Response Act (AHERA), a panel of asbestos microscopists was convened to provide a method streamlined to that task. The result, the Interim Transmission Electron Microscopy Analytical Method [*8*] , known as the AHERA TEM method is a simplified version of the draft EPA air method. With it, the analyst counts in terms of asbestos structures: fibers, bundles, clusters, and matrices. Interestingly, abatement contractors, who had no problem passing PCM air tests after abatement, initially failed AHERA clearance. After adopting more rigorous cleaning procedures, however, clearance by TEM was found to be routinely achievable.

In general, AHERA counts and OSHA monitoring counts are not comparable. It has happened that a sample which resulted in high PCM counts, over 1 fiber per cubic centimeter (f/cc), has been found to contain no detectable asbestos structures when analyzed according to AHERA because the fibers on the filter were all non-asbestos fibers. Conversely, samples that resulted in no detectable fiber concentration values by PCM count have resulted in high AHERA asbestos structure counts because the asbestos fibers on the filter were all thinner than 0.25 mm in diameter or shorter than 5 mm in length. Although the AHERA counting rules require classifying structures into those larger than 5 mm and those less than 5 mm, a 5 mm AHERA structure may contain no PCM countable fibers or it may contain several. A complex asbestos structure may also occur on an air filter, where under the PCM method counting rules, it would be counted as many fibers, whereas under AHERA counting rules it would be counted as one structure [*9*].

In 1995, The International Standards Organization (ISO) method entitled: "Ambient Air - Determination of Asbestos Fibres - Direct-transfer Transmission Electron Microscopy Procedure" was published as ISO 10312 [*10*]. It was developed to provide a clearer picture of the type, size and composition of airborne asbestos fibers. Using a carefully constructed set of counting rules, the analyst records information not only about the overall size of the asbestos structure identified but also about the component parts of the structure. The method differs considerably from the AHERA air clearance analysis method in that the ISO counting rules provide a more complete characterization of the sizes of structures and structural components of asbestos-containing particles in the air. With the information derived from the full analysis, it is possible to determine the total asbestos structure count, the total fiber count, the PCM equivalent fiber count, the AHERA equivalent structure count or other count values based on size criteria.

In the 1990s the ASTM committee D22.07 decided to discontinue development of an ambient air method that had been begun in the 1980s by Dr. Arthur Langer and instead begin the task of converting the ISO method to ASTM format. Long time ASTM member, Carl Hummel, performed the initial conversion and the format (as well as some of the content) has been refined through several subcommittee ballots of D22.07. The major change currently under consideration from the Method ISO 10312 is the clarification of the concept of 'detection limits'. Rather than detection limits, the analyst is encouraged to construct two-sided confidence intervals from the Poisson distribution from a Table with values for counts from zero to 470. When a count is zero, however, a one-sided 95% confidence interval based on the upper confidence limit is used. The 95% confidence limit in this case is from 0 structures/ liter of air to (2.99 times the analytical sensitivity) structures/ liter of air. The method notes that 2.99 is the upper 95% confidence limit for the Poisson distribution when the observed sample count is zero. According to the discussion on the ASTM Air Method (Z7077Z), the "detection limit" has been set at 2.99 structures counted in any area of any filter because of concerns that false positives (counting a structure when none exists) may occur in both blanks and sample filters. Based on the assumption of a Poisson distribution of false positives, the detection limit of 2.99 would protect against a false positive rate as high as 5% (5 false positive structures per 100 blank filters counted). This a very conservative level, since the actual false positive rate is believed to be 2% or lower. Thus, many of the samples reported as being below the "detection limit" (less than 3 structures counted) will actually contain true positives.

The information for determining upper and lower confidence limits for structure counts from 0 to 50 is given in Table 1. For example a count of 5 structures which corresponds to 0.8 structures per cubic centimeter (s/cc) would have a 95% confidence interval from 0.3 to 1.9 s/cc, whereas a count of 50 structures which corresponds to the same 0.8 s/cc would have a confidence interval from 0.6 to 1.05 s/cc.

Interferences

As described in other ASTM asbestos TEM methods such as the Asbestos Dust Methods: D5755 and D5756, minerals that have properties (chemical or crystalline structure) which are very similar to asbestos minerals may interfere with the analysis by causing a false positive to be recorded during the test. When using the methods, information must be maintained by the laboratory so that non-asbestos minerals are not misidentified as asbestos minerals. The interferences listed are:

1. Antigorite
2. Palygorskite (Attapulgite)
3. Halloysite
4. Pyroxenes
5. Sepiolite

Table 1.
Upper and lower limits of the Poissonian 95% confidence interval of a count from 0 to 50 structures

Count	Lower	Upper	Count	Lower	Upper
0	0	3.689*	26	16.983	38.097
1	0.025	5.572	27	17.793	39.284
2	0.242	7.225	28	18.606	40.468
3	0.619	8.767	29	19.422	41.649
4	1.090	10.242	30	20.241	42.827
5	1.624	11.669	31	21.063	44.002
6	2.202	13.060	32	21.888	45.175
7	2.814	14.423	33	22.715	46.345
8	3.454	15.764	34	23.545	47.512
9	4.115	17.085	35	24.378	48.677
10	4.795	18.391	36	25.213	49.840
11	5.491	19.683	37	26.050	51.000
12	6.201	20.962	38	26.890	52.158
13	6.922	22.231	39	27.732	53.315
14	7.654	23.490	40	28.575	54.469
15	8.396	24.741	41	29.421	55.622
16	9.146	25.983	42	30.269	56.772
17	9.904	27.219	43	31.119	57.921
18	10.668	28.448	44	31.970	59.068
19	11.440	29.671	45	32.823	60.214
20	12.217	30.889	46	33.678	61.358
21	13.000	32.101	47	34.534	62.501
22	13.788	33.309	48	35.392	63.642
23	14.581	34.512	49	36.251	64.781
24	15.378	35.711	50	37.112	65.919
25	16.178	36.905			

**The one-sided upper 95% confidence limit for zero structures is 2.99*

6. Vermiculite Scrolls
7. Fibrous Talc
8. Horneblende.

Mineralogical texts are helpful in providing information about the potentially interfering minerals [*11*]. One article has been published which describes the differences between chrysotile, halloysite and palygorskite [*12*]. Another shows images of talc fibers [*13*]. Other information, although not published, has been presented at meetings [*14*]. A recent technical monograph (without photographs) has reviewed the characteristics of a number of the potentially interfering minerals [*15*]. However, there

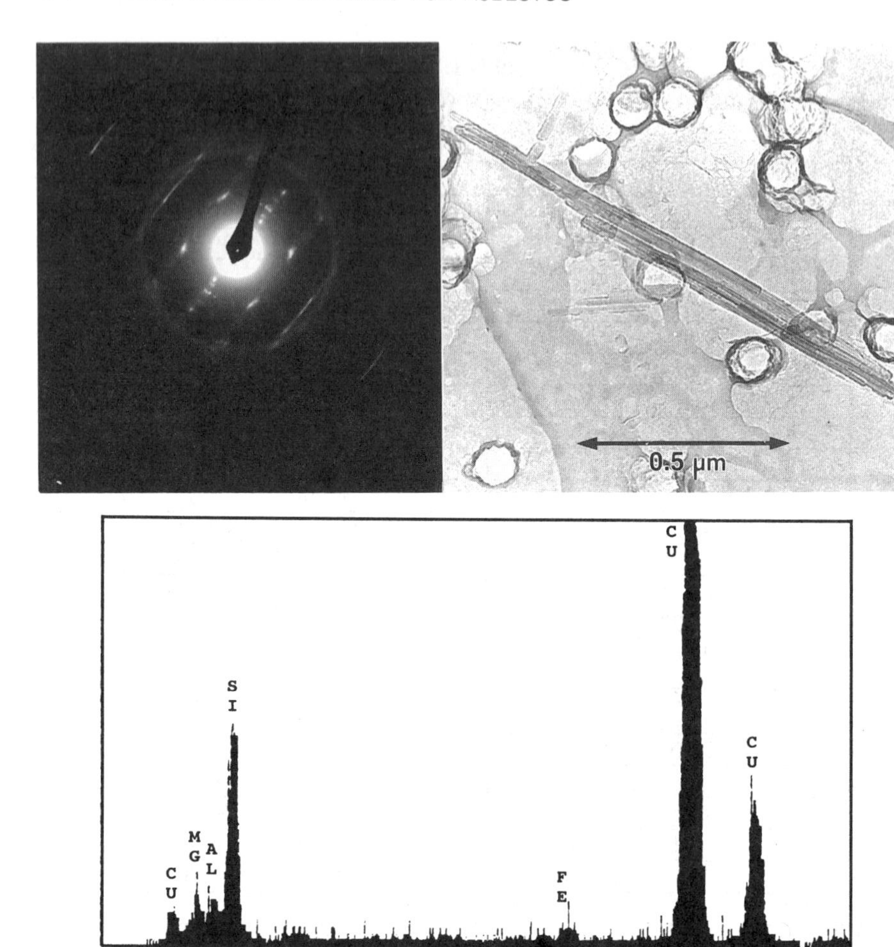

Figure 1. Attapulgite

is no one article to which an analyst can go for micrographs, diffraction patterns and x-ray spectra of these minerals. The following is a brief description of differentiating characteristics with micrographs and spectra for some of the more common potentially interfering fibers.

Palygorskite (also known as Attapulgite) is a fibrous clay (Figure 1). It is a thin lath mineral which can appear to have a central canal when laths are lying parallel. The row spacing in the diffraction pattern shown in Figure 1 is 5.36 angstroms. The elemental spectrum is distinctive from chrysotile as it has a significant amount of aluminum.

Halloysite is a fibrous clay (Figure 2). It is a cylindrical crystal like chrysotile

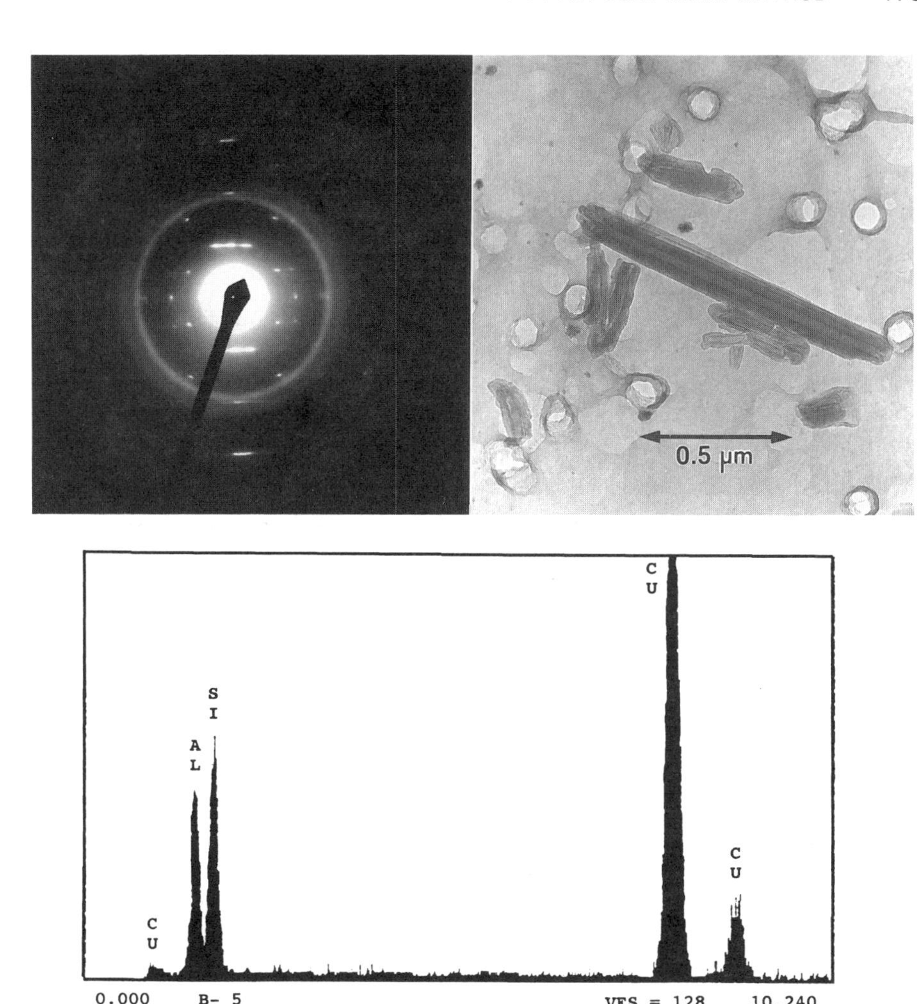

Figure 2. Halloysite

asbestos but is an aluminum silicate rather than a magnesium silicate and thus has a distinctively different x-ray spectrum. The row spacing in the diffraction pattern of halloysite shown in Figure 2 is 9.05 angstroms and is distinctly different from the 5.31 angstroms row spacing seen in diffraction patterns of chrysotile.

Sepiolite is a fibrous clay (Figure 3). It is a thin lath magnesium silicate mineral which can appear to have a central canal when laths are lying parallel. As a lath, sepiolite does not have a diffraction pattern like chrysotile, but several laths together can produce a streaked appearance and a row spacing of 5.24 angstroms compared to the row spacing of in a chrysotile diffraction pattern of 5.31 angstroms. The magnesium content is lower in sepiolite than in chrysotile.

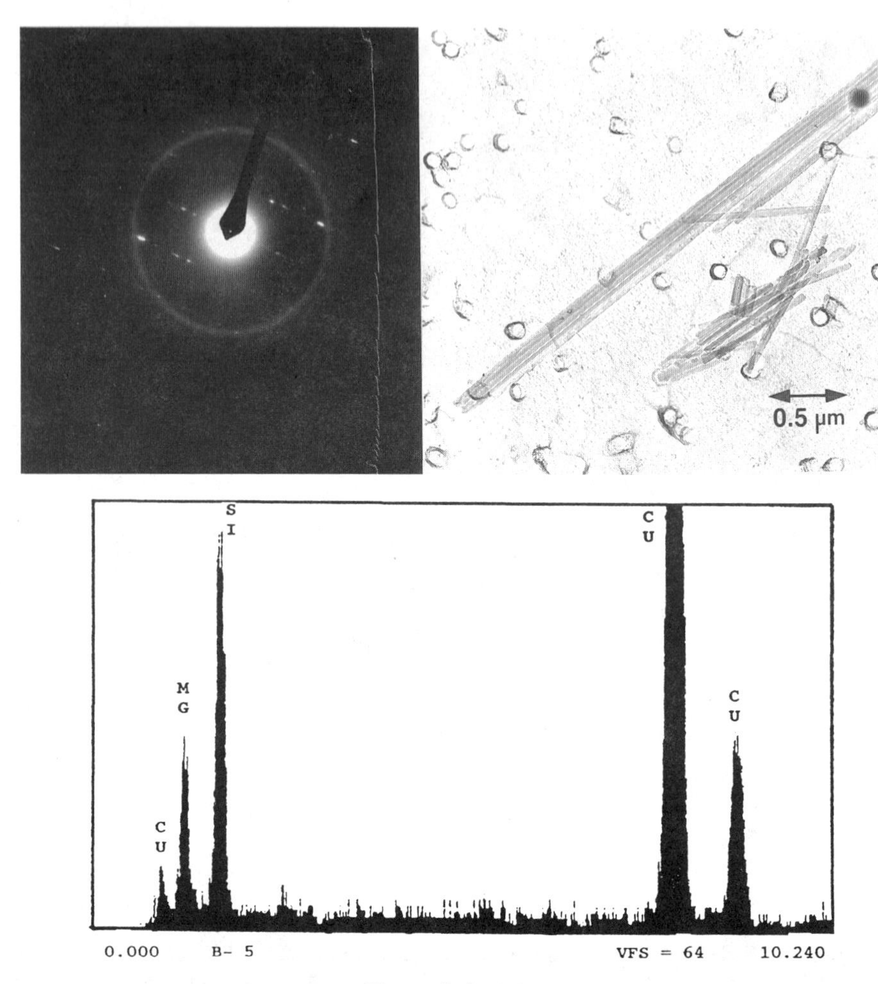

Figure 3. Sepiolite

Talc can occur as plates or as in a fibrous form (Figure 4). Often a portion of the fiber will have a hexagonal diffraction pattern, while another segment may show a pattern with rows of spots. The diffraction pattern shown in Figure 4 measures 5.33 angstroms. The elemental composition as shown in the x-ray spectrum is typical of talc with magnesium about half the height of silicon.

Antigorite is a fibrous serpentine mineral (Figure 5). Its elemental analysis is the same as chrysotile but it is a lath rather than a cylindrical crystal. The diffraction pattern is very distinctive for antigorite because of its 40 angstrom spacing as evidenced by very closely spaced dots on the diffraction pattern. The row spacing of the diffraction patterns shown in Figure 5 is 4.59 angstroms.

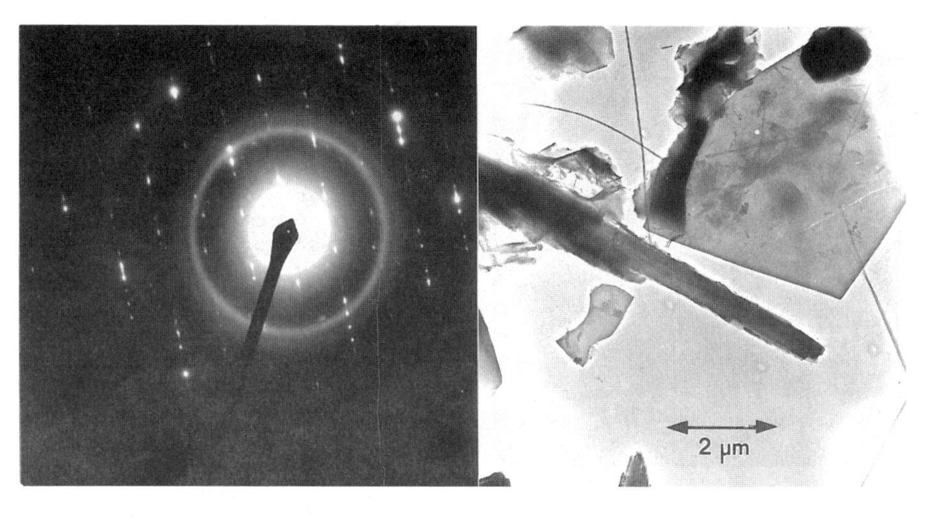

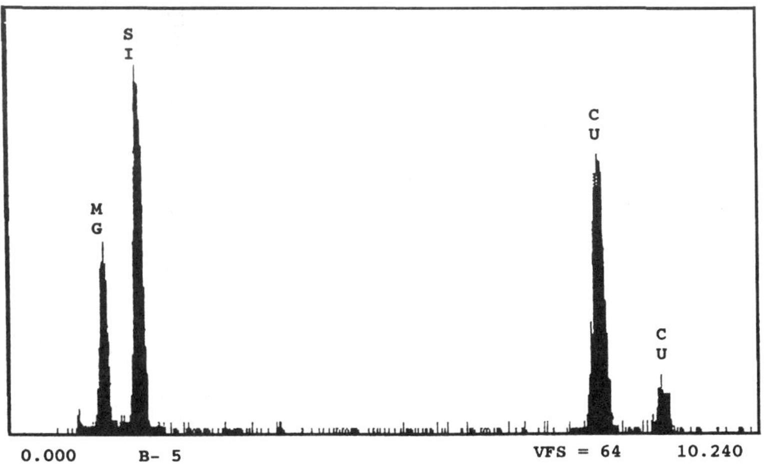

Figure 4. Talc

Glass fibers are generally larger than asbestos fibers and do not diffract (Figure 6). The elemental composition often contains elements such as aluminum which are not present in asbestos types.

The Horneblende shown in this paper (Figure 7) is from a sample from Tory Hill, Ont., Canada. It is a calcic amphibole containing Mg, Al, Si, K, Ca, and Fe. X-ray diffraction performed on the bulk sample most closely matched a hornblende. The x-ray spectrum is similar to tremolite except for the small but identifiable Al and K peaks. This sample is typical of amphibole cleavage fragments. The sample contained a

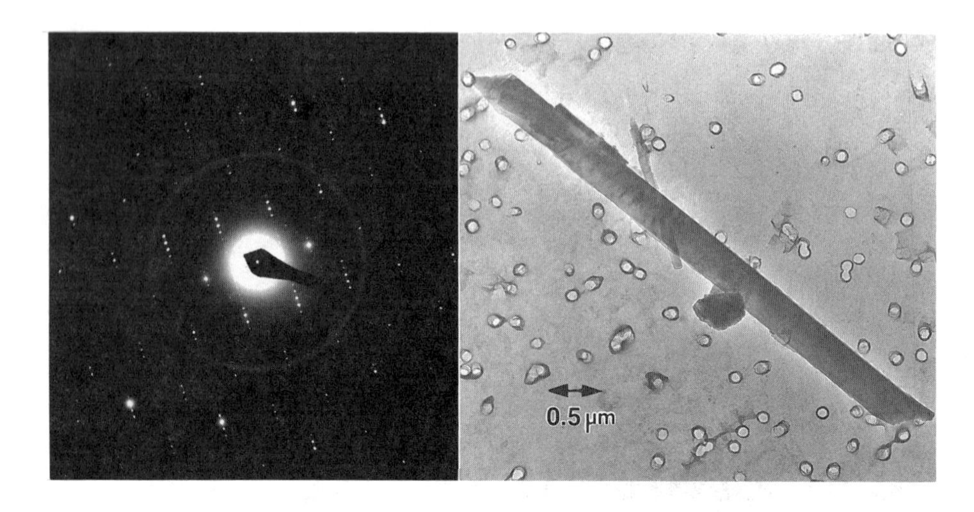

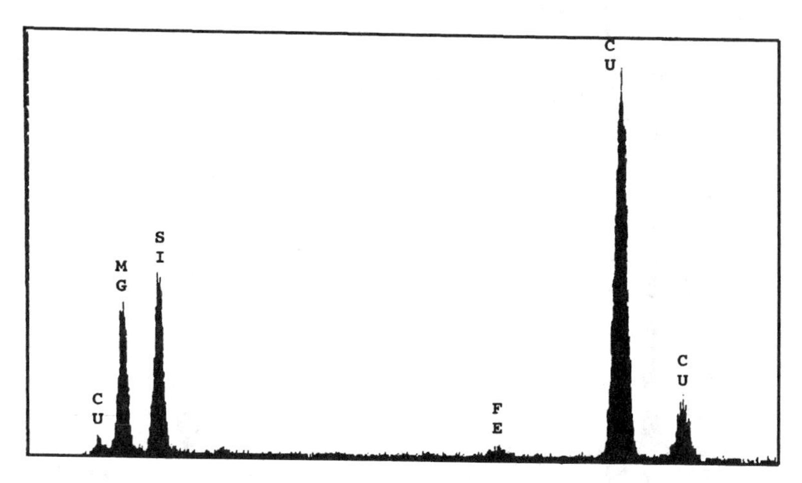

Figure 5. Antigorite

number of elongated particles that did not have parallel sides. The diffraction pattern shown in Figure 7 has a 5.07 angstrom row spacing.

Vermiculite scrolls are rolled-up edges of vermiculite sheets (Figure 8). They generally give an x-ray spectrum more consistent with vermiculite showing elements such as aluminum and potassium not consistent with asbestos minerals. The diffraction pattern shown in Figure 8 has a 5.27 angstrom row spacing.

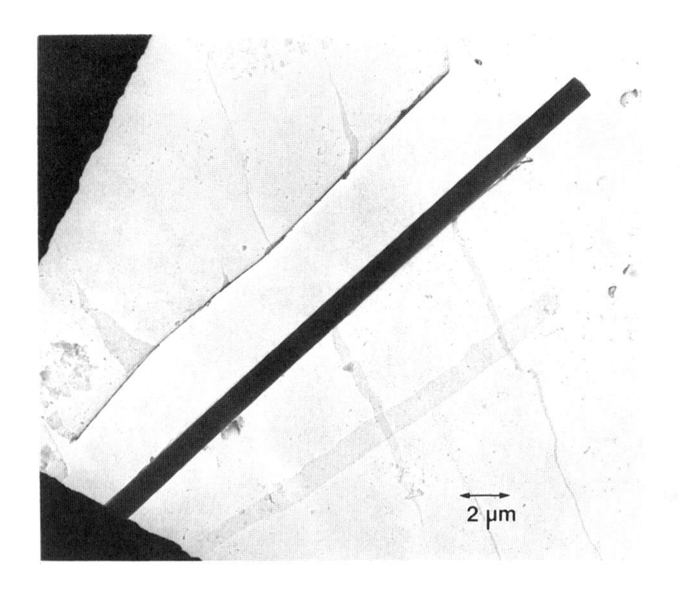

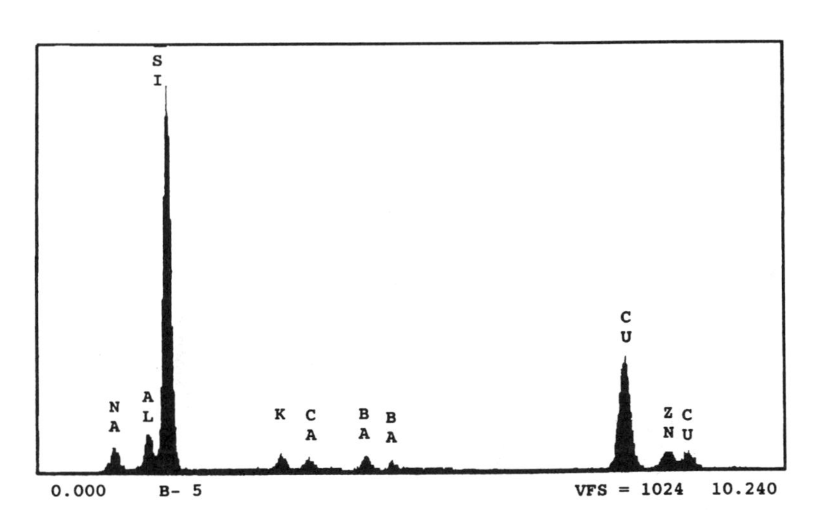

Figure 6. Glass fiber

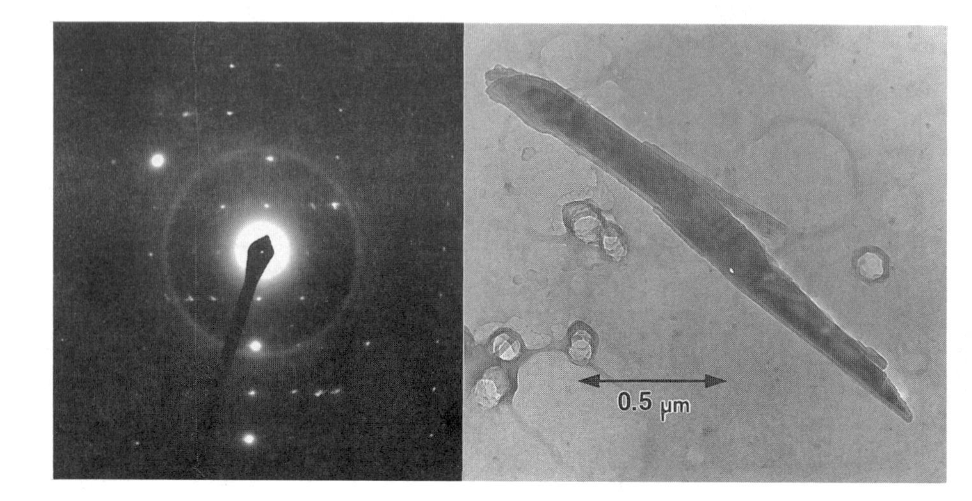

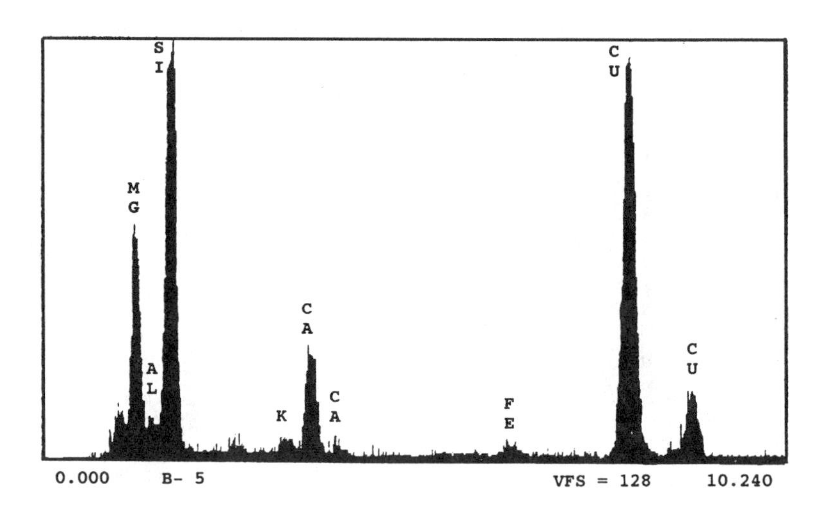

Figure 7. Horneblende

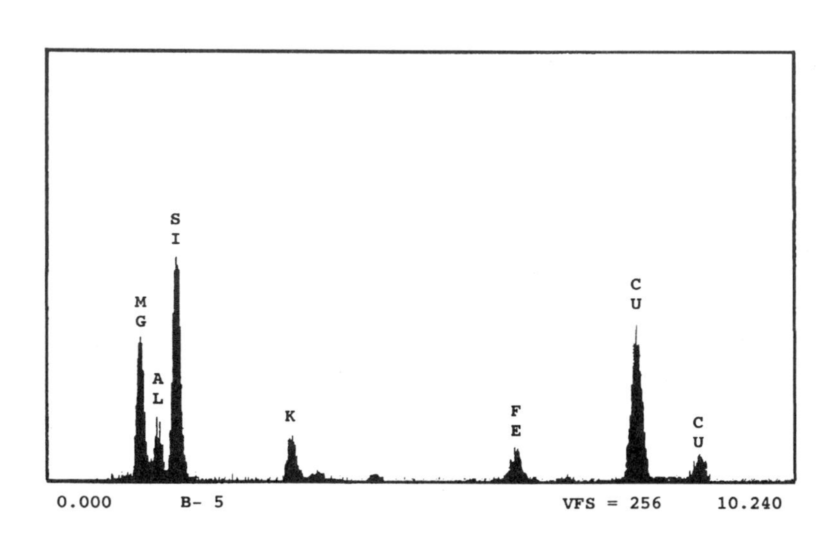

Figure 8. Vermiculite Scrolls

References

[1] Millette, J. R., Reference Methods for Asbestos Analysis, *Microscope*, 1997, Vol. 45, pp. 57-59,.

[2] NIOSH, "Physical and Chemical Analysis Method 239" Nat'l. Inst. Occup. Safety and Health, USPHS/NIOSH "Membrane Filter Method for Evaluating Airborne Asbestos Fibers", U.S. Dept. Health, Education and Welfare. 1979

[3] "Asbestos and Other Fibers by Phase Contrast Microscopy" - Method 7400, NIOSH Manual of Analytical Methods, 4th Ed., U.S. Department of HHS, NIOSH Publ. 94-113, 1994

[4] "Asbestos Fibers by Transmission Electron Microscopy" - Method 7402, NIOSH Manual of Analytical Methods, 4th Ed., U.S. Department of HHS, NIOSH Publ. 94-113, 1994

[5] "Method for the Determination of Airborne Asbestos Fibres and Other Inorganic Fibres by Scanning Electron Microscopy", *AIA Health and Safety Publication Recommended Technical Method No.2*, Asbestos International Association, AIA London, England, 13 pp 1984

[6] "Electron Microscopy Measurement of Airborne Asbestos Concentrations", *A Provisional Methodology Manual*, EPA 600/2/77/178, Research Triangle Park, NC (Revised 1978)

[7] "Methodology for the Measurement of Airborne Asbestos Concentrations by Electron Microscope". Yamate, M., S.C. Agarwal, and R.D. Gibbons; *Draft Report*, Washington, DC: Office of Research and Development, USEPA Contract No. 68-02-3255 1984

[8] Appendix A to Subpart E - "Interim Transmission Electron Microscopy Analytical Methods", U.S. EPA, 40 CFR Part 763. Asbestos-Containing Materials in Schools, Final Rule and Notice. *Fed. Reg.* 52(210): 41857-41894 1987.

[9] Millette, J.R. and Boltin, W.R. Effectiveness of Asbestos Operations and Maintenance: Measurement Issues. *Appl. Occup. Environ. Hyg.* 9(11): 785-790; 1994

[10] "Ambient Air - Determination of Asbestos Fibres - Direct-transfer Transmission Electron Microscopy Procedure". International Standards Organization, 1995. ISO 10312.

[11] Deer, W.A., R.A. Howie, and J. Zussman, *Rock Forming Minerals*, Vol 2. Longman Group LTD, London, 1974

[12] Millette, J. R.; Twyman, J. D.; Hansen, E. C.; Clark, P. J.; and Pansing M. F. Chrysotile, Palygorskite, and Halloysite in Drinking Water, *Scanning Electron Microscopy* I: 579-586 1979

[13] Kremer, T. and Millette, J. R., A Standard TEM Procedure for Identification and Quantification of Asbestiform Minerals in Talc, *Microscope*, 38:457-468, 1990.

[14] Dagenhart, T., Fisher, J. Bishop, K, Dunmyre, G., Lee, R. Fibrous pyroxenes: Common rock-forming minerals which may be confused with amphiboles, Presentation at the NAC Conf. & Exhibit,. San Antonio, TX, 1990

[15] Montague, N. TEM Back to Basics: Distinguishing Asbestos from "Look-alike" Minerals. *EIA Technical Monograph 1*, The Environmental Information Association, 1997

James S. Webber,[1] Alex G. Czuhanich,[1] and Laurie J. Carhart [1]

SHORTCOMINGS IN AIRBORNE ASBESTOS ANALYSIS FILTERED FROM NEW YORK STATE'S PROFICIENCY-TESTING DATA

REFERENCE: Webber, J. S., Czuhanich, A.G., and Carhart, L. J., **"Shortcomings in Airborne Asbestos Analysis Filtered from New York State's Proficiency-Testing Data,"** *Advances in Environmental Measurement Methods for Asbestos, ASTM STP 1342,* M. E. Beard, H. L. Rook, Eds., American Society for Testing and Materials, 2000.

ABSTRACT: Since 1992, the Environmental Laboratory Approval Program of the New York State Department of Health has conducted semi-annual proficiency testing (PT) of laboratories analyzing airborne asbestos by transmission electron microscopy. While these PT rounds have included magnification measurements, diffraction measurements, zone-axis calculations, and identification of unknown mineral fibers, most samples have been dispersions of asbestos and other mineral fibers on MCE filters. Participating laboratories in each PT round were required to fill out a questionnaire detailing preparation and analytical methods. When performance of laboratories was correlated with their reported preparation and analytical methods, some significant trends were uncovered. Quality of grids was best for laboratories that 1) collapsed filters using a DMF solution, 2) used gravimetric reduction of collapsed filters to calibrate etching time, 3) etched filters for three to six minutes, and 4) used a combination of Jaffe-wick and condensation washer to dissolve filters. Reported structure concentrations were highest for laboratories that 1) etched filters for longer than six minutes, 2) evaporated carbon in a tilt/rotation geometry, and 3) performed analysis at magnifications between 15,000 and 20,000.

KEYWORDS: asbestos, proficiency testing, transmission electron microscopy

Introduction

According to the New York State legislative mandate of 1984 [*1*], all laboratories analyzing environmental samples must be certified by the New York State Department of Health's Environmental Laboratory Approval Program (ELAP). The passage of the federal Asbestos Hazard Emergency Response Act (AHERA) [2] in 1987 precipitated a

[1]Research scientist and laboratory technicians, respectively, Wadsworth Center, New York State Department of Health, P.O. Box 509, Albany, NY 12201-0509

dramatic increase in the number of laboratories analyzing airborne asbestos by transmission electron microscopy (TEM). A total of 52 laboratories from 26 states have enrolled at some time during the period 1988 to 1997. Maximum enrollment of 45 laboratories occurred early in 1992 and decreased to 32 laboratories by the end of 1996.

As part of ELAP certification, a laboratory must demonstrate analytical competence by successfully participating in a proficiency-testing (PT) program. For analysis of airborne asbestos, these PT rounds have included magnification measurements, diffraction measurements, zone-axis calculations, and identification of unknown mineral fibers. The majority of samples, however, have been dispersions of aqueous suspensions of asbestos and other mineral fibers on mixed-cellulose ester (MCE) filters. Laboratories were required to prepare these filters for TEM analysis according to the protocol specified in AHERA.

PT materials were screened by reference laboratories before distribution to participating laboratories. These reference laboratories were not enrolled in ELAP (usually for geographic purposes - most were west of the Mississippi River) but had demonstrated high quality during visits by one of the authors. These laboratories received the PT materials, prepared and analyzed them, and reported results in the same way that ELAP-participating laboratories would. PT samples that produced highly variable or inconsistent results when analyzed by the reference laboratories were excluded from PT materials sent to ELAP-participating laboratories. Reference laboratories also provided critical feedback on the quality of the test in general, e.g., packaging, clarity of instructions, etc. Competence of ELAP-participating laboratories was based on consensus statistics from ELAP-participating laboratories, not on data produced by the reference laboratories. Results from reference laboratories are not incorporated into the present study.

Proficiency-Testing Parameters

Four types of asbestos were used for the ten MCE PT samples distributed from 1992 to 1996. Three sets each of chrysotile, crocidolite, and actinolite were distributed with consensus-mean concentrations ranging from 330 to 1800 structures/mm^2 (s/mm^2). Two sets of filters with amosite (fibrous grunerite) with consensus-mean concentrations of 390 and 2500 s/mm^2 were also distributed.

As part of each PT round, laboratories were required to answer a detailed questionnaire concerning preparation and analytical methods. Specifically, laboratories were asked to describe filter collapsing-method, etching time, etching calibration method, type of evaporated carbon source, geometric configuration during carbon evaporation, filter-dissolution method, filter-removal solvent, and magnification used during analysis. Answers to these questions form the basis for this paper; preparation and analytical methods that seem to have affected either grid quality or analytical quantitation will be discussed.

A total of 351 analyses (and preparations) were accomplished on the ten sets of PT samples by 30 to 45 ELAP-participating laboratories; the data typically did not fit a normal distribution and because data sets were irregular in size, non-parametric statistical analyses were employed. Results from ELAP-participating laboratories typically were divided into two to four categories on the basis of differences in answers to a question from the questionnaire. For example, for filter-collapsing method, one category was composed of all laboratories using a hot block, a second category was composed of all laboratories using DMF, and a third category was composed of all laboratories using an acetone chamber. The different categories were evaluated two at a time by comparing their respective mean results for each of the ten samples. These differences were compared for significance as detected by the Wilcoxon paired-rank sum test [*3,4*]. This test evaluated both the number of times that one category exceeded the other as well as the magnitude of those differences. Comparisons that produced differences of $p<0.10$ are discussed.

TEM Grid Preparation

During the first three PT rounds, all laboratories were required to prepare MCE filters on a specific grid type that was distributed to them by ELAP. On the basis of comments by some laboratories that these grids were different from their routine grids and that this affected the quality of their preparations, ELAP had all laboratories in subsequent rounds prepare PT filters on their own grids. These, of course, had to be indexed (finder) 200-mesh grids.

All grids returned by laboratories were photographed and evaluated for suitability of preparation using phase-contrast microscopy. Following AHERA criteria, the carbon film covering each of the grid openings was judged acceptable only if (a) less than 5% of the carbon film was broken, (b) less than 5% of the carbon film was folded or doubled, and c) less than 5% of the filter was undissolved. Pseudo-color images of grid openings with broken film areas ranging from 2 to 10% were generated in an STEM/image-analysis system as references for comparison. A laboratory's number of acceptable grid openings was the average of the two grids submitted by the laboratory. Obviously, the greater this number, the better the quality of grid preparation.

Fiber Quantitation

Laboratories were required to analyze their grid preparations of PT samples for asbestos structures according to AHERA protocol. They were required to analyze sufficient areas on two grid preparations from one filter to achieve an analytical sensitivity of 16 s/mm^2. Laboratories were required to submit hard copies of electron-diffraction patterns collected from the first four asbestos structures counted per sample. Laboratories were graded on their proximity to a consensus mean, outliers removed. For this paper, each laboratory's result was normalized to the consensus mean for that PT sample. For example, a laboratory reporting a value of 1500 s/mm^2 for a PT sample with a consensus mean of 2000 s/mm^2 would have a normalized result of 0.75.

Results and Discussion

Filter-Collapse Method

The three commonly utilized methods for collapsing filters were 1) a hot block with injected acetone vapor (commonly used in the NIOSH 7400 method [5])(n=115), 2) the DMF/water/acetic acid solvent of Burdett and Ruud [6]) (n=91), and 3) a closed chamber with a passive acetone atmosphere (n=150).

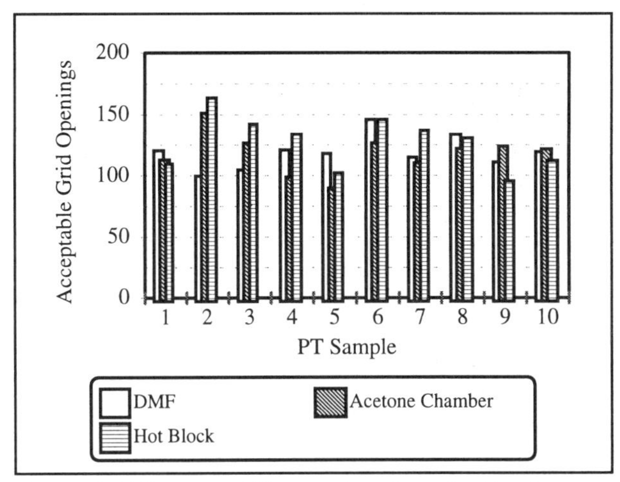

Figure 1. Grid acceptability as a function of collapsing mechanism.

No significant differences were detected among collapsing methods for the number of structures counted. However, use of the DMF method produced significantly (p=0.088) more acceptable grid openings than did the passive acetone atmosphere (Figure 1). The reason for this difference is not clear. One might have expected fewer acceptable grid openings from the turbulence of the hot-block method, but this was not observed.

Etching Duration

Etching times ranged from 20 seconds to 20 minutes. These times were divided into three periods: < 3 minutes (n=141), 3 to 6 minutes (n=115), and > 6 minutes (n=135).

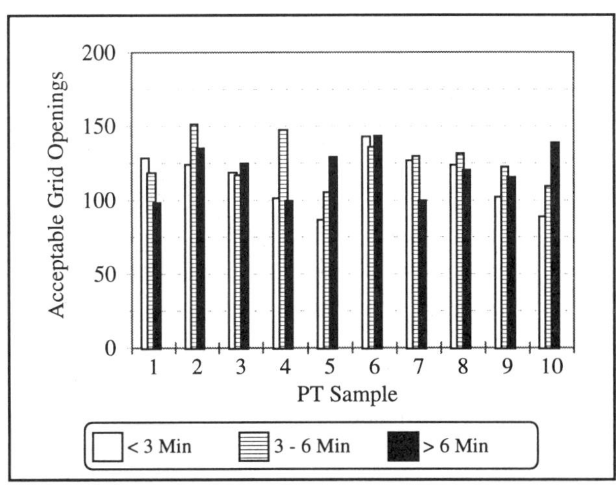

Figure 2. Grid acceptability as a function of etching duration.

Etching for the intermediate period (3 to 6 minutes) produced significantly (p=0.032) more acceptable grid openings than etching for less than 3 minutes (Figure

2). This might be due to greater flexibility of the lightly etched film from the intermediate group than of the glossier films of the less-etched group.

Etching for the long period (> 6 minutes) produced more structures than the short period ($p=0.014$) or the intermediate period ($p=0.081$) (Figure 3). Apparently the shorter fibers deposited in an aqueous medium on the 0.45-μm MCE filters penetrated deeply below the filter surface and were exposed and captured only with extensive etching (*7*). The sample with the greatest proportion of high-etch structure recoveries (Sample 5) was a chrysotile sample. Of all asbestos types, chrysotile generally yields the shortest and thinnest fibers and these likely penetrated deeply into the MCE filter, where they could be recovered only by extensive etching. Effects of etching duration on filters containing aerosol-deposited fibers is uncertain. In the absence of an aqueous medium, even short fibers may be intercepted higher on the filter surface due to electrostatic interactions.

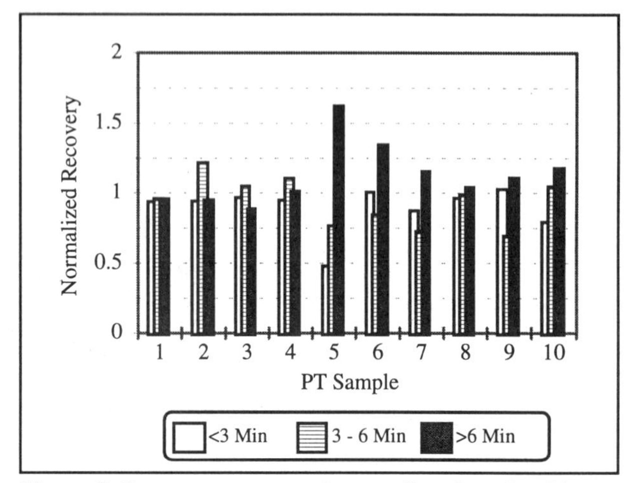

Figure 3. Structure concentration as a function of etching duration.

Etcher Calibration

Three methods widely used for calibrating etchers were 1) complete ashing of collapsed filter (n=44), 2) complete ashing of uncollapsed filter (n=90), and 3) gravimetric tracking of collapsed filter (n=236).

No significant differences were detected among calibration methods for the number of structures counted. Gravimetric tracking of

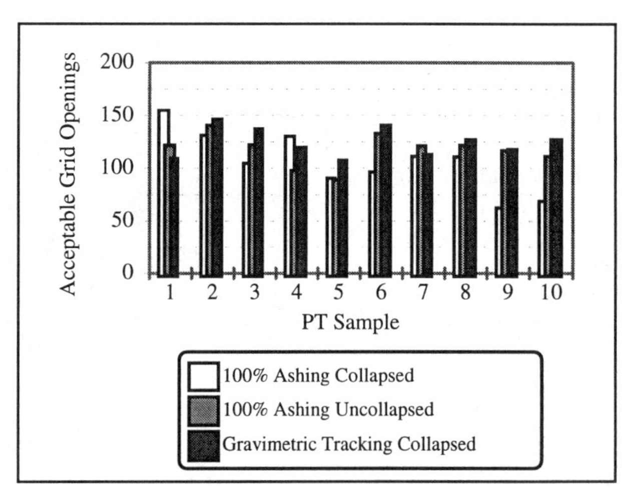

Figure 4. Grid acceptability as a function of etching calibration method.

collapsed filter was associated with significantly more acceptable grid openings than complete ashing of collapsed filter (p=0.042) (Figure 4). One explanation for this might be the difficulty in determining the endpoint (total loss) of a collapsed filter. The slide containing the collapsed filter must be repeatedly removed and carefully scrutinized to determine whether all clinging residue is gone. A second complicating factor is the non-linearity of etching rates during the etching of an entire filter. Either factor could lead to an off-target etching duration. As just reported, under-etching was associated with film breaking. And, although not statistically significant and therefore not reported above, intermediate etching (3 to 6 minutes) produced more acceptable grid openings (mean = 128) than extensive etching (> 6 minutes, mean = 118).

Carbon Source

Most laboratories used carbon rods with 1-mm necks (n=222) while others used thicker carbon rods with tapered points (n=121). Only a few preparations were made with carbon thread (n=25).

No significant differences in either acceptable grids or structure concentrations were associated with the type of carbon source.

Evaporator Geometry

Three geometric configurations were commonly employed during carbon evaporation: 1) rotate and tilt (n=262), 2) rotate with carbon source on rotation axis (n=42), and 3) rotate with carbon source off rotation axis (n=84).

No significant differences in grid acceptability were detected among geometric configurations.

Structure concentrations were significantly associated with evaporator geometry. Rotation-and-tilt geometry was consistently associated with higher concentrations than either rotation on-axis (p=0.001) or rotation off-axis (p=0.097). Rotation off-axis was associated with higher structure concentrations than rotation on-axis (p=0.005) (Figure 5). This is predictable in that fibers

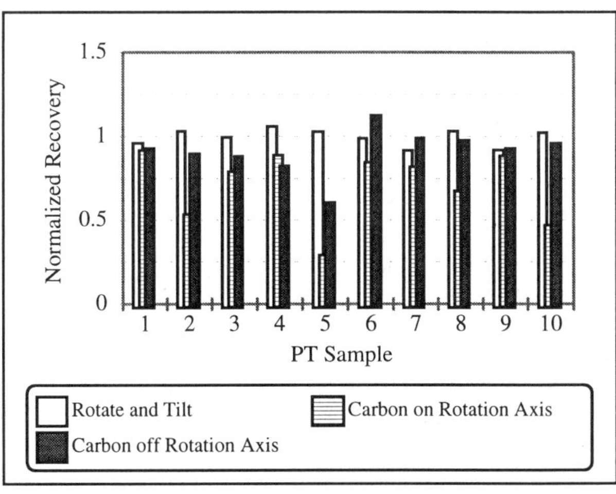

Figure 5. Structure concentration as a function of carbon-evaporation geometric configuration.

coated while the carbon source is on the rotation axis are being coated from a single direction. This leads to an incomplete carbon film holding the fiber and the fiber is subject to loss [*8*].

Solvent

Acetone was the most commonly used solvent (n=205), followed by DMF (n=86) and a combination of the two (n=62). Dimethyl sulfoxide was used 17 times.

No significant differences were detected for either grid-preparation quality or structure concentrations on the basis of solvents employed.

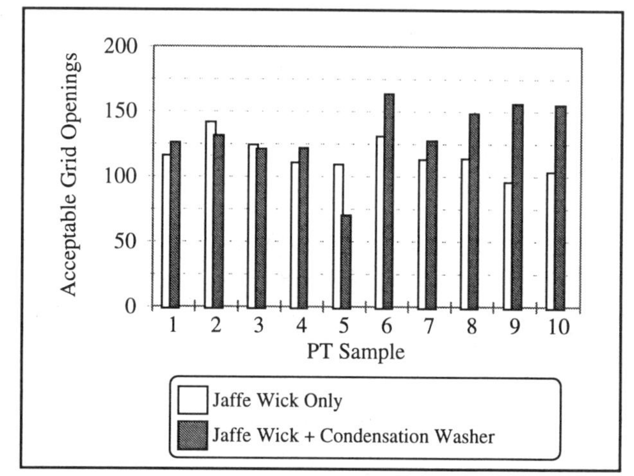

Figure 6. Grid acceptability as a function of filter-dissolution method.

Dissolution Methods

The two most common methods of dissolving filters were the Jaffe wick washer (n=244) and a combination of Jaffe wick and a condensation washer (n=89). Only 18 preparations were made with a condensation washer alone.

For grid preparation quality, the combination of Jaffe wick and condensation washer produced more intact grid openings than Jaffe wick alone (p=0.053) (Figure 6). Likewise, this combination was associated with a higher structure count (p=0.053) than Jaffe wick alone (Figure 7). The association of increased intact grid openings with a combination of Jaffe wick and condensation washer is

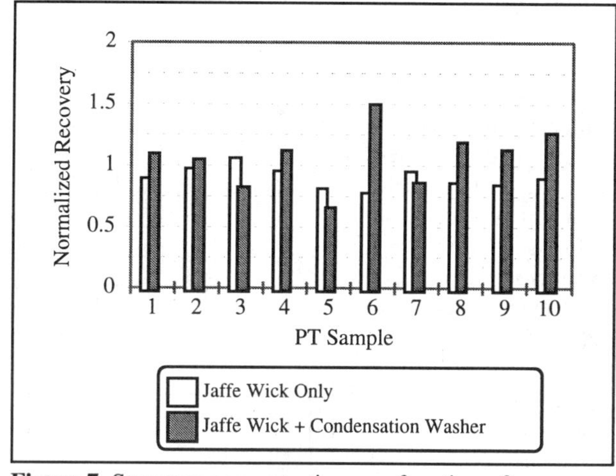

Figure 7. Structure concentration as a function of dissolution method.

unexpected in that the additional handling of grids between the two washes and the turbulence of the condensation washer would tend to damage carbon films.

Magnification

Results were divided into two categories on the basis of the reported nominal magnification used during analysis. Magnifications of 20,000 and higher were used during 117 analyses while magnifications of 15,000 to 20,000 were used during 119 analyses. Magnifications were not reported for Samples 2 or 10.

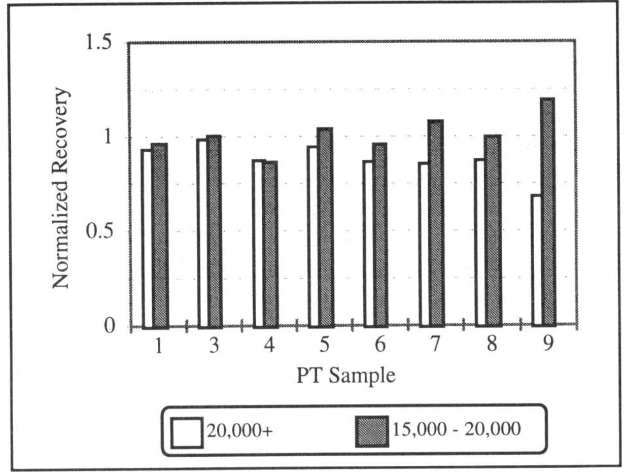

Figure 8. Structure concentration as a function of magnification during analysis.

No significant difference between sets was detected for grid quality (as expected), but laboratories analyzing at magnifications of 20,000 or higher yielded structure concentrations that were significantly ($p=0.0078$) *lower* than lower-magnification laboratories (Figure 8). For the typically minute chrysotile fibers, one might expect the opposite, i.e., higher counts at magnifications of 20,000 and greater. This was not the case: for samples containing chrysotile, reported concentrations were *higher* from analyses performed at the *lower* magnification. One explanation for this apparent anomaly might be the tolerance (or "slop") of a TEM's specimen translators. If a grid on the same microscope were scanned at magnifications of 15,000 and 20,000 with the same percentage overlap between scans, e.g., 25% overlap, the tolerance in the translator would be 33% greater at the higher magnification and could increase the likelihood that a scan might overpass a section.

Constraints and Conclusions

The significant associations detected during this study may or may not portray a cause and effect. The correlation of good preparations and moderate etching and the correlation of higher structure concentrations with rotation/tilt geometry and with extended etching are intuitively correct. On the other hand, the correlation of good preparations with dual dissolution and of elevated concentrations with lower magnification is perplexing. One confounding factor in this survey might be the possibility of covariance between parameters. For example, the laboratories that etched for greater than six minutes might also have analyzed samples at magnifications less than 20,000, thus creating an apparent correlation between magnification and structure concentration. Another source of uncertainty is the use of different grids by the laboratories in the later PT rounds. Laboratories that used grids with 10 by 10 (=100) grid openings would automatically have fewer potentially acceptable grid openings than laboratories using conventional 200-mesh grids with 200 or more grid openings.

References

[*1*] Section 502 Amended, New York State Public Health Law, 1 April 1984.

[*2*] 40 CFR 763, "Asbestos Hazard Emergency Response Act - Detailed Procedure for Asbestos Sampling and Analysis - Non-Mandatory" *Federal Register*, Vol. 52, No. 210, 30 October 1987, pp. 41845-41905.

[*3*] Lehmann, E.L.: *Nonparametrics: Statistical Methods Based on Ranks.* San Francisco, Calif.: Holden-Day, Inc., 1975. Table H.

[*4*] Beyer, W.H. (ed.): *Handbook of Tables for Probability and Statistics.* 2nd ed. Cleveland OH, The Chemical Rubber Company, p. 400, 1968.

[*5*] "Method 7400: Asbestos and Other Fibers by PCM," *NIOSH Manual of Analytical Methods,* Fourth Edition, 15 August 1994, pp. 1-15.

[*6*] Burdett, G. J., and Ruud, A. P., "Membrane-Filter, Direct-Transfer Technique for the Analysis of Asbestos Fibers or Other Inorganic Particles by Transmission Electron Microscopy," *Environmental Science and Technology,* Vol. 17, No. 11, 1983, pp. 643-648.

[*7*] Chatfield, E. J., "Measurements of Chrysotile Fiber Retention Efficiencies for Polycarbonate and Mixed Cellulose Ester Filters," *Advances in Environmental Measurement Methods for Asbestos, ASTM STP 1342,* M. E. Beard, H. L. Rook, Eds., American Society for Testing and Materials, 1998.

[*8*] Webber, J. S. , and Janulis, R .J., "Effects of Evaporation Configuration on Carbon Film Integrity," *Environmental Information Association Technical Journal,* Vol. 3, No. 1, 1995, pp.14-19.

Andrew F. Oberta[1] and Kenneth E. Fischer[2]

NEGATIVE EXPOSURE ASSESSMENTS FOR ASBESTOS FLOOR TILE WORK PRACTICES

REFERENCE: Oberta, A. F., and Fischer, K. E., **"Negative Exposure Assessments for Asbestos Floor Tile Work Practices,"** *Advances in Environmental Measurement Methods for Asbestos, ASTM STP 1342*, M. E. Beard, H. L. Rook, Eds., American Society for Testing and Materials, 2000.

ABSTRACT: Tests were conducted to evaluate fiber release and control methods from certain work practices used for installation of equipment on asbestos-containing resilient floor tile. Air samples collected during controlled operations showed that worker exposure can be expected to remain below the OSHA Permissible Exposure Limit of 0.10 fiber/cc for an 8-hour Time-Weighted Average, and below the Excursion Limit of 1.0 fiber/cc for a 30-minute period.

The criteria for a Negative Exposure Assessment under the OSHA asbestos standard are met, providing that the procedures and controls, including clean-up, described herein are employed by workers properly trained in their use. These procedures are shown to be the equivalent of the Resilient Floor Covering Institute Recommended Work Practices for resilient floor tile insofar as worker protection is concerned. These test results are not applicable to sheet vinyl linoleum with asbestos backing material.

KEYWORDS: asbestos, negative exposure assessment, floor tile, air sampling

Background

Resilient floor covering that contains asbestos is commonly found in buildings; it is present in 1,526,000 (or 42% of) public and commercial buildings, according to the Environmental Protection Agency (EPA) [1]. This material is found in the form of resilient floor tile, of which the 9 × 9-in. size, almost without exception, contains asbestos. The adhesive mastic used to install these

[1]The Environmental Consultancy, 107 Route 620 South, #35E, Austin, TX 78734.

[2]Principal, Dames & Moore, Inc., 7101 Wisconsin Avenue, Chevy Chase, MD 20814-4870.

tiles quite often contains asbestos as well. Asbestos is also found in the woven or matted backing of sheet vinyl linoleum.

Removal of vinyl asbestos floor tile has become a common abatement procedure, either using aggressive methods, including machinery, as described in NIBS Section 02087, Asbestos Abatement and Management in Buildings: Model Guide Specification [2], or using hand tools for smaller quantities according to the Recommended Work Practices (RWP) of the Resilient Floor Covering Institute (RFCI) [3]. The RWPs are based on tests that showed worker exposure below the Occupational Safety and Health Administration (OSHA) Permissible Exposure Limit (PEL) of 0.1 fibers per cubic centimeter of air (f/cc) [4]. The RWPs can be relied on as a Negative Exposure Assessment (NEA) for compliance with the OSHA standard as long as conditions, materials, work practices and training "closely resemble" those in the study on which the RWPs are based.

Asbestos floor tile is often present in areas where equipment must be secured to the floor underneath--usually a concrete slab that does not contain asbestos. This is typically done by drilling holes through the tile into the concrete and setting anchors for bolts. In many cases, removal of entire sections of flooring to avoid disturbing the asbestos-containing material (ACM) is not feasible. Besides drilling through the tile, other methods include removing a small piece of a tile, or punching a hole in the tile, before drilling into the concrete.

Previous studies have shown that drilling and punching holes in floor tile can produce exposures that either comply with or exceed the PEL, depending on the material, procedure, and control method used. Lawrence Livermore Laboratory (LLL) conducted tests to support an NEA by punching and breaking floor tiles and monitoring worker exposure levels [5]. The tests were conducted in a room-sized enclosure with High-Efficiency Particulate Air (HEPA)-filtered exhaust. The workers performing the tests had completed 40 hours of asbestos training. In one series of tests, eight holes were "cold-punched" (tiles were not heated) in each of three asbestos-containing tiles with asbestos in the mastic (percent not reported). The tiles were mounted on a wood substrate, and clear shaving gel was used to suppress debris and fiber release. Personal air samples showed exposure below the limit of detection of 0.017 f/cc, which is also below the OSHA PEL. No information is available on whether the removed material remained intact. In the other test, a worker broke a 12 × 12-in. tile into 72 pieces. His personal exposure was below 0.026 f/cc. No drilling or hole cutting tests were performed by LLL on floor tile, but such tests on Transite and asbestos siding showed exposures below the PEL.

Brackett et al. [6] used vacuum shrouds to capture airborne fibers; their test results showed fiber levels below the PEL. They also conducted tests without a shroud or other dust control measures, such as a shaving gel. Fiber levels exceeded the PEL in these control tests.

This paper reports on three different work practices for resilient floor tile: intact removal of a small piece of tile, drilling a hole through the tile, and punching a hole in the tile. Control methods included water, an encapsulant, a heat gun, and a vacuum shroud. No tests were conducted on sheet vinyl linoleum.

Methodology

The tests were conducted by a licensed asbestos abatement contractor under the direction of a licensed asbestos consultant. Air samples were collected by licensed air monitoring technicians and asbestos program managers. Personal and area samples were analyzed by a licensed laboratory using Phase Contrast Microscopy (PCM) [7]. Area samples were also analyzed by a licensed laboratory using Transmission Electron Microscopy (TEM) [8]. The same laboratory also analyzed bulk samples of floor tile and mastic using Polarized Light Microscopy (PLM) and TEM [9].

Description of Test Facility

The tests were conducted in an equipment room that was secured against unauthorized entry. One corner of the room has approximately 1,000 ft^2 of 9 × 9-in. floor tile, previously shown to contain asbestos.

Two test chambers were constructed by the abatement contractor to expedite completion of the large number of tests planned. The chambers were kept under negative pressure by a High Efficiency Particulate Air (HEPA) filtered exhaust unit to eliminate the possibility of fibers being released into adjacent occupied areas of the building and to prevent cross-contamination of the chambers. Each chamber had its own entrance and make-up air was drawn into the test chambers through the airlocks in the change rooms. A negative pressure of at least 0.02 inches of water relative to the space adjacent to the test chambers was maintained at all times and continuously recorded.

Because the nearest source of water was distant enough to require running a hose through the equipment room, and because of the small amounts of fiber release and debris expected, no showers were used. Personnel "double-suited" when entering a test chamber and removed the outer suit in the change room upon exiting. This procedure also helped to prevent cross-contamination by the individuals who went from one chamber to the other for inspections, air sampling, and other purposes.

The walls inside the test chambers were covered with fire-retardant plastic sheets, and a ceiling of plastic sheeting was suspended from overhead cable trays. The ceiling height was approximately 10 feet. All surfaces inside both test chambers were HEPA-vacuumed and wet-wiped before the first test. A waste storage chamber was constructed in Test Chamber #1 for temporary storage of bags of removed tile pieces and cleaning cloths. This chamber remained sealed except when bags were placed in it, and no air was drawn through it. Waste was disposed of by the abatement contractor after all tests were completed.

Tile and Mastic Analysis

Bulk samples of the tile and mastic were analyzed (Table 1). Results of a previous analysis, conducted in 1990, are shown in Table 1 for comparison. When PLM analysis of floor tile indicates no asbestos, EPA requires use of TEM to be sure there are none of the short fibers used in floor tile manufacture nor any interfering substances. That was not the purpose in this case; TEM was used to determine whether the PLM analysis had underestimated the asbestos content of the floor tile. The TEM results agreed very closely with the PLM results. The 1996 analysis of the mastic showed a lower asbestos content than reported in the 1990 analysis.

TABLE 1--Bulk sampling analytical results.

1996 Analysis		1990 Analysis	
Floor Tile	Mastic	Floor Tile	Mastic
14% chrysotile by PLM	8% chrysotile by PLM	5 - 15% chrysotile by PLM	30 - 50% chrysotile by PLM
13 - 16% chrysotile by TEM		...	...

Testing Operations

Two tests were conducted daily for 10 days. The following test procedures were used, with the number of tests for each shown in parentheses:

IR1 - Intact removal with water (3)
IR2 - Intact removal with encapsulant (2)
IR3 - Intact removal with heat gun (2)
HD1 - Hole drilling with water (3)
HD2 - Hole drilling with encapsulant (3)
HD3 - Hole drilling with vacuum shroud (1)
HP1 - Hole punching with water (3)
HP2 - Hole punching with encapsulant (3).

Test Procedure HD2 is shown in the box below.

Test Procedure HD2: Hole drilling with encapsulant.

1. This test procedure consists of drilling holes in 9 × 9-in. tiles. Each worker shall drill twenty-five holes at 4-min intervals.
2. The following items of equipment will be brought into Test Chamber #2: Debris encapsulant (Palmolive™ shaving cream), water sprayers, impact drill with ⅝-in. bit, cleaning cloths, and HEPA-filtered vacuum cleaner.
3. Each worker shall proceed as follows, starting with the first tile marked by the consultant and proceeding to the last.
   a. Cover a 1-in. diameter circular area of the tile with debris encapsulant.
   b. Position the drill in the encapsulant and turn the drill on low speed. Slowly drill through the tile into the concrete until the drill bit has made a hole completely through the tile. Make sure that the debris encapsulant covers the area being drilled, adding encapsulant if necessary.
   c. Turn off the drill and slowly retract the drill bit without disturbing the debris, concrete dust, and encapsulant around the hole. Wipe the debris, concrete dust, and encapsulant off the drill bit with a cloth.
4. The above operation will be repeated on the remaining tiles. After the consultant has inspected the tiles, any debris, concrete dust, and encapsulant will be removed from the tiles by using a HEPA vacuum and by wet-wiping.
5. When the drilling tests are completed, the consultant will inspect the floor for visible debris. Workers will then clean the walls and floor of the entire test chamber by using a HEPA vacuum and by wet-wiping.
6. All items of equipment will be cleaned and the contaminated cloths will be placed in disposal bags. Workers will exit the test chamber into the change room and remove disposable clothing.

The tests began each day at approximately 9:00 a.m. and were completed by noon. Before each test, the consultant marked the tiles to be removed, drilled, or punched each day, and the abatement contractor's supervisor explained the procedures to the workers.

Two workers in each test chamber removed pieces from, or drilled or punched holes in, 25 tiles at approximately 4-min intervals, alternating by 2 min. (For example, Worker #1 drilled a hole at 9:00, Worker #2 at 9:02, Worker #1 at 9:04, Worker #2 at 9:06, etc.) Fifty tiles were removed, drilled, or punched in approximately 100 min. Afterward, the floor and walls were cleaned and the consultant conducted a visual inspection and directed re-cleaning if visible debris was found. After both chambers passed the visual inspections, air samples were taken for clearance.

The consultant remained in the test chamber during the initial series of tests to direct the workers in the test operations and to ensure that the 4-min schedule was maintained. He collected samples of debris and pieces of tile, took notes on his observations, and photographed the operations. For the remainder of the tests, a licensed asbestos program manager directed the test operations and performed the visual inspections.

Air Sampling

A total of 152 personal air samples were taken to provide exposure monitoring data for an NEA per 29CFR1926.1101(f)(2)(iii), Occupational Exposure to Asbestos: Construction Industry Standard. Sixty long-term samples, taken over the duration of the testing or clean-up, or both, were termed PEL samples and ranged from 40 to 240 min (average = 121 min). Ninety-two short-term samples of 30-min duration were taken to evaluate peak exposures and were termed Excursion Limit or EL samples. The workers wore two pumps during part of the operations to take advantage of available sampling time. All personal samples were taken at 2.0 litres/minute (Lpm) on 0.8 micrometre (μm) pore-size Mixed Cellulose Ester (MCE) filters.

Area samples were collected at fixed locations inside each test chamber. Two samples were taken with high flow-rate (8.0 and 9.5 Lpm) pumps during testing and cleaning: one for analysis by PCM and one by TEM. Two clearance samples in each chamber were taken aggressively, with the test chamber swept once with an electric leaf blower. These samples were also analyzed by PCM and TEM. Sample volumes were 1,040 to 2,080 L during testing and cleaning, and 1,273 to 1,473 L for clearance. A total of 80 area samples were taken in the test chambers. Area samples were also taken outside the test chamber with high flow-rate pumps to verify critical barrier integrity for building occupant protection.

Results and Discussion

Observations

Intact removal--During Procedure IR1 using water control, a small amount of debris was generated by the chisel or scraper blade as it was driven through the tile. The amount of debris produced was probably affected by the sharpness of the blade. During Procedure IR2, the encapsulant (Palmolive™ shaving gel) contained the debris. Most of the pieces were removed intact, with a corner occasionally breaking off. (Note that accidental breakage of tile does not render it friable, according to OSHA and EPA interpretations and the RFCI RWPs.) During Procedure IR3, debris and breakage were minimized by cutting the tile immediately after heating it and while it was still soft. Use of a heat gun with a blower, as shown in the RFCI RWPs, is not advised because of the possibility of dispersing debris and fibers.

Hole drilling--A low-speed impact drill was used; the debris from the tile was in the form of helical shavings rather than fine particles. The water

(Procedure HD1) caused some of the concrete dust to form a small semi-circular berm of soft paste around the hole that captured some of the tile debris. The remaining dust was ejected asymmetrically as much as 1 ft from the hole. The encapsulant (Procedure HD2) also formed a partial berm and showed this asymmetrical concrete dust ejection pattern. (Subsequent to this test program, it was shown that this dispersal of concrete dust can be eliminated by a more liberal application of shaving gel; i.e. a mound about ½-in. high and 2 in. diameter.)

The vacuum shroud (Procedure HD3) consisted of a bottomless, clear plastic 5-gal bucket with a sheet of heavy plastic across the top with a slit for the drill bit. The tile was sprayed with water for fiber control before the shroud was placed over the tile and the hole drilled. The HEPA-vacuum hose was attached to the bottom of the bucket. While it collected airborne fibers, the suction of the hose did not appear to disturb the concrete dust and tile debris around the hole.

Hole punching--Two sizes of industrial hole punches (¾-in. and 1-½-in. diameter) were used. Debris was produced by the punches, some of which was contained by the encapsulant in Procedure HP2. The punches caused over half of the tiles to crack. This cracking as well as the amount of debris were probably due in part to the sharpness of the punches. In one test, the impact caused the adjacent tile to lift.

Clean-up--The procedures tested reduced, but did not completely eliminate, the generation and dispersal of dust, debris, and fibers. Proper clean-up methods are essential to minimizing worker and occupant exposure from these operations, and thorough clean-up is an indispensable part of the procedures. For the tests, the test chambers were cleaned before a visual inspection and air sampling clearance (as the work area on an abatement project would be).

Thorough clean-up of the mastic is essential to minimizing post-removal fiber release after the intact removal and hole punching procedures, because dry residue can remain after the mastic remover evaporates. Using the hole drilling procedures, asbestos fibers may be contained in the concrete dust and thorough clean-up of the drill bit and the floor is essential. A more liberal application of the encapsulant than was used in the tests will also prevent dispersal of the debris and dust, as was shown subsequent to this program. For all procedures, precautions should be taken to avoid tracking debris and contaminated dust around and beyond the work area. Wearing disposable foot coverings or cleaning footwear is recommended.

Air Sampling

Personal samples for worker exposure--The results of the personal samples are summarized in Tables 2 and 3, and are also shown in Fig. 1. None of the PEL samples exceeded 0.1 f/cc, and none of the EL samples exceeded 1.0 f/cc.

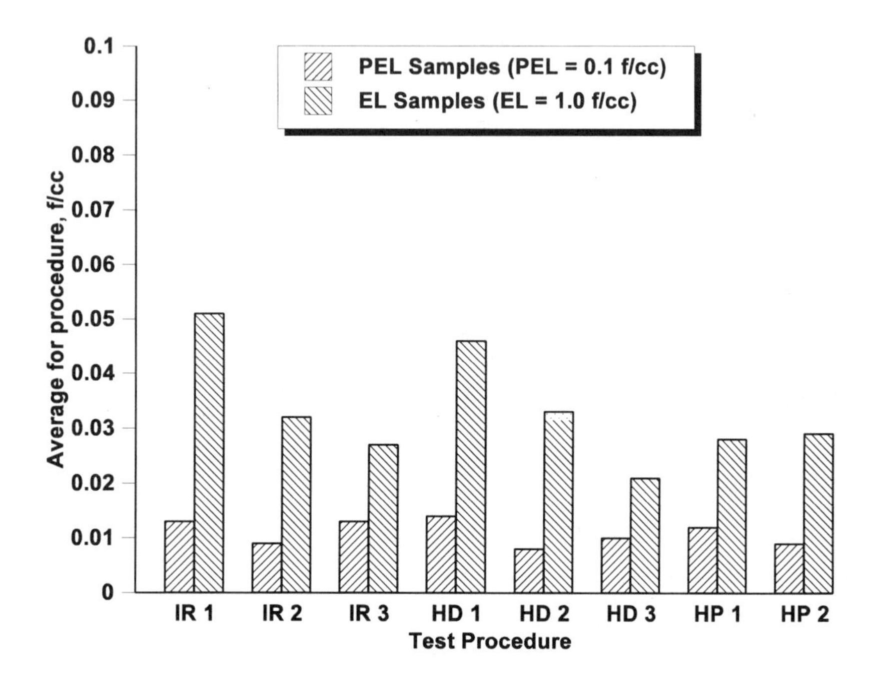

Figure 1. Personal Samples taken during Testing and Clean-up

The PEL values in Table 2 are for the duration of the sample and are not adjusted for an 8-hour Time-Weighted Average (TWA), which would lower the average exposure over a working day by as much as 50% due to the amount of non-exposed time. The actual number of fibers counted in 108 of the 152 personal samples (71%) was below the detection limit of four fibers/100 fields for PCM analysis as specified in Appendix B to 29CFR1926.1101. For those samples, four fibers/100 fields was used in Tables 2 and 3 for the calculation of the airborne fiber concentration, which is conservative relative to using the actual number of fibers counted.

TABLE 2--Summary of personal (PEL) samples during testing and cleaning.

Procedure	No. of tests	No. of samples	Range, f/cc	Average, f/cc	95% UCL, f/cc
IR1 - Intact removal with water	3	8	0.0067 - 0.0292	0.013	0.026
IR2 - Intact removal with encapsulant	2	6	0.0022 - 0.0175	0.009	0.019
IR3 - Intact removal with heat gun	2	6	0.0086 - 0.0344	0.013	0.029
HD1 - Hole drilling with water	3	10	0.005 - 0.028	0.014	0.025
HD2 - Hole drilling with encapsulant	3	8	0.0053 - 0.0114	0.008	0.012
HD3 - Hole drilling with vacuum shroud	1	4	0.0037 - 0.0146	0.010	0.017
HP1 - Hole punching with water	3	10	0.0053 - 0.022	0.012	0.021
HP2 - Hole punching with encapsulant	3	8	0.0041 - 0.0131	0.009	0.015
Total	20	60			

Also shown in Tables 2 and 3 are the 95% Upper Confidence Limits (UCL) for the PEL samples and EL samples for each procedure. The UCL is calculated from [10]

$$UCL = AC + (1.645)(RSD)(AC) \qquad (1)$$

where AC = average concentration (f/cc) and RSD = relative standard deviation, or SD/AC. None of the 95% UCLs for the PEL samples exceed 0.1 f/cc for any procedure, and none of the 95% UCLs for the EL samples exceed 1.0 f/cc for any procedure.

TABLE 3--Summary of personal (EL) samples during testing and cleaning.

Procedure	No. of tests	No. of samples	Range, f/cc	Average, f/cc	95% UCL, f/cc
IR1 - Intact removal with water	3	12	0.0218 - 0.2777	0.051	0.164
IR2 - Intact removal with encapsulant	2	12	0.0123 - 0.0531	0.032	0.048
IR3 - Intact removal with heat gun	2	12	0.0245 - 0.0327	0.027	0.034
HD1 - Hole drilling with water	3	10	0.0327 - 0.0694	0.046	0.070
HD2 - Hole drilling with encapsulant	3	16	0.0237 - 0.098	0.033	0.062
HD3 - Hole drilling with vacuum shroud	1	4	0.0204 - 0.0245	0.021	0.024
HP1 - Hole punching with water	3	12	0.0204 - 0.049	0.028	0.042
HP2 - Hole punching with encapsulant	3	14	0.0245 - 0.0327	0.029	0.035
Total	20	92			

The twelve highest total fiber counts for personal samples are shown in Table 4. With the exception of samples IR1-2-08 and IR3-1-06, the actual number of fibers counted by PCM on the personal samples was quite low, and apparently independent of the operation being performed. One EL sample (IR1-2-08) exceeded and one EL sample (HD2-1-15) effectively equalled the PEL of 0.1 f/cc. Work practices and air sampling procedures followed in Test No. IR1-2 were reviewed with the contractor and air monitoring technician, and no explanation was found for the high EL value.

Area samples analyzed by PCM are summarized in Table 5. The fiber levels detected during testing and clean-up are well below the personal samples, possibly because there appeared to be little mixing of the air within the test chambers, and the area samples were taken from 5 to 20 ft away from the location of the work. Table 5 also shows that clearance samples were below 0.01 f/cc by PCM, with the exception of Test HD1-2. Although these are referred to as "clearance" samples, it should be remembered that no building occupants were allowed in the test chambers until all the tests were complete. Table 6 shows the results of samples analyzed by TEM. Asbestos structures were found on 9 of the 20 samples taken during testing and cleaning, and on only one clearance sample.

TABLE 4--Twelve highest fiber counts for personal samples.

Sample No. *	Type **	Volume, litres	f/100 fields	f / cc	Operation
IR1-2-08	EL	60	34	0.2777	Removal
IR3-1-06	PEL	370	27	0.0344	Removal
HD1-1-06	PEL	210	12	0.028	Drilling
IR1-1-06	PEL	168	11	0.0292	Removal
IR1-3-03	PEL	220	11	0.0223	Removal & cleaning
HD1-1-03	PEL	210	9	0.0210	Drilling
HD2-1-15	EL	40	9	0.098	Cleaning
HD1-1-10	EL	60	8.5	0.0694	Cleaning
HP2-1-03	PEL	276	8	0.0124	Punching
IR3-1-03	PEL	370	7.5	0.0086	Removal
IR2-2-07	EL	60	7.5	0.0531	Removal
HD1-1-09	EL	60	7	0.0572	Cleaning

* Sample No.: Procedure - Test No. - XX, e.g. IR1-2-08.
** PEL - Permissible Exposure Limit; EL - Excursion Limit.

Samples taken at four locations outside the test chambers for verification of critical barrier integrity were all below 0.01 f/cc by PCM analysis.

NEA

The test results show that the work practices evaluated comply with 29CFR1926.1101 and can be used to support an NEA when these operations are performed in the future. The work practices "closely resemble" those in 29CFR1926.1101(g)(8)(i) for Class II vinyl asbestos flooring and are also considered Class III operations under 29CFR1926.1101(g)(9) insofar as their purpose is not removal or abatement of ACM, but repair and maintenance that may involve the disturbance of ACM. It is noted that the removal of intact pieces of floor tile, as performed in Procedures IR1, IR2, and IR3, is the equivalent of "incidental breakage" that does not render the material friable according to OSHA.

The OSHA *Asbestos Advisor* [11] contains the following Question and Answer:

Q. If floor tiles are broken during removal, are they no longer "intact?"

A. Not necessarily. Some incidental breakage of floor tiles is to be expected. Under the standard, material is not intact only if it has crumbled, been pulverized, or has otherwise deteriorated so that the asbestos fibers are not longer likely to be bound with their matrix. Therefore, the incidental breakage of tiles does not by itself mean that the material is not intact.

TABLE 5--Summary of PCM area samples.

Procedure	Test No.	During testing and cleaning		Clearance	
		f/cc	f/100 fields	f/cc	f/100 fields
IR1 - Intact removal with water	1	0.0026	9.0	0.0025	7.5
	2	0.0056	17.5	0.0023	6.0
	3	0.0023	6.0	0.0057	16.0
IR2 - Intact removal with encapsulant	1	0.0044	15.5	**0.0015**	**<4.0**
	2	0.0025	10.0	0.0011	4.0
IR3 - Intact removal with heat gun	1	0.0033	13.0	0.0011	4.0
	2	**0.0009**	**2.0**	**0.0015**	**1.0**
HD1 - Hole drilling with water	1	0.0058	18.5	0.0031	8.0
	2	0.0013	4.5	0.0222	58.0
	3	**0.0014**	**1.0**	**0.0011**	**3.5**
HD2 - Hole drilling with encapsulant	1	0.0013	5.5	0.0033	11.0
	2	0.0020	7.5	0.0050	13.0
	3	0.0046	14.0	0.0038	10.5
HD3 - Hole drilling with vacuum shroud	1	0.0041	15.5	0.0040	13.5
HP1 - Hole punching with water	1	0.0036	12.5	0.0027	7.5
	2	0.0026	7.0	0.0027	8.5
	3	0.0032	7.0	0.0050	13.0
HP2 - Hole punching with encapsulant	1	0.0020	7.0	0.0023	7.5
	2	**0.0009**	**2.0**	**0.0015**	**1.0**
	3	0.0037	11.0	0.0071	18.5

Numbers in bold face are detection limits for the analytical method used.

TABLE 6--Summary of TEM area samples.

Procedure	Test No.	During testing and cleaning		Clearance	
		s/cc	s/100 fields	s/cc	s/100 fields
IR1 - Intact removal with water	1	0.0041	0	0.0042	0
	2	0.0042	1	0.0042	0
	3	0.0150	3	0.0042	0
IR2 - Intact removal with encapsulant	1	0.0189	4	0.0042	0
	2	0.0041	0	0.0041	0
IR3 - Intact removal with heat gun	1	0.0046	1	0.0042	0
	2	0.0026	0	0.0042	0
HD1 - Hole drilling with water	1	0.0042	0	0.0042	0
	2	0.0289	6	0.0042	0
	3	0.0044	0	0.0042	0
HD2 - Hole drilling with encapsulant	1	0.0139	3	0.0044	0
	2	0.0041	0	0.0042	0
	3	0.0045	0	0.0042	0
HD3 - Hole drilling with vacuum shroud	1	0.0046	0	0.0044	0
HP1 - Hole punching with water	1	0.0344	7	0.0126	3
	2	0.0100	2	0.0042	0
	3	0.0050	0	0.0042	0
HP2 - Hole punching with encapsulant	1	0.0043	1	0.0042	0
	2	0.0026	0	0.0042	0
	3	0.0045	0	0.0042	0

Results are asbestos structures > 0.5 μm long per cubic centimetre of air. Concentrations in bold face are detection limits for the analytical method used.

In 29CFR1926.1101, OSHA uses the term "substantially intact," which could also apply to condition of the floor tile following the use of these procedures.

These procedures may be considered equivalent to the RFCI RWPs, specifically the procedures entitled "Complete Removal of Existing Resilient Tile Floor Covering" on pp. 23-27 of the publication cited earlier. It is noted that the procedures for drilling and punching holes do not result in worker exposures in excess of the PEL and EL, even with the generation of visible debris, because the control methods used are effective in minimizing the release of airborne fibers. Therefore, the drilling and punching procedures provide a degree of worker protection equivalent to the intact removal procedures and to the related RFCI RWPs.

Because no prior monitoring had been performed using these procedures, these test results also constitute an Initial Exposure Assessment (IEA) per 29CFR1926.1101(f)(2)(ii), as well as an NEA per 29CFR1926.1101(f)(2)(iii). To rely on the NEA for future work, the employer must document compliance with the provisions of 29CFR1926.1101(f)(2)(iii)(B):

1. The employer has monitored prior asbestos jobs for the PEL and the excursion limit within 12 months of the current or projected job;
2. The monitoring and analysis were performed in compliance with the asbestos standard in effect;
3. The data were obtained during work operations conducted under workplace conditions "closely resembling" the processes, type of material, control methods, work practices, and environmental conditions used and prevailing in the employer's current operations;
4. The operations were conducted by employees whose training and experience are no more extensive than that of employees performing the current job;
5. These data show that under the conditions prevailing and which will prevail in the current workplace there is a high degree of certainty that employee exposures will not exceed the TWA and excursion limit.

A task performed in the future for purposes of equipment installation will constitute the "current or projected job," and these tests constitute the "prior asbestos jobs" as expressed in Item 1. The data in the NEA are valid for work performed within 12 months of the date of these tests. In response to Item 2, the monitoring and analysis were performed in accordance with the current OSHA standard.

The employer must certify that future work "closely resembles" the tests on which the NEA was based (Item 3). Even though the tests were done in negative-pressure enclosures, the air sampling results showed that the use of such enclosures will not be necessary for future work. In response to Item 4, the training received by the abatement workers who performed the tests was of a longer duration than the training that will be given to equipment installers, but abatement worker training includes many work practices not applicable to work contemplated for equipment installation purposes. Furthermore, the equipment installers will receive training in the specific use of these procedures, including "hands-on" exercises and a 2-hour awareness course.

The "high degree of certainty" required by Item 5 that the PEL and EL will not be exceeded is demonstrated by the fact that the 95% UCL for all procedures is below 0.1 f/cc for the PEL samples and below 1.0 f/cc for the EL samples, as shown in Tables 2 and 3.

The NEA must be produced in the event of an OSHA compliance inspection, but need not be submitted to OSHA otherwise. It is intended to show that worker exposure will be below the PEL and EL and may be relied on for purposes of compliance with certain provisions of the standard, including exposure monitoring, respiratory protection, protective clothing, decontamination, and medical surveillance. Although the employer may still elect to do so, it will not be necessary to perform these measures on each job, as long as conditions conform to those in the NEA, to remain in compliance with 29CFR1926.1101.

Conclusions

A series of twenty tests on asbestos-containing resilient floor tile demonstrated that worker exposure can be kept below the OSHA PEL of 0.1 f/cc and the EL of 1.0 f/cc during intact removal of pieces of the tile and during drilling and punching holes in the tile, provided that certain methods are used to control debris and fiber release and that proper clean-up is performed. The tests also demonstrated that there is no significant exposure to other building occupants as a result of these operations. An NEA was prepared in compliance with the OSHA Construction Industry Standard.

REFERENCES

[1] U.S. Environmental Protection Agency, *EPA Study of Asbestos-Containing Materials in Public Buildings: A Report to Congress*, Appendix 1, Washington, DC, 1988.

[2] National Institute of Building Sciences, *Asbestos Abatement and Management in Buildings: Model Guide Specification*, Third Edition, Washington, DC, 1996.

[3] Resilient Floor Covering Institute, *Recommended Work Practices for the Removal of Resilient Floor Coverings*, Rockville, MD, 1995.

[4] Occupational Safety and Health Administration, *Occupational Exposure to Asbestos: Construction Industry Standard*, 29CFR1926.1101, 1994.

[5] Zalk, D., CIH, *Minuscule Asbestos Procedures*, Lawrence Livermore National Laboratory, Presented at the EIA '96 Conference of the Environmental Information Association, Las Vegas, NV, March 2, 1996.

[6] Brackett, K. A., Hanna, J. L., Vincent, H. S., and Clarke, P. J., "Engineering Control Practices for Reducing Emissions During Drilling of Asbestos-Containing Flooring Materials," *Technical Monograph Series*, Environmental Information Association, Bethesda, MD, 1997.

[7] National Institute for Occupational Safety and Health, Centers for Disease Control, "Asbestos and Other Fibers by PCM," *NIOSH Manual of Analytical Methods*, Method 7400, Issue 2, Fourth Edition, Cincinnati, OH, August 15, 1994.

[8] U.S. Environmental Protection Agency, "Asbestos-Containing Materials in Schools," *Federal Register*, 40 CFR Part 763, Appendix A to Subpart E. Washington, DC, October 30, 1987.

[9] U.S. Environmental Protection Agency, *Method for the Determination of Asbestos in Bulk Building Materials*, EPA/600/R-93/116, Research Triangle Park, NC, 1993.

[10] Jordan, R. C., "Reporting the Results of Final Air Sampling: 1. Consideration of Variability," *NAC Journal*, National Asbestos Council, Inc., Atlanta, GA, 1988.

[11] Occupational Safety and Health Administration, *Asbestos Advisor for Building Owners,* Release 2.0. Section 6. Flooring Operations. http://www.osha.gov/oshasoft/asbestos/asbxwin3.html, 1997.

G. J. Burdett,[1] G. Archenhold,[2] A. R. Clarke,[2] D. M. Hunter[2]

THE USE OF SCANNING CONFOCAL MICROSCOPY TO MEASURE THE PENETRATION OF ASBESTOS INTO MEMBRANE FILTERS

REFERENCE: Burdett, G. J., Archenhold, G., Clark, A. R. and Hunter, D. M., **"The Use of Scanning Confocal Microscopy to Measure the Penetration of Asbestos into Membrane Filters."** *Advances in Environmental Measurement Methods for Asbestos, ASTM STP 1342,* M. E. Beard, H. L. Rook, Eds. American Society for Testing and Materials, West Conshohocken, PA, 2000.

ABSTRACT: Nearly all current methods for asbestos analysis rely on the use of a filter to separate particulate from the aerosol, or in the case of indirect preparations, particulate from a liquid. The filter that is most commonly used is formed from cellulose acetate or cellulose nitrate or a mixture of both. The resulting membrane has a sponge - like structure forming a series of tortuous pores which have dimensions larger than the nominal pore size of the filter. This allows particulates to penetrate into the filter before capture so that for many types of microscopy analysis [e.g. phase contrast microscopy (PCM), scanning electron microscopy (SEM) and transmission electron microscopy (TEM)], it is important how far the fibre penetrates into (or through) the membrane, as it may not be recoverable for analysis.

Confocal microscopy has the advantage over conventional light microscopy of being able, through means of an adjustable slit, to discard light from above and below the focal plane, to makes it possible to observe a precise point within a transparent film. Computer controlled quantitative scanning confocal microscopy may be used to carry out precise measurements of depth profiles of entrapped particulates by scanning a vertical series of 2D planes into the membrane. The use of pixel connectivity routines also allow the 3D orientation of fibres to be determined so foreshortening effects on fibre length can be measured. The technique has been applied to 0.8μm nominal pore size filters, which had been loaded with crocidolite asbestos, to determine the fibre penetration and orientation.

KEYWORDS: confocal, microscopy, membrane filter, PCM, TEM, asbestos, penetration

---

[1] Health and Safety Laboratory, Broad Lane, Sheffield, S3 7HQ, U.K.

[2] Department of Physics and Astronomy, University of Leeds, Leeds LS2 9JT, U.K.
© British Crown Copyright 1997.

Introduction

Membrane filter PCM analysis

The collection of airborne fibres, by drawing air through a membrane filter, forms the basis of several regulatory measurements for asbestos [1,2,3] and man-made mineral fibres [4,5] by light microscopy. In general these fibres are collected on membrane filters, which have a sponge like interior of 'tortuous pores' formed from the cold casting of cellulose acetate or cellulose nitrate or more often a mixture of both, known as mixed esters of cellulose (MEC or MCE) filters. The surface of these filters consists of an open filament structure, which has spaces larger than the nominal pore size assigned to the overall filtration efficiency. This means that fibres and particles have an opportunity to migrate some way into the filter before being captured. When the filter is subsequently 'cleared' and analysed by PCM, there is a probability that fibres have penetrated some way into the filter, so as to be outside the depth of focus of the PCM objective. The fibres may also be captured with a high angle of incidence to the filter so that when viewed in the microscope the length is foreshortened and either appears to be less than 5μm long or to have an aspect ratio (length at least 3x greater than the width) <3:1 and would not be counted as a fibre.

The ability to focus through a transparent matrix is common to all transmitted light microscopy imaging methods. However, at the magnifications used for asbestos analysis, the restricted depth of focus makes it important that the plane in which the fibres are present is both horizontal and relatively narrow. For membrane filter analysis at x 400 - x500 magnification with a x40 phase contrast objective, of numerical aperture (N.A) = O.66, the depth of focus is approximately 1 μm with a working distance of 700 μm. This means that if fibres are spread over a depth of more than a few micrometers they will be out of focus, have poor contrast and will be missed by the counter, unless they rigorously scan through the focus.

When the membrane filter - PCM method was first developed [6,7,8] it was common practice to clear the filter by filling the pore space with a liquid of a similar refractive index as the filter material (e.g. triacetin or a mixture of dimethyl pthalate and diethyl oxalate). This changes the light scattering properties between the air in the voids (RI= 1.00) and the membrane filter material (RI = 1.43), so that the white opaque membrane filter will become transparent. As most 'tortuous pore' membrane filters have a quoted porosity of about 90% and a thickness of 150-200 μm, particles and fibres which progress some way into the filter could only be found by changing the sample to objective distance, to observe a range of focal planes above and below that being observed, for every field viewed. The amount of depth to be scanned was never quantified and largely became intuitive to the microscopist, who when counting many hundreds of fields of view per day may adjust the fine focus only within a range of what is a comfortable and quick, giving adjustments of the order of $\pm$ 10 μm. If fibres penetrated up to 10% into the filter this would represent a depth of 20 μm.

Improved methods of filter preparation [9,10] involved collapsing the open pore structure of the membrane filter by the use of solvents (dimethyl formamide) or vapours (acetone). This produced a thin film of plastic-like material, from which the air had been removed, and therefore appears transparent to light. Two or three drops of a matching refractive index mount is then added to the surface so that a glass coverslip can be attached to the surface of the filter to protect it. However, as the collapsing procedure is only partially complete (collapsed filters have widths of 40-50 μm) the refractive index mount must also have the ability to penetrate into the matrix (Euparal - a thermoset resin was used with DMF collapsed filters to give a permanent mount and triactin was used with acetone to give a semi-permanent mount). That the filters are only partially collapsed can be seen by the poor phase contrast image from a newly collapsed filter and normally several hours are required at room temperature or heating of the slide for 30 minutes or more at 60 °C before a clear background is obtained.

The advantages of the chemical collapsing of the pore structure of the filter is that the depth where the fibres reside is reduced by a factor of 4 and that the collapsing process helps align fibres in a horizontal plane to the filter surface so their true length is likely to be seen during the microscopic evaluation. However, no investigation of this effect or the orientation of fibres collected in membrane has been reported.

Membrane filter - SEM and TEM analysis

The problems with depth penetration become ever more important for scanning electron microscopy (SEM) and transmission electron microscopy (TEM) analysis of fibres. SEM analysis depends on the secondary electrons to form an image. As electrons have very little ability to penetrate into or through a solid matrix, the image formed is only from the surface layer and is akin to reflected light from a sample surface in optical microscopy. Any penetration of fibres into the matrix will render the fibre or part of the fibre invisible, so for SEM samples it is necessary to etch the surface layer of the filter, to reveal any buried fibre [11]. For TEM, a similar procedure of etching the surface must also be used to ensure the fibres are exposed so they can efficiently attach to the evaporated carbon film. The rest of the filter plastic is then dissolved away chemically leaving the fibres in a thin carbon film, which is sufficiently electron transparent for electrons to pass through and form a transmitted electron microscopy image. The filter etching is normally carried out in a low temperature asher with an oxygen plasma and the rate of etching was investigated [12] by applying increasing etching times until the fibre count became stable. This was found to be equivalent to about 5 μm on a 0.8 μm pore size filter, but later methods (e.g. USEPA, AHERA) [13] asked for much longer etching periods, resulting in ash artifacts being produced which complicated the analysis.

The potential problems with filter penetration are even more significant when the filters are used for liquid filtration, either during indirect preparation or waterborne fibre analyses. The membrane filter capture efficiency in air is particularly enhanced by the electrostatic attraction of the particles to the membrane filaments. This is absent in liquid filtration and furthermore the much denser medium will decrease the effectiveness of the interception and impaction processes, as the fibres will be more able to follow the flow lines through the filter. Therefore interception is likely to be the main method for capture of fibres during water filtration. The efficiency of water filtration was investigated by Burdett and Rood [12] who found that a 0.1 μm pore size filter should be used and that the penetration depth increased with increased nominal pore size so that recovery losses would occur, even with etching to remove the top few micrometers of the surface plastic.

Confocal laser scanning microscopy

Confocal microscopy differs from conventional light microscopy in that it can be used to form an image at precise depths in a sample rather than from a relatively broad focal plane. This is achieved by using an adjustable pin-hole or slit to discard light from above and below the focal plane, taking only the image of a very thin section within the object. This is essentially a two dimensional slice, whose thickness is controlled by the width of the slit used and depth within an object is controlled by changing the distance of the path length by moving a set of mirrors. The one disadvantage is that the optics only allow for this to be done over a small area if a good image is to be obtained. In confocal laser scanning microscopy the conventional illumination system is replaced by a laser to provide a monochromatic light source and the system scans the object area to give a series of two dimensional (2D) slices into the object. Therefore scanning allows the optics of the microscope to be optimised, and with computer control, CCD cameras and data capture facilities, it is possible to make accurate three dimensional measurements of the fibre orientation, projected length and to calculate the actual length from the stack of 2D slices of the sample under investigation. Furthermore the digital image can be analysed by software for particle recognition and sizing, and reconstruction into a true three dimensional (3D) image. Confocal laser scanning microscopy is therefore an excellent tool to investigate the depth of penetration and orientation of fibres in a transparent matrix [14].

The microscope can be used to measure fibres with lengths down to 2 μm long. The main restriction is the measurement of the diameter of the fibre so that the 3:1 aspect ratio is exceeded. In this paper, for reasons of space, only the 5 μm fibre information is given (except for the 3D study).

Objectives

The objectives of this work were to investigate the depth of penetration of >5 μm long airborne and waterborne crocidolite fibres into uncollapsed and collapsed 0.8 μm nominal pore size membrane filters. Also preliminary studies of the 3D orientation of the fibres were carried out.

Experimental method

Filter loading: airborne asbestos

A number of membrane filters were exposed to asbestos in HSL's 1 m^2 cross sectional area dust box, using a fluidised bed generator to produce an aerosol of textile grade crocidolite.

The fibres were injected into the top of the dust box where they were neutralised in an àir stream of positive and negative ions and mixed before being allowed to settle in a downward air velocity of approximately 8 cm / second. This downward air flow was designed to reduce the build up of very fine fibres and was produced by applying extraction to the bottom of the dust box, which also produced a negative pressure inside, preventing any escape of fibres. The airborne asbestos fibres were sampled onto 0.8 μm nominal pore size mixed ester of cellulose membrane filters which were held in downward facing conductive plastic cowled holders. A measured flow rate of approximately 2 litres/ minute was used for each filter, over a two hour sampling period.

Filter loading: waterborne asbestos

A small amount of UICC crocidolite was dispersed in filtered distilled water by shaking and treatment in a low powered ultrasonic bath using 0.1% v/v of Decon 90 as a surfactant. Aliquots of the suspension were filtered through a number of Millipore MEC filters of different pore sizes. The filtration was carried out using a Millipore Type 25 mm diameter glass filtration funnel. A pre-wetted backing filter (5.0 μm pore size MEC filter) was placed on top of the glass frit along with the pre-wetted test filter before clamping the funnel on top. 10 ml of filtered distilled water was then placed into the funnel to which a 0.5 ml aliquot of the asbestos suspension was added. Vacuum was applied using a small mains powered pump. After filtration the filters were removed and placed in labelled containers to dry. The method is similar to that described in ISO 13794:1998 [15].

TEM sample preparation and size analysis

Air and water filtered membranes were prepared for TEM using ISO 10312:1995 method [16]. This involves chemically collapsing the filter with a mixture of 35% di-methyl formamide, 15% acetic acid and 50% distilled water. After 10 minutes of heating at 65 °C the filter is then etched in an oxygen plasma at 100W power for 3 minutes. The exposed fibres are placed in a carbon evaporator to coat the surface with about 30 nm of carbon. Small sections are cut from the filter and placed onto 200 mesh TEM index grids in a Jaffe washer using a small amount of dimethyl formamide to wick away the filter plastic to leave the fibres in the thin carbon film supported on the TEM grid. The grids were analysed in a TEM at 11,800 magnification for all fibres > 5 μm long and at 20,000 magnification for all fibres > 0.5 μm long. A CCD camera with software was used to collect the image which was calibrated against a 2160 cross-grating traceable to a national standard. Approximately 300 fibres of each size category were counted and sized.

Confocal microscopy analysis

The confocal system used in this work used an argon ion laser emitting light at 514 nm and 488 nm with a total power of 25 mW. The instrument was used in the reflected mode where only the light of the same wavelength of the laser was used to construct the image, all other wavelengths being removed by optical filters. The objective used in the investigation was an oil immersion planapochromat X60 (NA = 1.40). The final magnification that was produced depended on the size of the viewing screen. The magnification and area scanned was compared to a manufacturer supplied standard cross-grating which was checked against a standard grating traceable to national standards as having rulings of 51 µm. A quarter filter was used for the confocal microscope investigation. The filters examined were either viewed uncollapsed or collapsed in acetone onto a glass slide. An oil immersion mount was formed using immersion oil of RI = 1.518, this gave a resolution of about 0.2 µm. A length cut -off of 2 µm was used for sizing, which would allow most inclined fibres fibres above this length to be detected and measured (given that 1 µm slices were taken).

Results and Discussion

TEM Results

The TEM was used to collect detailed size information from the airborne fibres on the filter samples prepared. The airborne crocidolite fibres were of a textile grade bulk material and the fibre length, diameter and aspect ratio distributions were sized for all fibres >5µm long. Typically 300 fibres were sized and an example of the > 5 µm size distribution from a water filtered sample of UICC crocidolite is given in figures 1-3, with the size statistics from both airborne and waterborne crocidolite summarised in Tables 1 and 2.

Table 1: TEM size distribution of air filtered textile grade crocidolite asbestos: fibres > 5µm long (sample ref. RL/829274546/96): 320 fibres sized.

Length (µm)		Diameter (µm)		Aspect Ratio	
Geometric Mean	8.48	Geometric Mean	0.20	Geometric Mean	42.73
Mean	9.66	Mean	0.24	Mean	35.84
Standard Error	0.39	Standard Error	0.01	Standard Error	1.54
Median	7.56	Median	0.19	Median	42.19
Standard Deviation	7.05	Standard Deviation	0.21	Standard Deviation	56.49
Sample Variance	49.69	Sample Variance	0.046	Sample Variance	3194
Minimum	4.5	Minimum	0.048	Minimum	7.96
Maximum	85.9	Maximum	2.69	Maximum	303
Sum	3091.6	Sum	76.35	Sum	11468.8

Table 2: TEM size distribution of water filtered UICC crocidolite asbestos: fibres > 5µm long (sample ref. HSL/84546/96): 301 fibres sized.

Length (µm)		Diameter (µm)		Aspect Ratio	
Geometric Mean	8.17	Geometric Mean	0.33	Geometric Mean	24.95
Mean	9.27	Mean	0.41	Mean	30.80
Standard Error	0.40	Standard Error	0.02	Standard Error	1.37
Median	7.17	Median	0.32	Median	24.95
Standard Deviation	6.97	Standard Deviation	0.38	Standard Deviation	23.75
Sample Variance	48.57	Sample Variance	0.14	Sample Variance	563.98
Minimum	5.01	Minimum	0.07	Minimum	5.23
Maximum	87.50	Maximum	4.59	Maximum	258.23
Sum	2790.3	Sum	124.9	Sum	9271.1

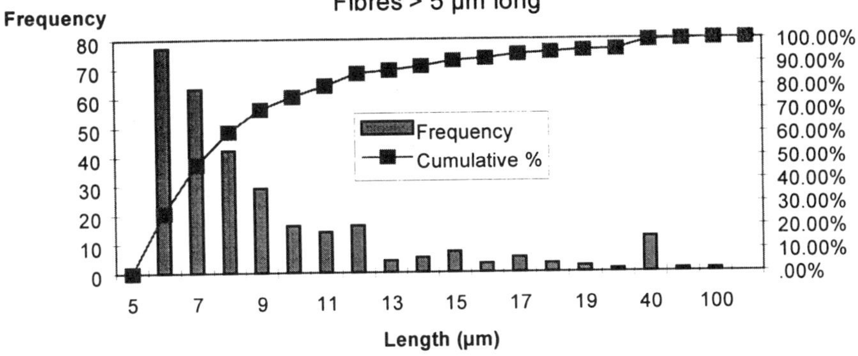

Figure 1: Fibre length distribution of UICC Crocidolite fibres. Fibres > 5 μm long

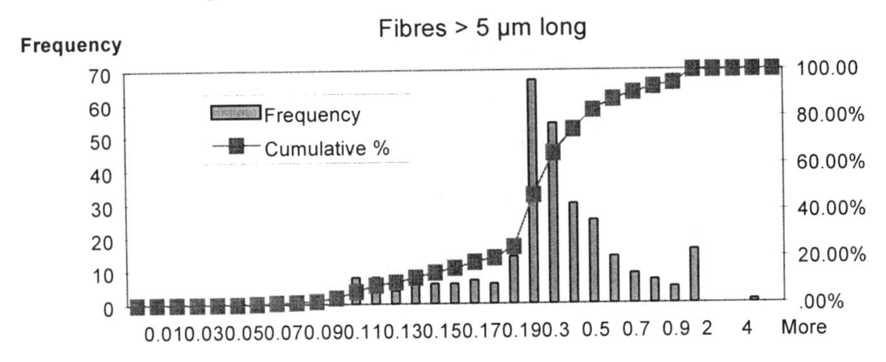

Figure 2: Fibre width distribution of UICC Crocidolite Fibres > 5 μm long

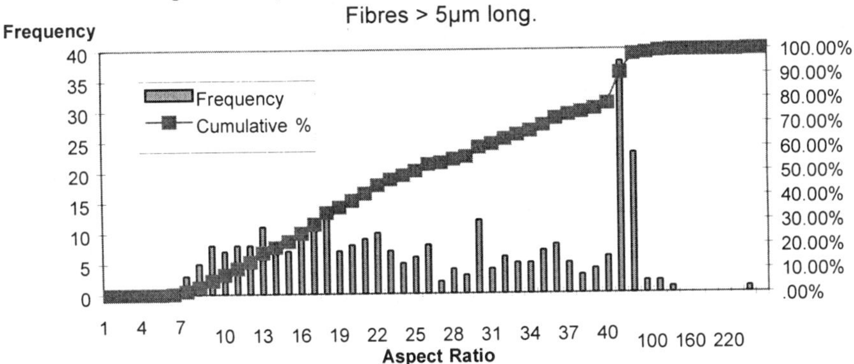

Figure 3: Aspect ratio distribution of UICC Crocidolite Fibres > 5μm long.

Confocal microscopy results

Water filtered UICC crocidolite on uncollapsed membrane filters: The 0.8μm filter produced a maximum measured fibre penetration depth of 20μm (see Figure 4) for >5 μm long fibres. The peak capture of longer than 5μm fibres appears at 4 μm depth with ~85% of fibres over 5μm measured becoming trapped within the first 6μm. After 6 μm the numbers of fibres decreases rapidly towards zero at a depth of 10μm with only three fibre like objects seen below this depth.

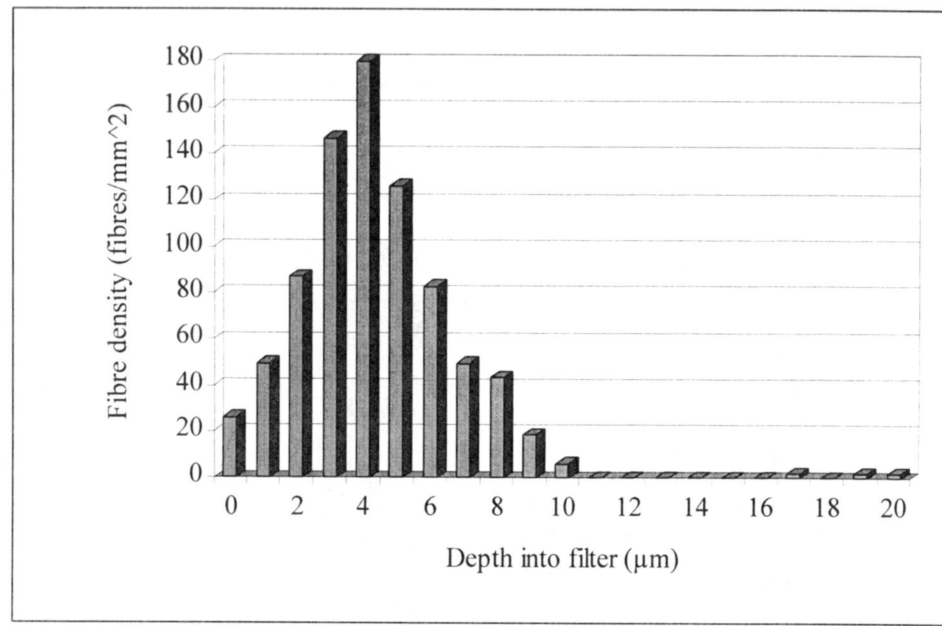

Figure 4: Fibre number density at varying depths into the uncollapsed 0.8 μm pore size filter - loaded with UICC crocidolite in water.

Air filtered crocidolite on a 0.8 μm uncollapsed membrane filter: The penetration of >5 μm long fibres into the filter is shown in figure 5. Again the peak capture was at 4 μm depth and appeared to fall rapidly towards zero at 10 μm depth, except for a few fibres which penetrated up to a maximum depth of 14 μm.

Crocidolite on a 0.8 μm acetone collapsed membrane filter: The effect of collapsing the filter is readily apparent from the maximum depth of penetration for the water filtered UICC crocidolite decreasing from 20 μm to 4μm, (figure 6) with the peak capture depth decreasing from 4 μm to about 1 μm depth. This is entirely in line with the overall reduction of the membrane filter thickness by collapsing (reduces from around 180 μm to about 40 - 50 μm). The air filtered textile crocidolite sample gave a similar result except the maximum penetration depth was 3 μm.

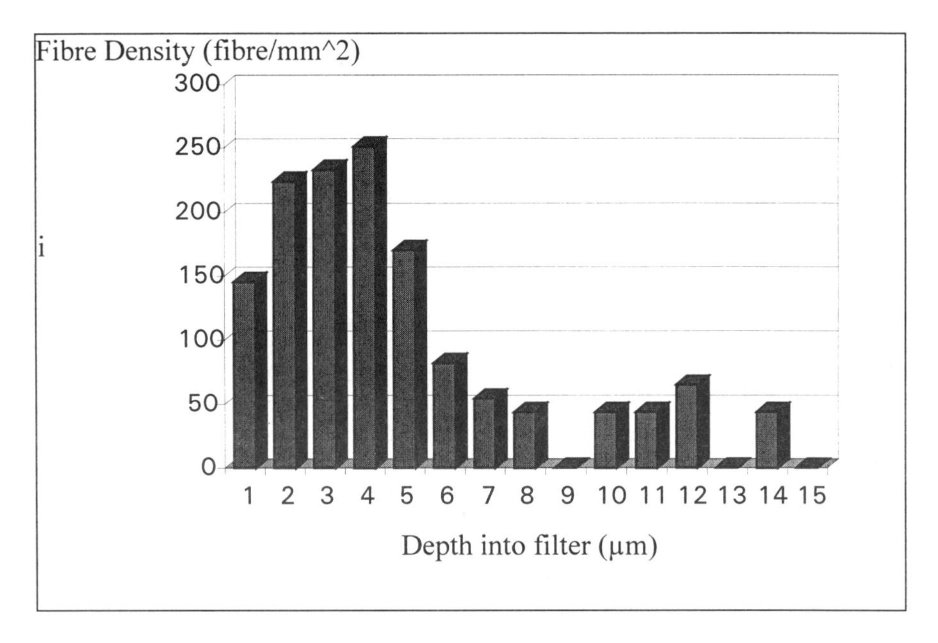

Figure 5: Fibre number density at varying depths into the uncollapsed 0.8 μm pore size filter - loaded with crocidolite in air.

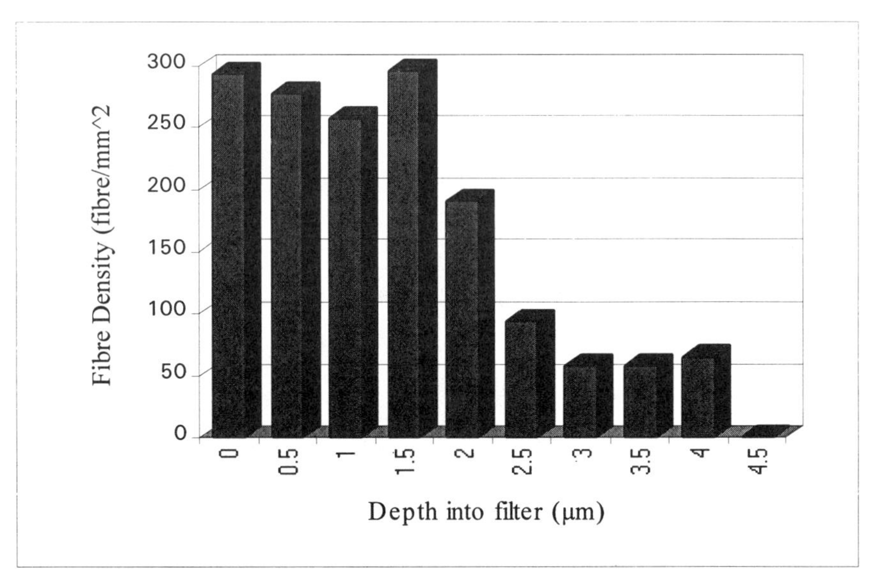

Figure 6: Fibre number density at varying depths into the uncollapsed 0.8 μm pore size filter - loaded with UICC crocidolite in water.

Three dimensional orientation of > 2 μm UICC crocidolite fibres in an uncollapsed 0.8 μm pore size membrane filter.

The orientation of fibres in an uncollapsed filter was studied, as the collapsed filters restricted most fibre penetration to ± 1 μm depth, restricting the out of plane angle to the filter surface. A similar maximum depth of penetration of 20 μm was found as reported in Figure 4. A 3D plot of the penetration distance v fibre length and frequency is given in Figure 7. This shows clearly that the majority of >5 μm long fibres are efficiently captured in the upper few micrometers of the filter and that shorter >2 μm long fibres have longer penetration distances, but are not more than a factor of 2 greater.

The out of plane angles for all fibres > 2 μm fibres is given in figure 8. The large number of in-plane fibres at 0 degrees is an artifact due to the depth profile being taken in 1 μm slices, making many short fibres appear as in-plane as they did not reach the next slice down. However, taking this into account there appears to be a normal distribution of orientations around 0 degrees with few exceeding 25 degrees. To determine whether there were any significant biases, the orientation of the fibres was plotted against both fibre length (Figure 9) and fibre depth (Figure 10). At first sight it is surprising that long fibres appear to have shallower out of plane angles and that fibres which penetrate further, do not have significantly higher out of plane angles. However, as long fibres are unlikely to penetrate into the filter, they will rest on the horizontal surface of the filter and will be in-plane. Similarly, when fibres are captured in the filter by interception, the fibre is most likely to be caught as it changes from a vertical angle of incidence to the filter surface to a horizontal orientation. Fibres extending vertically across two or more slices were considered to have penetrated to the deepest slice.

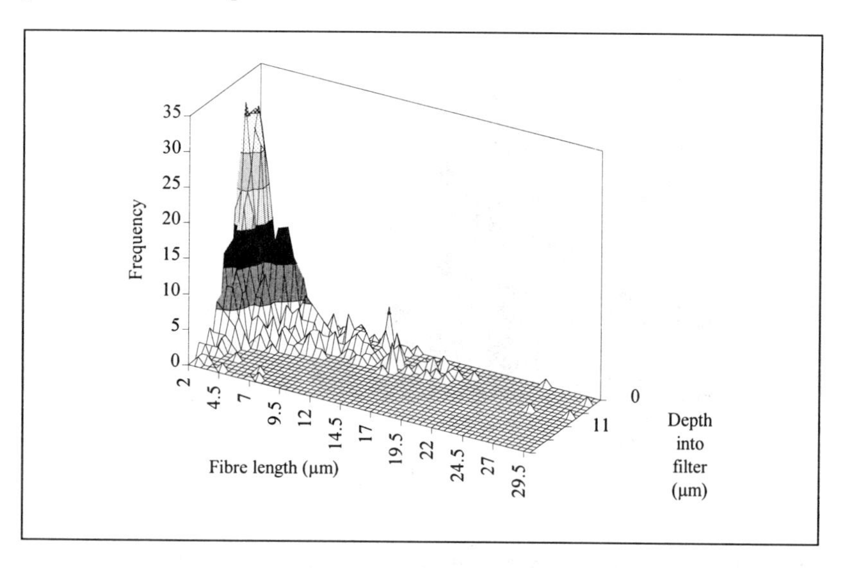

Figure 7: Fibre length frequency distribution variation with depth for an uncollapsed 0.8 μm pore size filter - loaded with UICC crocidolite in water.

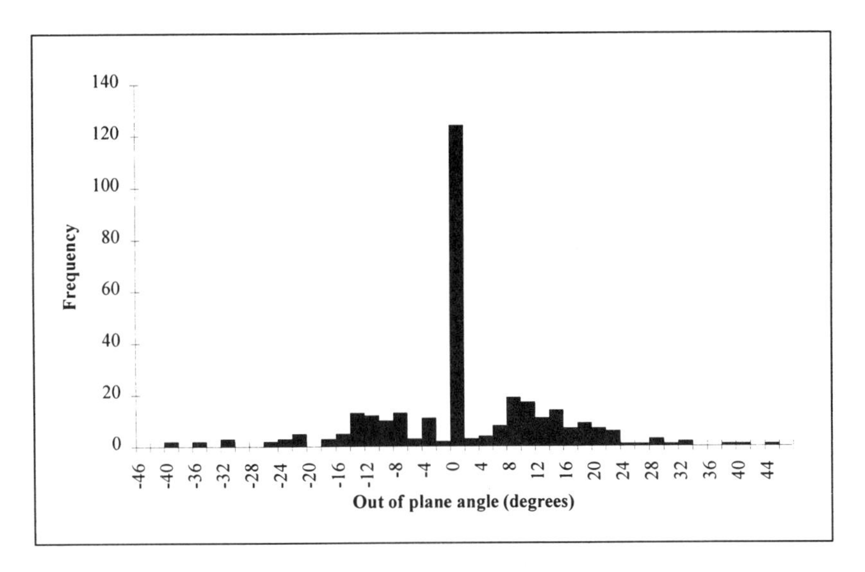

Figure 8: Variation of out of plane angle for UICC crocidolite fibres in an uncollapsed 0.8 μm pore size filter.

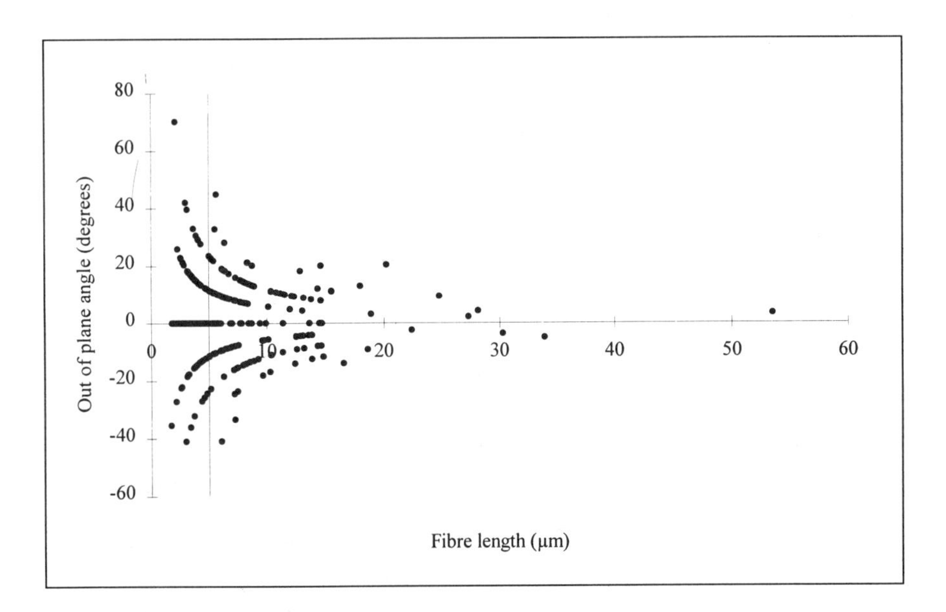

Figure 9: Variation of out of plane orientation with fibre length.

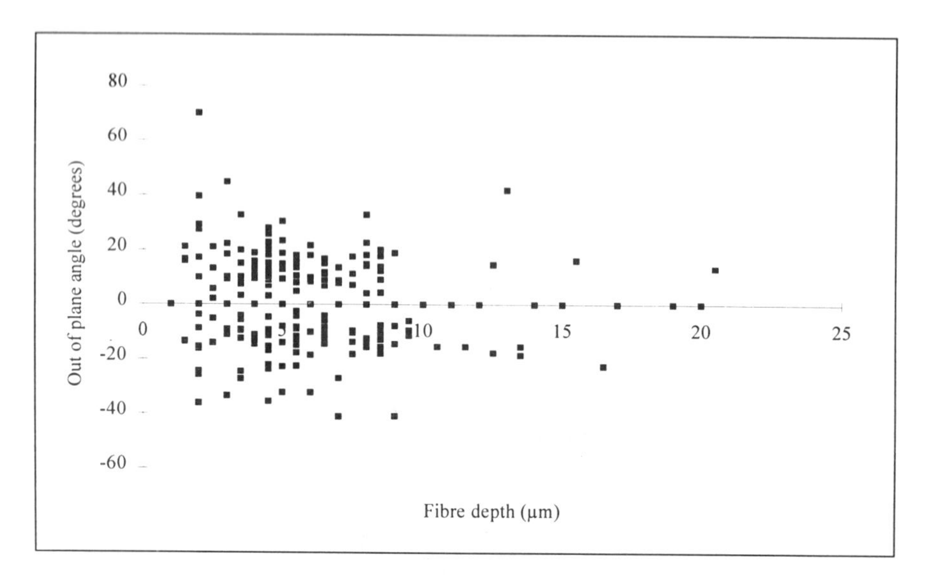

Figure 10: Variation of out of plane angle with fibre depth.

Three dimensional fibre visualisation

A 3D reconstruction of the orientation, position and depth of the out of plane fibres is given in Figure 11. The Z scale has been exaggerated by a factor of 2.50 and the X and Y dimensions have been reduced by a factor of 2.56 to give a Z dimension that is exaggerated by 640% in comparison to the X and Y scales.

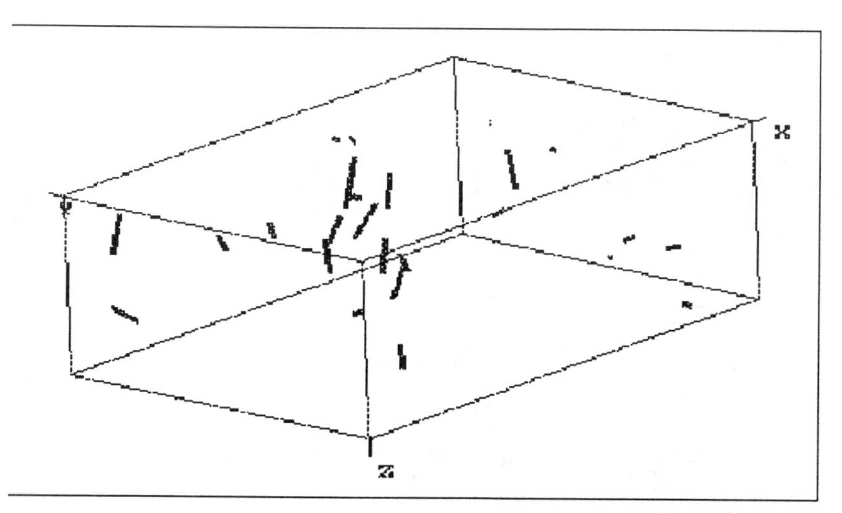

Figure 11: Three dimensional reconstruction of the fibres in a typical extended focus image (the XYZ space corresponds to 186μm x 124μm x 9μm).

Conclusions

The maximum penetration of >5 um long fibres into a 0.8 μm pore size uncollapsed MEC membrane filter is limited to about 20 μm depth for water filtration and about 13 μm for air filtration, but both have peak capture of fibres at 4 μm depth.

Collapsing the filter in acetone reduces the peak and maximum depth of penetration by about a factor of 4, the same as the overall reduction in the filter thickness. The peak fibre distribution of >5 μm long fibres was therefore about 1 μm below the surface of a 0.8 μm pore size collapsed membrane filter when used for air sampling and between 1-2 μm when used for water sampling. The maximum penetration depths were 3 and 4.5 μm respectively.

This means that an air sampled, 0.8 μm pore size, acetone collapsed filter will have some 85 % of the >5 μm long fibres within the ± 1 μm depth of focus for a X40 phase contrast objective, assuming the image plane is 1 μm below the filter surface. Therefore this study supports the current advice for PCM analysis in NIOSH 7400 and MDHS 39/4 that the sample to objective distance must be changed to observe focal planes above and below, but a relatively small movement is required and is unlikely to be a cause for substantial error.

The out of plane angles for most >5 μm fibres is within ± 25 degrees for uncollapsed filters and will have produced relatively small foreshortening effects of fibre lengths in historic mounting methods. Although out of plane measurements were not made on filters which had been collapsed, it would seem from the limited depth of penetration, the collapsing procedure is very efficient for aligning the fibres in a horizontal plane and any bias due to foreshortening effects are likely to be minimal.

SEM and TEM analysis of air samples on 0.8 μm pore size membranes requires the top few micrometers of the collapsed filter surface is etched away to fully expose the fibres. This appears to be sufficient to capture all fibres down to 2 μm in length monitored in this study. Given that electrostatic effects are important in capturing short fibres it is unlikely that short airborne fibres penetrate much further into the uncollapsed membrane filter. Short asbestos fibres filtered from liquid suspensions will penetrate into and through a membrane filter unless a much smaller pore size is used, 0.1 μm is recommended (12).

Crocidolite fibres were chosen for this study as they are thought, due to their shape and size distribution, to be the type of asbestos which will penetrate the furthest into the membrane filter and provide the worst case.

Further work to extend the range of pore sizes and types of fibres studied would add to the current knowledge base on membrane filter sampling and recovery efficiencies for the microscopical analysis of fibres.

The success of the method in being able to reconstruct a 3D representation shows the method can also be used to study fibre orientation, fibre size and packing densities. Therefore it should be possible to obtain these types of data from filtration materials, as well as directly measuring where particles are captured, to further understand and model filter efficiency and performance.

References

[1] WHO, "Recommended method for the determination of airborne fibre number concentrations by phase contrast optical microscopy". World Health Organisation, Geneva, (in Press). 1997.

[2] MDHS 39/4 "Asbestos Fibres in Air: Light Microscope Methods for use with the Control of Asbestos at Work Regulations", Health and Safety Executive, HSE Books. 1995. ISBN 0 7176 0913 8.

[3] NOSH Method 7400, " Light microscopy, Phase contrast", NOSH Manual of Analytical Methods, National Institute for Occupational Safety and Health, Cincinnati, USA. Revision 4, 1992.

[4] WHO/EURO, "Reference method for measuring airborne man-made mineral fibres (MMMF)". World Health Organisation, Copenhagen. 1985.

[5] MDHS 59 "Man Made Mineral Fibre: Airborne Number Concentration by Phase Contrast Light Microscopy", Health and Safety Executive, HSE Books, 1988. [ISBN 0 7176 0319 9].

[6] ARC Asbestosis research Council, "The measurement of asbestos dust by the membrane filter method". Technical note 1, 1968.

[7] Ayer, H. E., Lynch, J. R. and Fanney, J. H. "A comparison of impinger and membrane filter techniques for evaluating air samples in asbestos plants". *Annals of the New York Academy of Sciences*., vol. 132, pp.274-279, 1965.

[8] P&CAM 239 'Asbestos fibres in air', National Institute for Occupational Safety and Health, Cincinatti, USA. 3/30, 1977.

[9] Leguen, J.M.M. and Galvin, S. 'Clearing and mounting techniques for the evaluation of asbestos fibres by the membrane filter method'. *Annals of Occupational Hygiene*, vol. 24, pp. 273-280. 1981.

[10] AIA-RTM1, "Reference method for the determination of airborne asbestos fibre concentrations at workplace by light microscopy (membrane filter method)". Asbestos International Association. London. 1979.

[11] Rooker, S.J., Vaughan, N.P. and LeGuen, J.M.M., "On the visibility of fibres by phase contrast microscopy", *American Industrial Hygiene Association Journal*. Vol. 43, number 7, pp. 505-515, 1982.

[12] Burdett G.J. and Rood A.P. "Membrane-filter, direct-transfer technique forthe analysis of asbestos fibres or other inorganic particles by transmission electron microscopy". *Environmental Sciences & Technology*, vol. 17, No.11, pp.643 -648. 1983.

[13] AHERA,Asbestos containing materials in schools: Final rule and notice.40 CFR Pt 763, Oct. 30th, 1987.

[14] Clarke, A.R., Archenhold, G., & Davidson, N.C., 'A novel technique for determining the 3D spatial distribution of glass fibres in polymer composites', *Composites Science and Technology*, vol. 55, pp. 75-91, 1995.

[15] ISO 13794:1997 "Ambient Air - Determination of Asbestos Fibres - Indirect Transfer Transmission Electron Microscopy Method". International Standards Organisation. Geneva.

[16] ISO 10312:1995 "Ambient Air - Determination of Asbestos Fibres - Direct Transfer Transmission Electron Microscopy Method". International Standards Organisation

Measurement Methods for Asbestos in Water

James R. Millette[1], Pronda Few[1], and Joseph A. Krewer[2]

ASBESTOS IN WATER METHODS: EPA'S 100.1 & 100.2 AND AWWA'S STANDARD METHOD 2570

REFERENCE: Millette, J .R., Few, P., and Krewer, J. A., "**Asbestos in Water Methods: EPA's 100.1 & 100.2 and AWWA's Standard Method 2570,**" *Advances in Environmental Measurement Methods for Asbestos, ASTM STP 1342,* M. E. Beard and H. L. Rook, Eds., American Society for Testing and Materials, 2000.

ABSTRACT: Although the three methods are based on common procedures, a number of differences in them provide useful variations for specific types of analyses. In 1992 the U.S. Environmental Protection Agency (EPA) specified the 1983 EPA research report of Chatfield and Dillon as EPA Method 100.1. It differs from the other methods in that it contains information about filtering samples only through polycarbonate (PC) filters. A March 1993 EPA memorandum by Mr. Joseph Conlon provides information about how the EPA Method 100.1 can be used to analyze water samples to meet the EPA Drinking Water Standards for asbestos. EPA Method 100.2 was written in 1994 to be specific to the EPA Maximum Contaminant Level (MCL) for asbestos in drinking water - 7 million asbestos fibers over 10 µm in length per liter. It requires counting only those long fibers, provides information about preparing mixed cellulose ester (MCE) type filters, and includes the quality assurance information that had been developed under the Asbestos Hazard Emergency Response Act (AHERA) of 1987. The American Water Works Association Standard Method 2570 was approved in 1993. It contains information about preparing either MCE or PC filters and provides for counting asbestos fibers greater than 0.5 µm in length (the AHERA fiber definition). An interlaboratory comparison of a prepared sample grid showed good correlation between the laboratories for fibers greater than 10 µm.

KEYWORDS: Water, EPA 100.1, EPA 100.2, AWWA 2570, Filter Contamination

---

The development of methods to determine the asbestos content in water began in earnest in the early 1970s [*1-9*]. The American Society for Testing and Materials (ASTM) committee E-4 on drinking water was active during this time and provided a

[1] Executive Director and Research Scientist, respectively, MVA, Inc., 5500 Oakbrook Pkwy, No. 200, Norcross, GA 30093

[2] Environmental Specialist, Georgia Department of Natural Resources, 205 Butler St. SE, Atlanta, GA 30334

unique forum for discussion. Although no standard method was ever agreed upon, the discussions of this group were very influential in the asbestos methods (water, air and dust) that were later published. The interlaboratory data derived from this group (published in 1978) remains the best set of interlaboratory comparison data for water analyses yet published [*10*] . Two in-depth reviews, one in 1984 and the other in 1991, provide a good overview of the research on asbestos in drinking water prior to 1990 [*11,12*].

On July 17, 1992, the EPA in the Federal Register (57 FR 3184039) under paragraph 141.23(k)(1) specified that the analysis for asbestos in water was to be conducted using transmission electron microscopy by EPA Method "Analytical Method for Determination of Asbestos Fibers in Water", EPA-600/4-83, September 1983 [*13*]. This method, developed under EPA contract by Chatfield and Dillon, became EPA Method 100.1. It provides information about preparing samples on polycarbonate (PC) filters.

In March, 1993 a Memorandum was issued by James M. Conlon, Director of the Drinking Water Standards Division, Office of Ground Water and Drinking Water, United States Environmental Protection Agency on the subject of "Guidance and Clarification on the Analysis of Asbestos" [*14*]. The memorandum of clarification and guidance was needed because the 1983 method was not designed specifically to meet the 1992 EPA drinking water regulation on asbestos in water which pertained only to fibers greater than 10 μm in length. The memorandum was also needed to clarify the use of the ozone/UV equipment prescribed in the 1983 method because it was no longer generally available in laboratories. The memorandum was based on the findings of an official peer review, conducted in January 1993, of the test method for asbestos in drinking water [*15*].

In June 1994, the EPA Method 100.2 became available [*16*]. It was written specifically to be responsive to the drinking water standard for asbestos. This method is for the determination of asbestos structures over 10 μm in length in drinking water. It also provides information about preparing mixed cellulose ester (MCE) type filters, and includes the quality assurance information that had been developed under the Asbestos Hazard Emergency Response Act (AHERA) of 1987.

The American Water Works Association Standard Method 2570 was approved in 1993 and published in 1994 in the supplement to the 18th edition of Standard Methods for the Examination of Water and Wastewater [*17*]. It contains information about preparing either MCE or PC filters and provides for counting two size categories: those fibers greater than 10 μm and those in the range between 0.5 μm and 10 μm in length.

COMPARISON OF THE METHODS

Although the methods for the analysis of water for asbestos by TEM are based on the same general procedures, the methods differ in several areas. These differences involve such items as the filter materials used to capture the fibers and the filter preparation techniques. Fiber identification, sizing and enumeration protocols also vary between the methods, as do quality assurance requirements and statistical procedures used to validate the analyses. The differences are primarily related to the intended use of each method. For example, EPA 100.1 is a refinement of an earlier interim method for determining asbestos fibers in water samples for research projects. The method provides detailed identification, sizing and counting information which may be essential for research studies but unnecessary for more routine analyses. As additional knowledge was gained about the occurrence and significance of asbestos fibers in water, more streamlined methods were developed to meet the needs of the regulated and analytical communities.

Sample Container

For EPA 100.1, glass sampling containers are preferred, but low density polyethylene is acceptable. There is no difference in the recommended sampling container in 1993 Memo, EPA 100.2, or AWWA 2570.

Sample Collection

With EPA 100.1, if sampling a piped system, let water run until sample is representative of fresh water. If sampling a tank or a water body, try to sample a representative horizontal and vertical distribution. The directions for sample collection are essentially the same for the 1993 Memo, EPA 100.2, and AWWA 2570, with a note that a temperature change may indicate that a piped system has been run for a sufficient length of time.

Sample Storage

For EPA 100.1, no preservation with acids, store in dark at about 5 ° C, add mercuric chloride solution to preserve sample if it cannot be treated within 48 hours. Treatment consists of ultraviolet light/ozone (UV/Ozone) bubbling and sonication before the sample is filtered.

The 1993 Memo clarifies the procedure. It does not require the UV/Ozone treatment. However, the sample must be filtered within 48 hours or UV/Ozone treatment must be performed before filtering sample.

For EPA 100.2, if the sample cannot be filtered within 48 hours or high levels of organic contamination are suspected, UV/Ozone treatment is required. EPA 100.2 also requires that the sample be shipped on ice, but not frozen.

Method AWWA 2570 has the same sample storage requirements as EPA 100.2.

Filter type(s) used (25 or 47 mm diameter filters allowed in all methods)

The method EPA 100.1 describes the filtration procedure using a 0.1 µm pore size polycarbonate (PC) filter only.

The 1993 Memo states that only 0.1 µm pore size PC filter membranes may be used.

The 1994 method EPA 100.2 established that 0.1 to 0.22 μm pore size mixed cellulose ester (MCE) filters could also be used for water analysis. The method EPA 100.2 only describes the procedures for filtration with the MCE filters. The solvents, 1-methyl-2-pyrrolidone and chloroform, are listed in method EPA 100.2 but no procedures involving these solvents for the dissolution of MCE filters are described.

The method AWWA 2570 allows for the use of either 0.4 μm pore size or smaller PC filters and 0.45 μm pore size or smaller MCE filters.

Filter collapsing method (MCE filter only)

For Method EPA 100.1 or the 1993 memo, this step is not required because a PC filter has been used.

For the method EPA 100.2, DMF/acetic acid or acetone is used to collapse the filter.

For the method AWWA 2570, the same solvents as for the EPA 100.2 are used.

Filter dissolution method

For method EPA 100.1, the Jaffe wick or condensation washer is used with chloroform as the solvent.

The 1993 Memo uses the same equipment and solvent as EPA 100.1.

For method EPA 100.2, the Jaffe wick is used with dimethyl formamide (DMF) or acetone. The solvents, 1-methyl-2-pyrrolidone and chloroform, are mentioned in method EPA 100.2 but no procedures involving these solvents for the dissolution of MCE filters are described. These solvents are generally used with PC filter dissolution.

For the AWWA 2570 method, the Jaffe wick is used with DMF or acetone for MCE filters and with chloroform or 1-methyl-2-pyrrolidone for PC filters.

Plasma etching requirements (for MCE filters only)

Plasma etching is not required for the method EPA 100.1 or the 1993 Memo.

For the method EPA 100.2, plasma etching is required for 0.22 μm pore size MCE filters and is optional (but recommended) for 0.1 μm pore size MCE filters.

Under AWWA 2570 the plasma etching is required for all MCE filters.

Plasma etcher calibration requirements

The plasma etching requirements are not applicable to the method EPA 100.1 or the guidance provided by the 1993 Memo.

For method EPA 100.2, the plasma etching is to be calibrated to remove approximately 10% of the filter mass as described in AHERA or NISTIR references.

The method AWWA 2570 refers to the AHERA or NISTIR references for plasma etching calibration.

Countable fiber definition

For method EPA 100.1, the analyst is to count any fiber greater than 0.5 µm long with parallel or stepped sides and an aspect ratio of 3:1 or greater.

The guidance in the 1993 Memo uses the same counting definitions as EPA 100.1, but the analyst is to count fibers greater than 10 µm long only.

Under method EPA 100.2, the analyst is to count any fiber greater than 10 µm long with substantially parallel sides, without rounded ends and an aspect ratio of 3:1 or greater (counting fibers less than or equal to 10 µm long is optional).

With the method AWWA 2570, the analyst counts fibers with substantially parallel sides and an aspect ratio of 5:1 or greater in two size categories (if both categories are needed): 0.5 µm to less than or equal to 10 µm long, and greater than 10 µm long.

Structure types and sizes recorded

For the method EPA 100.1, information about fibers is recorded as length and width if over 0.5 µm long. For bundles the mean length and width is recorded if over 0.5 µm long. For fiber aggregates (cluster) the size and number of individual fibers is recorded. When fibers are attached to non-fibrous debris (matrix) the length is recorded as twice the visible portion or to the end of the matrix. Then the analyst is to count all fibers in a matrix individually. There are special counting and sizing rules for fibers on grid bars.

With the 1993 Memo clarification, the length and width is to be recorded for fibers over 10 µm long. For bundles the analyst is to record the maximum length and mean width if over 10 µm long. With clusters and matrices the analyst is to record all fibers over 10 µm long. The same rules for counting fibers on grid bars apply as in EPA 100.1.

Under the method EPA 100.2, the analyst is to record the sizes of fibers and bundles if over 10 µm long and record the maximum length and mean width if the fiber has stepped sides. With matrices the analyst is to record it as a single fiber if it contains a fiber or fibers over 10 µm long, but if structure is too complex, the analyst is to record the size but not count it toward the total count. The rules for measuring concealed portions of a fiber in a matrix or on a grid bar are the same as in EPA 100.1. There are additional rules regarding free ends and fibers on grid bars. For clusters the analyst is to record individual fibers if over 10 µm long, but if a structure is too complex, the analyst is to record the overall size but not count it toward the total count.

With the AWWA 2570, the analyst records the appropriate size category (no width recorded) for fibers and bundles. For clusters and matrices the analyst records the size category of the longest visible fiber or bundle.

Stopping rules

All methods require the examination of at least four grid openings, regardless of the other stopping rules.

The method EPA 100.1 specifies that the analysis may be stopped after 100 fibers have been counted or 20 grid openings have been examined.
The 1993 Memo allows the analysis to be stopped after an analytical sensitivity of 0.2 million fibers per liter has been reached, or 100 fibers have been counted.

The method EPA 100.2 specifies the same stopping rules as the 1993 Memo.

The method AWWA 2570 allows the analysis to be stopped after a specified detection limit is reached, 100 fibers have been counted or 10 grid openings have been examined.

Required magnification

The method EPA 100.1 requires that the analysis take place at about 20,000x.

The 1993 Memo provides the guidance that for fibers longer than 10 μm the required magnification needs to be greater than or equal to 10,000x.

With the method EPA 100.2 the magnification requirement is between 10,000 and 20,000x.

The method AWWA 2570 requires a magnification of 15,000-20,000x.

SAED and EDXA identification procedures

While all the methods require identification of asbestos fibers by morphology, selected area electron diffraction and energy dispersive x-ray analysis, there are significant differences regarding the amount of identification information collected and recorded.

The method EPA 100.1 contains a detailed classification system which uses a letter code to record the identification procedure applied to each individual fiber. Various levels of analysis (three for chrysotile and four for amphiboles) provide increasingly rigorous identification protocols. At least one typical SAED photograph and EDXA spectrum printout or storage on computer disk is required for each asbestos type found in every sample. The higher analysis may require two zone axis SAED photographs to identify certain amphibole fibers.

The 1993 Memo summarizes the identification system used in EPA 100.1 and specifies a certain acceptable identification protocol, with appropriate letter codes. No additional information is provided regarding SAED photographs or EDXA spectrum printouts or disk storage.

The method EPA 100.2 provides a general identification protocol and requires a count sheet record of the identification protocol for each fiber, but not a letter code record. At least one typical SAED photograph and EDXA spectrum printout or disk storage is required for each asbestos type found in every set of samples from the same source, or at least every fifth sample.

The method AWWA 2570 does not specify a procedure for recording the identification of each fiber and references other sources for additional information on asbestos identification. At least one typical SAED photograph is required for each group of samples or at least every fifth sample. At least one EDXA spectrum must be printed out or stored on disk for every tenth chrysotile structure and every fifth amphibole structure.

Quality Assurance Programs and Quality Control Procedures

The method EPA 100.1 does not contain a separate quality assurance program, but does provide a variety of quality control requirements for ensuring proper equipment and instrument performance and sample preparation acceptability. Quality control requirements for blanks, control samples and standards are also described. Statistical procedures are provided for calculating the mean and confidence intervals of the fiber concentration, uniformity of fiber deposits on the grids, fiber size distributions and other information. Method precision and accuracy is also discussed.

The 1993 Memo does not provide any additional quality assurance information for method EPA 100.1.

The method EPA 100.2 contains a quality control section, which states that the required quality control checks follow the TEM method described in the Asbestos Hazard Emergency Response Act (AHERA) [*18*] and NISTIR [*19*] guidance documents. The method also provides a section on calibration and standardization procedures and a table of data quality objectives. Information on method performance and data analysis and calculations is also provided.

The method AWWA 2570 contains a quality control/quality assurance section which includes references to the AHERA and NISTIR procedures. The method AWWA 2570 also provides information on precision and bias based on analyses using 0.1 μm pore size PC filters. No data are provided in AWWA 2570 specifically relating to other types of filters or filters with other pore sizes.

Discussion

It should be noted that method EPA 100.1 is a more self-contained method than the other three documents discussed here, in the sense that it contains extensive discussion and guidance on the identification of asbestos minerals, statistical procedures and other information. The 1993 Memo is a clarification of certain aspects of the 1983 document, EPA 100.1, for the purpose of making it more applicable to a specific regulatory situation. Both method EPA 100.2 and AWWA 2570 reference a number of other documents, including EPA 100.1, as sources of additional information for specific situations. Both of these methods also reference analytical and quality assurance procedures developed for the analysis of asbestos fibers in air, under the Asbestos Hazard Emergency Response Act (AHERA) of 1987. This is intentional, since most TEM asbestos laboratories analyze far more air samples than water samples. The AHERA analytical protocols and quality assurance procedures are therefore already in use and familiar to most TEM laboratories.

BLANK CONTAMINATION

In the 1970's one of the authors (JRM) found evidence that one type of membrane filter (polycarbonate) was occasionally contaminated with both chrysotile and crocidolite asbestos. Contamination of polycarbonate filters was also reported by Chatfield [*20*] and Ring [*21*] . Chatfield also found filter contamination on MCE filters [*22*]. Ring reported that "... there clearly is some contamination on the Nuclepore (polycarbonate) filters themselves". Chatfield found that the filter contamination was not at a constant level; some filters in a batch were unaffected, but some were contaminated with sufficient numbers of fibers to be of concern when measuring samples with low levels of asbestos fibers. Concern over the contamination of polycarbonate filters led the 1986 committee developing the TEM analytical methods for the clearance of asbestos abatement actions under AHERA to establish an acceptable filter blank average of less than 18 structures/mm^2 and an acceptable blank level for a single preparation of a maximum of 53 s/mm^2 [*18*].

In 1996 and 1997, during routine transmission electron microscopy analysis at MVA, Inc., asbestos contamination was detected on blank filters from two batches of 47 mm, 0.2 μm pore size polycarbonate filters: Lots R5KM64689 & R5KM64690 labeled: Millipore, August, 1995. An aluminum coating was detected on most of the chrysotile and crocidolite fibers found on the blank filters (Fig. 1 - 4). Contamination levels ranged from 10.4 str/mm^2 to 125 str/mm^2 (Table 1).

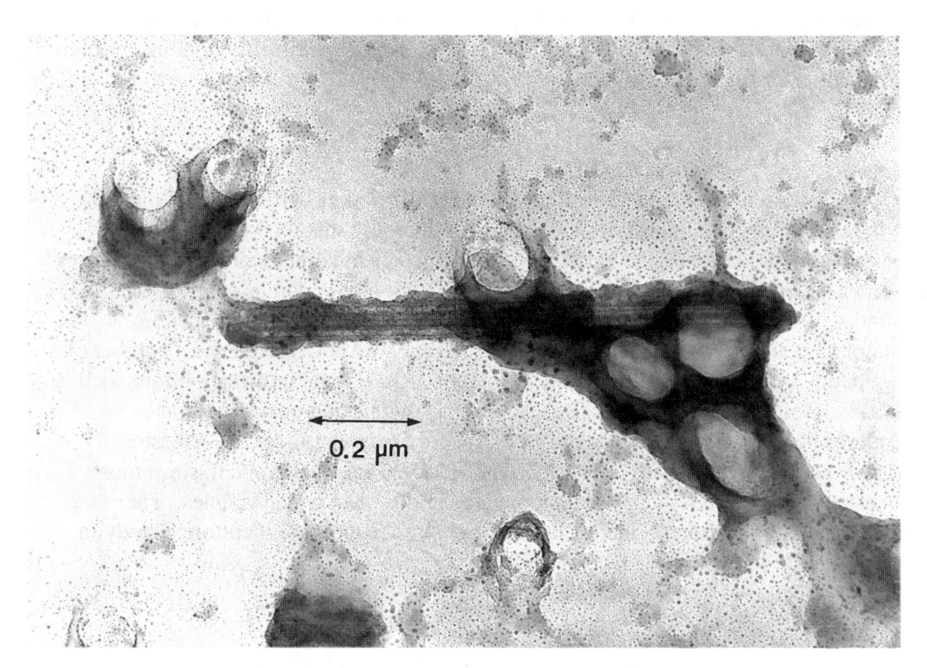

Figure 1. Chrysotile associated with a polycarbonate filter.

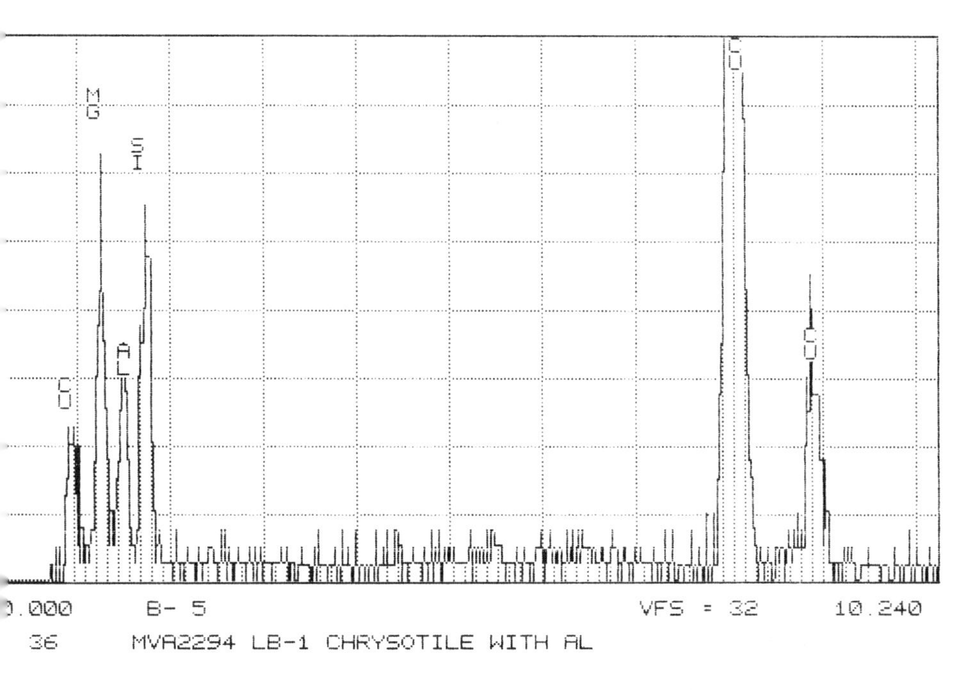

Figure 2. X-ray spectrum of Chrysotile in polycarbonate filter. Aluminum is present coating the fiber.

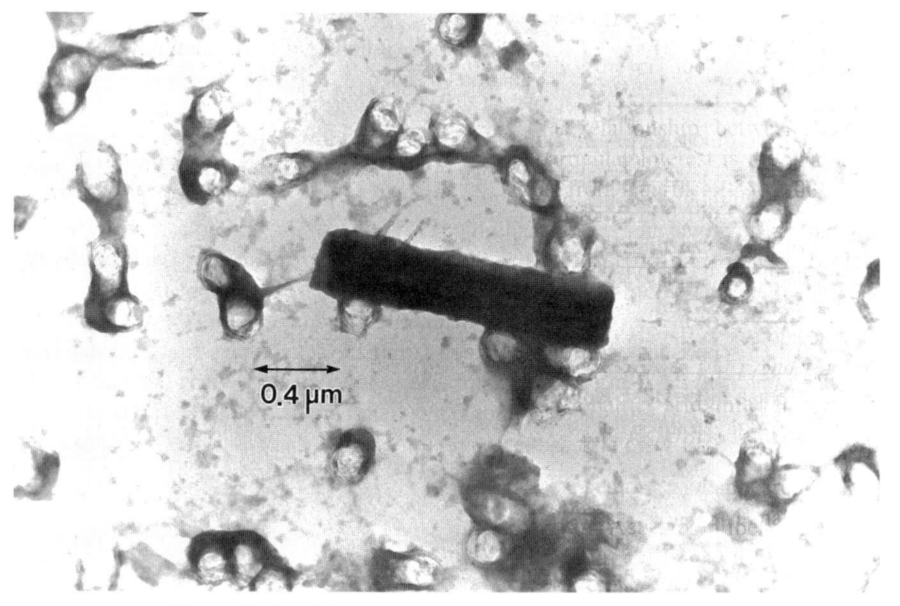

Figure 3. Crocidolite associated with a polycarbonate filter.

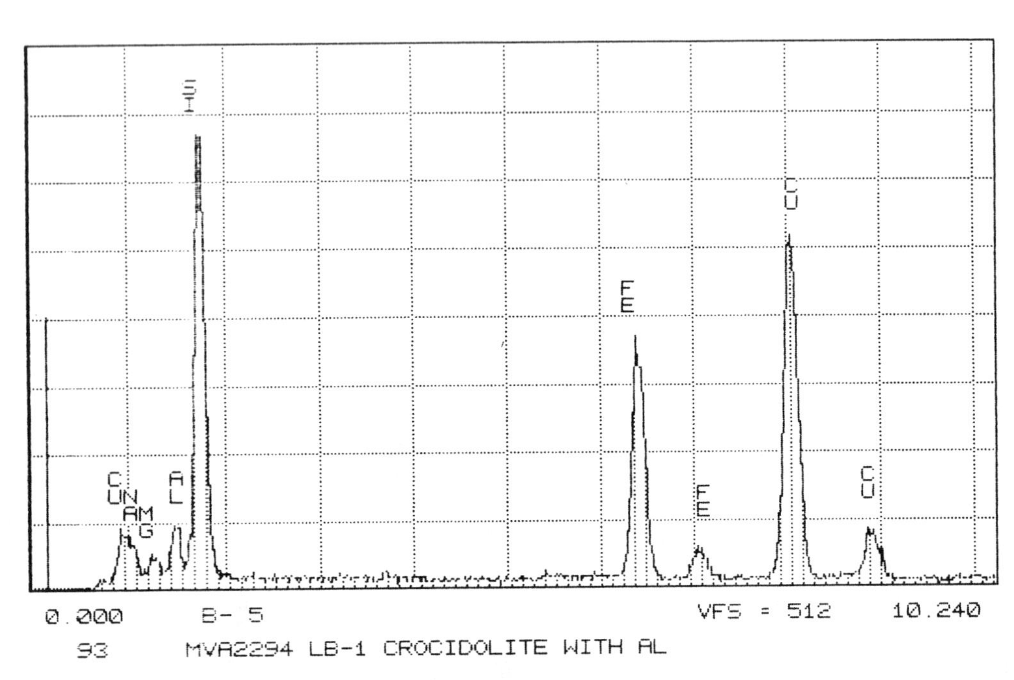

Figure 4. X-ray spectrum of Crocidolite in polycarbonate filter. Aluminum is present coating the fiber.

Two other lots of polycarbonate filters (R5SM69776, labeled: December 1995 and R6SM23752, labeled: December 1996) of the same product sold by Millipore were also analyzed without evidence of contamination. Two filters from lot R6SM23752 were analyzed with an analytical sensitivity of 12.5 fibers/mm^2 and no asbestos was detected. Five filters from lot R5SM69776 were analyzed with an analytical sensitivity of 10.4 - 12.5 fibers/mm^2 and no asbestos was detected.

Several filters from batches of Millipore mixed cellulose ester (MCE) filters, 47 mm, 0.22 µm pore size were also prepared during the same time period and analyzed using the same method. Chrysotile asbestos fibers were detected during the analysis of one filter of 17 MCE filters. The asbestos fibers found on the blank PC filters and the one blank MCE filter were all less than 10 µm in length.

TABLE 1. - - Study of PC and MCE filter contamination

Filter Lot #	Str/Grid Opening	Asbestos Str/mm^2	Analytical Sensitivity Str/mm^2
PC Filters			
R5KM64689	10/10	125	12.5
"	1/10	13	12.5
"	2/19	13	6.6
"	6/10	75	12.5
"	0/10	0	12.5
R5KM64690	1/12	10	12.5
"	2/12	21	10.4
"	2/10	25	12.5
"	2/12	21	10.4
R5SM69776	0/12	0	10.4
"	0/10	0	12.5
"	0/12	0	10.4
"	0/11	0	11.4
"	0/10	0	12.5
R6SM23752	0/10	0	12.5
"	0/10	0	12.5
MCE Filters			
H1BM92256C	0/10	0	12.5
"	0/10	0	12.5
"	0/10	0	12.5
"	0/10	0	12.5
"	0/10	0	12.5
"	0/10	0	12.5
"	0/10	0	12.5
H3NM00530	2/10	25	12.5
"	0/10	0	12.5
"	0/10	0	12.5
"	0/10	0	12.5
"	0/10	0	12.5
H4NM02985	0/10	0	12.5
"	0/10	0	12.5
"	0/11	0	11.4
"	0/10	0	12.5
"	0/10	0	12.5

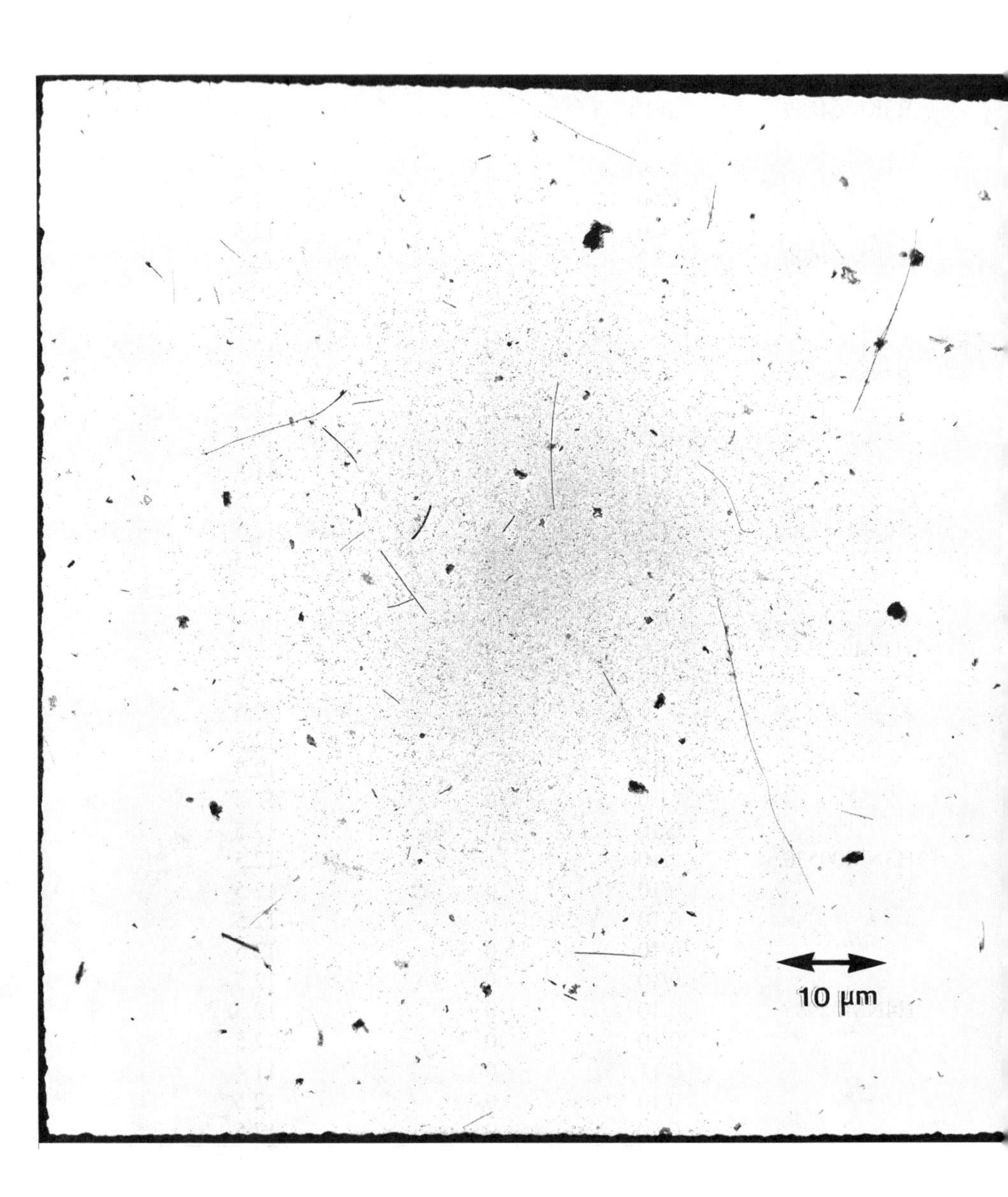

Figure 5. TEM view of the distribution of fibers on a Test grid for counting chrysotile fibers longer than 10 micrometers.

TABLE 2. - - Comparison of inter-laboratory analyses for fibers > 10 μm

Laboratory	Analyst	Asbestos Conc., MFL
001	A	16.5
001	B	15.7
002	A	16.5
003	A	18.5
003	B	18.7

APPLICATIONS OF THE DRINKING WATER METHODS

All the drinking water methods are applicable to clean, low turbidity water. For some turbid samples, the UV/Ozone treatment is successful in eliminating the turbidity when it is caused by biological growth or certain organic compounds. Difficulties arise with turbid waters that contain fine clay particles. With these samples it is not always possible to reach a desired analytical sensitivity.

The drinking water methods have been applied to the analysis of asbestos in a variety of liquids. Samples of beer, wine, urine, and pharmaceuticals have been tested using the steps described in the liquid filtration procedures and electron microscope of the water methods.

CONCLUSION

Three standard methods are available for the analysis of water for asbestos by TEM. All are based on the same general procedures and should provide comparable results when used to quantify the concentrations of asbestos fibers longer than 10 μm in length.

REFERENCES

[1] Biles, B. and Emerson, T. R.,. "Examination of fibres in beer," *Nature* , 219:93-94. 1968

[2] Cunningham, H. M. and Pontefract, R., "Asbestos fibres in beverages and drinking water", *Nature* 232:332-333, 1971.

[3] Kay, G., "Ontario intensifies search for asbestos in drinking water, " *Water Pollution Control*. 9:33-35. 1973

[4] Chatfield, E. J., "Quantitative analysis of asbestos minerals in air and water" In: *32nd Ann. Proc. Electron Microscopy Soc. Amer.*, St. Louis MO, 1974.

[5] Cook, P. M., Glass, G. E. and Tucker, J. H., "Asbestiform amphibole minerals detection and measurement of high concentrations in municipal water supplies," *Science* 185:853-855, 1974

[6] Elzenga, C. H. J., Meyer, R. B. and Stumphius, J., "Asbestos in drinking water," *Overdruk Vit H2O*, Sevende Jaargang, Nummer 19, Pagina, Netherlands 406 t/m 410, 1974

[7] Liska, R. J., Millette, J. R. and McFarren, E .F., "Asbestos analysis by electron microscope" *2nd Ann. Water Quality Technology Conf. of Amer. Water Works Assoc.* Dallas, TX Dec. 1-4. 1974

[8] McCrone, W. C. and Stewart, I. M., "Asbestos," *Amer. Lab.* 6(4):10-18. 1974

[9] Olsen, H. L., "Asbestos in Potable Water Supplies," *J. Amer. Water Works Assoc.* 66:515-518, 1974

[10] Chopra, K. S., "Inter-laboratory Measurements of Amphibole and Chrysotile Fiber Concentration in Water," NBS Spec. Publ. 506, Proceedings of the Workshop on Asbestos: Definitions and Measurement Methods., U.S. Gov. Printing Office, Washington, DC. 1978 377-380

[11] Toft, P., Meek, Wigle, M.E. and Meranger, D.T., "Asbestos in Drinking Water," *Crit.Rev. Environ. Control* 14(2) 151 1984

[12] Webber, J. S. and Covery, J. R., "Asbestos in Water," *Crit. Rev. Environ. Control*, 21(3,4):331-371 1991

[13] EPA Method 100.1, "Analytical Method for the Determination of Asbestos in Water," E. J. Chatfield and M. J. Dillon. (EPA-600-4-83-043), September 1983 Available from NTIS, Order Number PB83-260471

[14] EPA Memorandum: "Guidance and Clarification on the Analysis of Asbestos". J.M. Conlon. March 9, 1993. Available from the U.S. EPA Regional Drinking Water Offices.

[15] Feige, M. A., Clark, P.J. and Brackett, K. A.; "Guidance and Clarification for the Current U.S. EPA Test Method for Asbestos in Drinking Water," *Environ. Choices Techn. Suppl.* Fall 1993 13-14.

[16] EPA Method 100.2, "Determination of Asbestos Structures Over 10 µm in Length In Drinking Water," K.A. Brackett, Clark, P.J. and Millette, J. R. (EPA-600-R-94-134), June 1994. Available from NTIS, Order Number PB94-201902

[17] AWWA, "ASBESTOS", Standard Methods for the Examination of Water and Wastewater, American Public Health Association, 18th Ed Supplement, 1994, Section 2570, p10-15

[18] U.S. Environmental Protection Agency. "Asbestos-containing materials in schools: Final rule and notice," *Federal Register* 40 CFR Part 763, Appendix A to Sub-part E. October 30, 1987.

[19] National Institute of Standards and Technology & National Voluntary Laboratory Accreditation Program. Program Handbook for Airborne Asbestos Analysis. NISTIR 89-4137. U.S. Dept. of Commerce, Gaithersburg, Md. 1989.

[20] Chatfield, E. J., "Asbestos background levels in three filter media used for environmental monitoring," in: *Proc. 33rd Annual Meeting, Electron Microscopy Society of America*, Las Vegas, NV. C.J. Arceneaux, Ed., Claitor's Publishing Division, Baton Rouge, LA 1975

[21] Ring, S. and Suchanek, R. J. "Fiber Identification and Blank contamination Problems in the EPA Provisional Method for Asbestos Analysis," *NBS Spec. Pub. 619*. US Department of Commerce/ National Bureau of Standards, PB82-209628, 1982

[22] Chatfield, E. J., Glass, R.W. and Dillon, M.J., "Preparation of water samples for asbestos fiber counting by electron microscopy," *EPA Report EPA-600/4-78-011*, Office of Research and Development, Environmental Protection Agency, Athens, GA, 1978

Eric J. Chatfield[1]

A RAPID PROCEDURE FOR PREPARATION OF TRANSMISSION ELECTRON MICROSCOPY SPECIMENS FROM POLYCARBONATE FILTERS

REFERENCE: Chatfield, E. J., **"A Rapid Procedure for Preparation of Transmission Electron Microscopy Specimens from Polycarbonate Filters,"** *Advances in Environmental Measurement Methods for Asbestos, ASTM STP 1342,* M. E. Beard and H. L. Rook, Eds., American Society for Testing and Materials, West Conshohocken, PA, 2000.

ABSTRACT: A new method for rapid preparation of transmission electron microscope (TEM) specimen grids from capillary-pore polycarbonate (PC) filters has been developed. This procedure permits reliable and complete dissolution of the PC filter medium to be achieved in 10 minutes. Using this procedure, specimen grids can be available for TEM examination less than 30 minutes after carbon coating of the filter. The procedure is a simple solution to the problems of residual filter polymer and cracking of the carbon film that many laboratories have experienced in preparation of TEM specimen grids from PC filters. A method by which all silicate minerals, including asbestos contamination, can be removed from unused PC filters, has also been developed.

KEYWORDS: asbestos analysis, polycarbonate, membrane filter, dissolution, transmission electron microscopy

Introduction

Over the last 15 years, polycarbonate (PC) filters have varied in their resistance to complete dissolution in chlorinated hydrocarbon solvents, such as chloroform or dichloromethane. Consequently, almost all transmission electron microscope (TEM) specimens prepared by extraction replication of PC filters using chlorinated hydrocarbon solvents show some degree of residual filter polymer, the amount depending on the carbon coating technique and the procedures used for filter dissolution. These films of residual polymer reduce particle image contrast and also limit the visibility of electron diffraction patterns. Both effects increase the strain on the TEM operator and introduce the potential for a negative bias in the analysis.

Excessive heating of the surface of PC filters during carbon deposition has been found to render the surface layers of the filter polymer insoluble in chlorinated

---

[1]President, Chatfield Technical Consulting Limited, 2071 Dickson Road, Mississauga, Ontario, Canada L5B 1Y8.

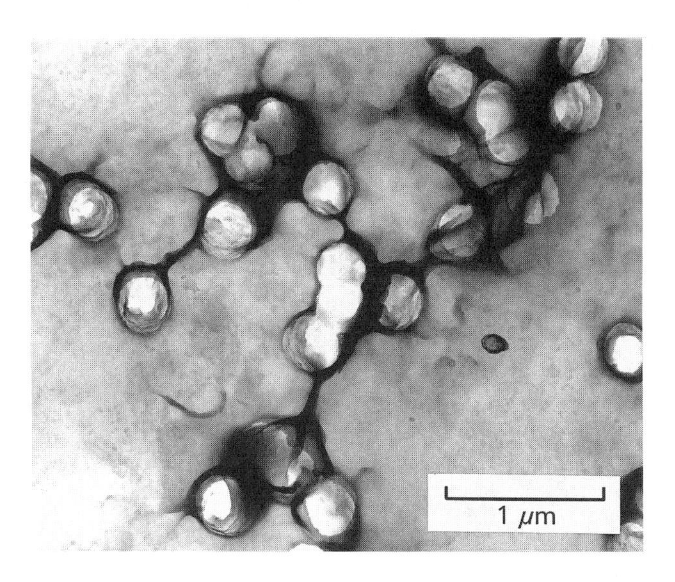

Figure 1--*TEM micrograph of polycarbonate filter, showing undissolved filter polymer around pores*

hydrocarbon solvents. However, if no excessive heating of the filter surface has occurred during carbon evaporation, use of chloroform in a Jaffe washer [*1*] for 2 hours, followed by a 20 minute treatment with chloroform in a condensation washer, has usually been found to produce TEM grids of marginally acceptable quality. After these procedures, very little additional removal of filter polymer occurs even if the condensation washing is extended significantly. TEM specimens prepared in this way almost always exhibit the "connected pore" appearance shown in Figure 1, caused by the presence of residual undissolved polymer which for some reason is more concentrated around the carbon replica of each pore. In some specimens, the regions of undissolved polymer extend further from the pores, and an extreme example of this is shown in Figure 2. Most of the chrysotile fibers on this specimen cannot be seen. Although rejection criteria, based on the proportion of the specimen area obscured by undissolved polymer, are specified by the National Voluntary Laboratory Accreditation Program (NVLAP), analysts have different interpretations of the degree of opacity which constitutes “obscured”. The NVLAP-accredited laboratory which prepared the specimen shown in Figure 2 proceeded with the analysis and reported the results without any reservations concerning specimen quality.

The difficulties experienced by analysts in preparing satisfactory TEM specimens from PC filters, and the continued presence of asbestos contamination on unused PC filters, have discouraged their use in analyses for asbestos in both water and air samples. This is unfortunate, because the PC preparation method is the closest to a truly direct-transfer, non-modifying, preparation method that is currently available. Preparation methods for mixed esters of cellulose (MCE) filters involve collapsing of the filter structure using liquids, often resulting in disruption and separation of matrix components,

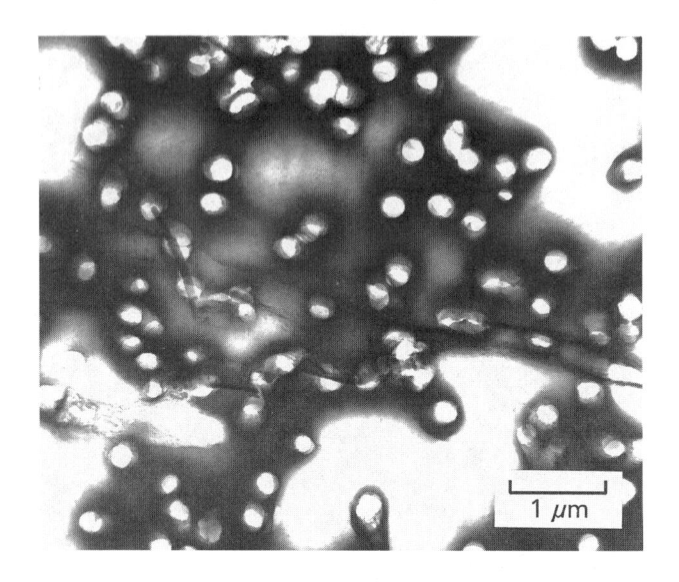

Figure 2--*TEM micrograph of polycarbonate filter, showing an extreme example of undissolved filter polymer*

and also involve plasma etching, which further modifies the sample by removing organic materials. Both of these sample-modifying procedures occur before the carbon replication process, whereas carbon replication is the first step in the preparation of TEM specimens from PC filters.

A new dissolution procedure, for preparation of TEM specimens from PC filters, has been developed which is based on the use of a mixture of 20% 1-2-diaminoethane and 80% 1-methyl-2-pyrrolidone in a Jaffe washer of the conventional wire mesh bridge design. This solvent mixture completely removes the PC filter polymer in a period of 10 minutes. It yields ideal specimen grids, even from filters that have been excessively heated during carbon evaporation. This TEM specimen preparation procedure allows the use of much thinner carbon films, than is the case for the conventional chloroform dissolution procedure, because the distortion of the PC filter observed during dissolution using chloroform does not occur. The new procedure has been incorporated in the International Organization for Standardization Methods ISO 10312 [*2*] and ISO 13794 [*3*] for determination of asbestos in air, and in a method for determination of asbestos fibers in parenteral medicines [*4*]. Figure 3 shows an example of a TEM specimen prepared from a PC filter by the new procedure, and it can be seen that chrysotile fibrils are readily visible.

Dissolution Procedure

1. prepare a Jaffe washer consisting of a stainless steel mesh bridge in a glass petri-dish;

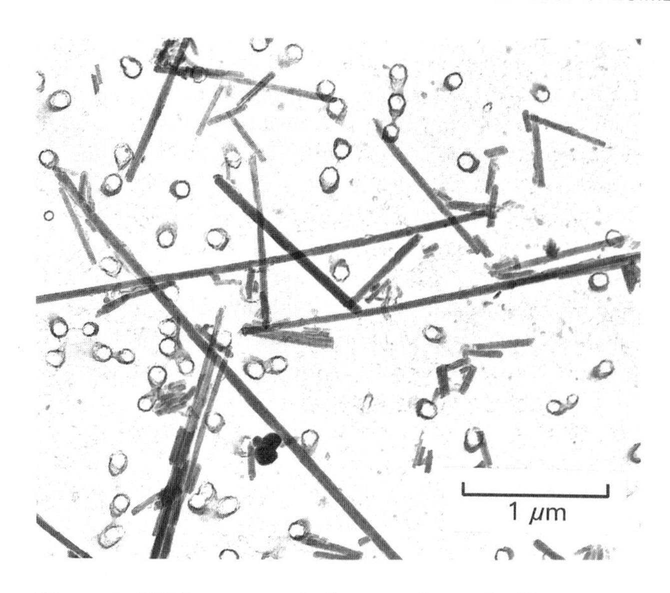

Figure 3--*TEM micrograph showing chrysotile fibers on a polycarbonate filter prepared by the new method*

2. prepare a mixture of 20% 1-2-diaminoethane and 80% 1-methyl-2-pyrrolidone and add a sufficient volume of the mixture to the Jaffe washer such that the meniscus touches the horizontal underside of the stainless steel bridge;

3. place each TEM grid with a portion of carbon-coated PC filter, carbon side facing upwards, on to the Jaffe washer mesh. Cover the dish and allow to stand for approximately 10-15 minutes;

4. transfer the stainless steel mesh with the grids to a second, empty petri-dish;

5. to the second petri-dish, add either distilled water or reagent alcohol (ethanol) until the meniscus touches the horizontal underside of the stainless steel bridge. Allow to stand for approximately 10 minutes;

6. remove the stainless steel bridge with the grids from the Jaffe washer, place it on a paper towel, and allow to dry. The drying process can be accelerated by absorbing the excess water or ethanol from the underside of the mesh using a paper towel.

The choice of distilled water or reagent alcohol as the final washing solvent depends on the application or the preference of the analyst. For water sample analysis, or for indirect TEM specimen preparations from air samples in which water has been the dispersal medium, either water or reagent alcohol may be used. If it is required to retain water-soluble particle species, such as gypsum, on the final TEM specimens from direct-transfer preparation of air sample filters, then ethanol must be used for the final wash. Even when ethanol is used, it will be found that some dissolution of gypsum will occur, but most of the gypsum in each particle will remain and the original forms of the gypsum particles will still be seen as carbon replicas. For direct-transfer air sample preparations in which it is desired to remove gypsum fibers from the preparation, water should be used for the final washing.

Some types of copper grid are attacked by the solvent mixture. It is not recommended to allow grids to remain in contact with this solvent mixture for more than approximately 15 minutes. Substantially less dissolution time may be sufficient in some cases. Solvent attack has not presented any analysis problems when water is used as the final wash. When using copper grids, a slight blue color in the solvent or in the water wash after use indicates that some chemical attack on the grids has occurred. If this chemical attack is of concern, or if for operational reasons the grids must remain in contact with the solvent for periods of time longer than approximately 15 minutes, gold grids may be used instead of copper and these are not attacked by the solvent. Nickel grids have also been found to resist attack by the solvent (J. S. Webber, New York State Department of Health, Personal Communication).

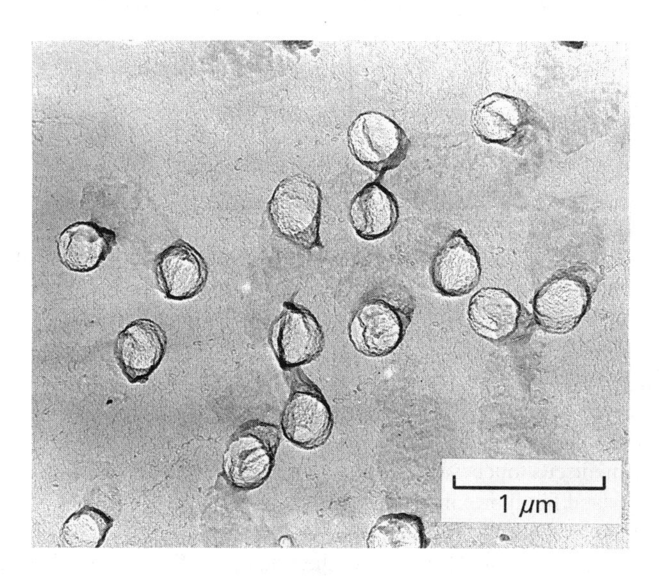

Figure 4--TEM micrograph showing detail of replicated pores in a specimen prepared by the new method

Figure 4 shows a TEM micrograph of a specimen prepared from a PC filter using the new method, in which it can be seen that the area of each pore appears to be covered by carbon replica. This is normal, and the process by which this occurs is illustrated in Figure 5. During carbon coating, the filter is rotated while being held at a fixed angle to

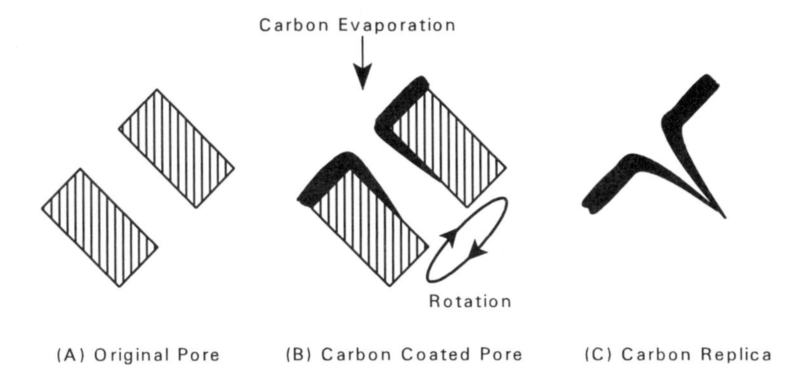

Figure 5--*Carbon replication of pores in a polycarbonate filter*

the direction of carbon evaporation. The resulting carbon replica of the interior surfaces of the pores becomes thinner at increasing depth in the pores. When the filter medium is completely dissolved, the fragile cylindrical carbon replica of the interior of the pore collapses on to itself by capillary action during evaporation of the solvent, leaving a hollow conical structure.

Using the new procedure, dissolution of PC filters occurs so rapidly that interference colors, created by the rapidly thinning polymer layer, can be observed only a few seconds after dissolution commences. No distortion of the shape of the carbon-coated PC filter portion is seen during dissolution. Starting from a PC filter, ideal TEM specimens can be available for TEM analysis in a period of less than 30 minutes. This is a significant improvement in preparation time over the conventional methods for preparation of TEM specimens from PC filters. It also represents considerable savings of time over the MCE preparation methods, bearing in mind the additional time requirements for collapsing and plasma etching of the MCE filter. In the analysis of water samples, use of this rapid specimen preparation method has the advantage that the particulate and fiber loading of filters can be assessed rapidly, and if necessary, new filters can then be prepared from the sample before it degrades.

Removal of Asbestos Contamination from Polycarbonate Filters

Contamination of filter media by asbestos fibers has been observed in both MCE and PC filters [*5, 6, 7*]. However, in recent years it has been rare to detect any asbestos contamination on TEM specimens prepared from unused MCE filters. Asbestos contamination on PC filters, however, is still frequently observed, and usually consists of short fibers of chrysotile and crocidolite. Claims are made by manufacturers that particular lots have been tested by TEM and are "asbestos-free", but asbestos contamination has been detected even on some of these filters. The sporadic nature of the contamination is such that any reasonable amount of testing by the supplier, or incorporation of a reasonable number of unused filters as blank controls, still does not give sufficient confidence for PC filters to be used for low background measurements.

It is possible to clean PC filters of all silicate mineral contamination by immersion of the filters for a period of 15 minutes in a mixture of 80% concentrated hydrofluoric acid (HF) and 20% concentrated hydrochloric acid (HCl). The filters are then given two successive washes in filtered distilled water. Although the filters may be dried after cleaning if they are to be used for air sampling, for filtration of liquids, the cleaned filters **must** remain wet until they are used. Immersion in water during the cleaning removes the wetting agent (polyvinylpyrrolidone) applied by the manufacturer, and, after drying, the filters are then hydrophobic. It is not possible to filter liquids through a cleaned 0.1 μm pore size polycarbonate filter if the filter has been allowed to dry. Transfer of a wet PC filter from the final distilled water wash to the filtration unit base is achieved by holding the edge of the filter with tweezers, while it remains under the water surface, and moving it to the surface of a 75 mm x 25 mm microscope slide. The slide can then be withdrawn from the liquid while the filter is held flat in position on it. The filter is then held by its edge close to the correct position on the base of the filtration unit, and the microscope slide is removed, allowing the filter to be transferred to the base of the filtration unit. This transfer procedure is somewhat easier than handling dry PC filters. The cleaning procedure for PC filters has been incorporated in the Verband der Chemischen Industrie method for determination of asbestos fibers in parenteral medicines [*4*].

Safety Considerations

It is recommended that the solvents for this dissolution method be used in a fume hood, and that the materials safety data sheets be reviewed before using the solvents. Hydrofluoric acid, used in the filter cleaning procedure, is a particularly dangerous acid and burns to as little as 2.5% of body area by this acid can produce hypocalcaemia resulting in death [*8*]. It is strongly recommended that the materials safety data sheet for hydrofluoric acid, and other available literature on handling precautions, be consulted before use.

References

[*1*] Jaffe, M. S., "Handling and Washing Fragile Replicas", *Journal of Applied Physics*, 1948, 19:1187.

[*2*] International Organization for Standardization, "Ambient air - Determination of asbestos fibres - Direct-transfer transmission electron microscopy method", ISO 10312, 1995.

[*3*] International Organization for Standardization, "Ambient air - Determination of asbestos fibres - Indirect-transfer transmission electron microscopy method", ISO 13794, 1999.

[4] Der Verband der Chemischen Industrie "Proposed Analytical Method (Draft) Parenteral Medicines - Determination of asbestos fibres - Direct-transfer transmission electron microscopy procedure", Verband der Chemischen Industrie, e.V., Karlstraβe 21, D-60329 Frankfurt, Germany, 1994.

[5] Chatfield, E. J., "Asbestos Background Levels in Three Filter Media Used For Environmental Monitoring", In: *Proceedings, Electron Microscopy Society of America, 33rd Annual Meeting*, G. W. Bailey, Ed., 1975, 276-277.

[6] U. S. Environmental Protection Agency, "Filter Blank Contamination in Asbestos Abatement Monitoring Procedures: Proceedings of a Peer Review Workshop", Water Engineering Research Laboratory, Contract No. 68-03-3264, United States Environmental Protection Agency, Cincinnati, Ohio 45268, 1986.

[7] Chatfield, E. J., Glass, R. W., and Dillon, M. J., "Preparation of Water Samples for Asbestos Fiber Counting by Electron Microscopy", Report EPA-600/4-78-011, National Technical Information Service, Springfield, Virginia 22161, 1978.

[8] Muriale, L., Lee, E., Genovese, J., and Trend, S., "Fatality Due to Acute Fluoride Poisoning Following Dermal Contact With Hydrofluoric Acid in a Palynology Laboratory", *Journal of the British Occupational Hygiene Society*, Vol. 40, No. 6, pp. 705-710, 1996.

Eric J. Chatfield[1]

Measurements of Chrysotile Fiber Retention Efficiencies for Polycarbonate and Mixed Cellulose Ester Filters

---

REFERENCE: Chatfield, E. J., "**Measurements of Chrysotile Fiber Retention Efficiencies for Polycarbonate and Mixed Cellulose Ester Filters**", *Advances in Environmental Measurement Methods for Asbestos, ASTM STP 1342*, M. E. Beard and H. L. Rook, Eds., American Society for Testing and Materials, West Conshohocken, PA, 2000.

ABSTRACT: The efficiencies with which chrysotile fibers in water are retained by polycarbonate (PC) and mixed cellulose ester (MCE) filters have been measured. The studies demonstrate that, for filtration of water samples, PC filters of 0.2 μm pore size or less and MCE filters of 0.22 μm porosity or less are satisfactory filters for determination of the concentrations of asbestos fibers of all lengths in water samples, but that the presence of surfactant may introduce a negative bias in the shorter fiber sizes. The results show that the observed fiber densities on 0.22 μm porosity MCE filters increase with plasma etching, and also exhibit a positive bias due to a variable degree of shrinkage which occurs during the filter collapsing procedure. In a second series of measurements, using a procedure similar to that used by the pharmaceutical industry for testing the retention of bacteria by membrane filters, it was found that all of the filters tested were capable of yielding reduction factors of better than approximately 10^7. However, for 0.22 μm porosity MCE filters, sporadic reduction factors as low as 10^2 occurred, and when surfactant was added to the water, these low reduction factors were observed consistently, and measurable proportions of the fibers up to approximately 2 μm long passed through the filters.

KEYWORDS: Asbestos analysis, fiber, water, polycarbonate, mixed cellulose ester, membrane filter, filtration efficiency, retention, surface tension, parenteral

Background

The U.S. Environmental Protection Agency (EPA) announcement permitting the use of 0.22 μm porosity mixed cellulose ester (MCE) filters [*1*], as an alternative to 0.1 μm pore size polycarbonate (PC) filters specified in EPA Method 100.1 [*2*], in the transmission electron microscopy (TEM) analysis of water samples for asbestos, raises

---

[1]President, Chatfield Technical Consulting Limited, 2071 Dickson Road, Mississauga, Ontario, Canada L5B 1Y8

questions concerning the comparability of results from analyses using the two filter types. A limited series of measurements reported by Brackett et al. [*3*], which formed the basis of the EPA response to comments on the proposal to permit the use of MCE filters, resulted in the conclusion that the PC procedure incurred serious losses of fibers, even those longer than 10 μm. It was also concluded that chrysotile fibers up to 4 μm in length were capable of passing through 0.1 μm pore size PC filters, but that transmission of fibers was not observed through 0.22 μm porosity MCE filters. The data were considered to validate the use of 0.22 μm porosity MCE filters in the measurement of the concentration of asbestos fibers longer than 10 μm, the size range on which the water quality regulation [*4*] is based.

A method for determination of asbestos in water has also been published by the American Water Works Association (AWWA) [*5*]. This method permits the use of PC filters up to 0.4 μm pore size, and MCE filters up to 0.45 μm porosity. However, in addition to measurement of the concentration of asbestos fibers longer than 10 μm, the AWWA method contains procedures to determine the concentration of asbestos fibers down to 0.5 μm in length. The only validation data to which the AWWA method refers are based on the use of 0.1 μm pore size PC filters [*6*], and on the results of an inter-laboratory study, reported in 1977, in which the participants used 0.1 μm pore size PC filters, 0.1 μm porosity MCE filters and 0.22 μm porosity MCE filters [*7*]. The AWWA method does not provide any data to validate the use of filters with larger pore sizes.

It is important to recognize that pore size specification for a membrane filter is an absolute specification **only** for capillary-pore type filters. In the case of capillary-pore filters, such as the PC filter, the pores have a very narrow distribution of diameters. For tortuous path filters, such as the MCE filters, the porosity is specified on the basis that a sterile filtrate is obtained when a filter is challenged with an aqueous suspension of a specific size of bacteria. The pore size rating for the 0.22 μm porosity MCE filter, for example, is determined using a challenge suspension of *pseudomonas diminuta*. The pore size rating for tortuous path filters, such as the MCE filters, is **not** a specification that particles exceeding that size are retained by the filter. Simonetti et al. [*8*] have demonstrated that, in aqueous suspension, latex spheres of diameters significantly larger than the nominal filter porosity can pass through MCE filters, and the presence of surfactant in the suspension further reduces the filtration efficiency.

Introduction

The efficiencies with which chrysotile fibers in aqueous suspension are transferred to TEM specimens were compared for various types of filter by measurement of the fiber densities on TEM specimens prepared from the different types of filters through which identical aliquots of a chrysotile suspension had been filtered. A series of filter challenge experiments were also performed, in which the different types of filter were used to filter identical aliquots of an aqueous suspension of chrysotile fibers, and the concentrations of chrysotile fibers passing through the filters were measured. An additional study of the effects of surfactant in the chrysotile suspension was also conducted.

Fiber Density Comparisons Using Chrysotile in Distilled Water

Experimental Procedure

Constant volume aliquots of a stable suspension of chrysotile fibers in distilled water were filtered through 0.1 μm and 0.2 μm pore size PC filters, and also through 0.1 μm and 0.22 μm porosity MCE filters. In order to guard against the introduction of bias resulting from any change in the fiber concentration of the suspension during the time required to perform the filtrations, the filter types were systematically alternated throughout the filtration session. TEM grids were prepared from the PC filters using the method reported in these proceedings [*9*]. TEM grids were prepared from the 0.1 μm porosity MCE filters by direct carbon-coating replication and also by the collapsing procedure of Burdett and Rood [*10*] incorporating removal of 10% of the filter weight by plasma etching. TEM grids were prepared from the 0.22 μm porosity filters by the collapsing procedure of Burdett and Rood incorporating no etching, removal of 1% of the filter weight by plasma etching, and removal of 10% of the filter weight by plasma etching. For each filter type and preparation method combination, TEM specimen grids were prepared from 8 individual filters. For each filter, approximately 200 fibers longer than 0.5 μm were counted and the fiber dimensions recorded. Each reported mean fiber density for fibers longer than 0.5 μm is based on a total of between 1657 and 1694 fibers. Each reported mean fiber density for fibers longer than 5 μm is based on a total of between 69 and 123 fibers. The fiber densities on the TEM specimens were compared and it was found that the fiber densities on the 0.22 μm porosity filters prepared by collapsing and 10% plasma etching exceeded the fiber densities observed for all other filter types and preparation methods. It was then noticed that the collapsed filter sectors for the 0.22 μm porosity MCE filters were visibly smaller than the portions originally removed from the filters. Measurements were made by projecting the areas of the filter sectors before and after collapsing, using 8 filters of each type, and the reduction in area was calculated for each of 0.1 μm, 0.22 μm, 0.45 μm and 0.8 μm porosity MCE filters. The calculated reductions in area were used to correct the fiber densities reported.

Results

The results of the study of area reduction during filter collapsing are given in Figure 1. The magnitude of the area reduction increases with the porosity of the filter. A mean area reduction of 4.1% was observed for 0.1 μm porosity filters, 6.8% for 0.22 μm porosity filters, 7.4% for 0.45 μm porosity filters, and 13.4% for 0.8 μm porosity filters.

The results of the fiber density experiments using suspensions of chrysotile in distilled water are shown in Figures 2 and 3.

For fibers longer than 0.5 μm (Figure 2), the fiber densities calculated for TEM specimens prepared from 0.1 μm pore size PC filters, 0.2 μm pore size PC filters, 0.1 μm porosity MCE filters prepared without collapsing or etching and 0.1 μm porosity MCE filters prepared by collapsing and etching were found to be similar. The fiber density on 0.22 μm porosity MCE filters prepared by collapsing and incorporating no etching was lower than either of the PC filter pore sizes or either of the 0.1 μm porosity MCE filter

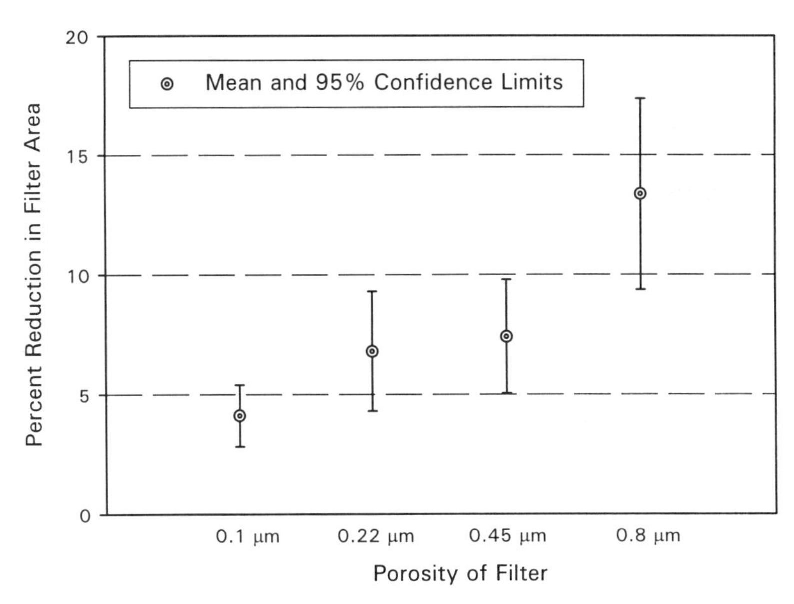

Figure 1--*Reduction in area of MCE filters during collapsing procedure*

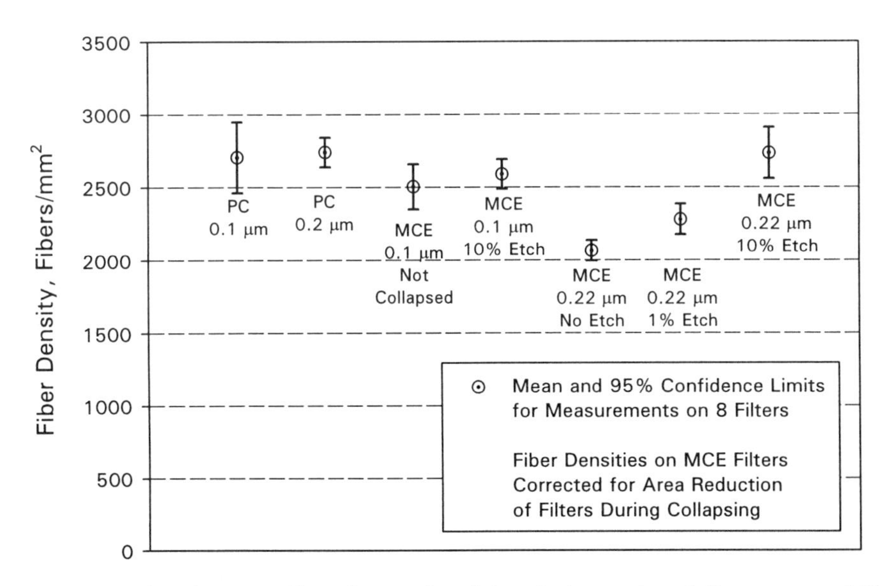

Figure 2--*Fiber densities (fibers longer than 0.5 μm) observed on different types of filter*

preparations. Using an etching procedure which removed 1% of the filter weight, the fiber density increased somewhat. Using an etching procedure which removed 10% of the filter weight, the fiber density on the 0.22 μm porosity MCE filters, after correction for shrinkage, was found to be similar to the fiber densities obtained for the 0.1 μm and 0.2 μm PC filters and both preparations of the 0.1 μm porosity MCE filters. At the 1% significance level, the differences between the mean fiber densities for these five preparations were not statistically-significant.

For fibers longer than 5 μm, the calculated fiber densities are shown in Figure 3. It can be seen that the fiber densities obtained using either 0.1 μm pore size PC filters or 0.1 μm porosity MCE filters appear to be lower than those obtained using PC filters of 0.2 μm pore size or MCE filters of 0.22 μm porosity. The reason for this trend may be that longer chrysotile fibers tend to aggregate during the longer filtration times which occur when using filters of small pore sizes, leading to fewer individual long fibers but a small number of localized fiber aggregates. The probability of observing one of these large aggregates on TEM specimens prepared from the filter is quite small, and the result is that the observed concentration of the long fibers will exhibit a negative bias.

In Figure 3, it can be seen that the fiber densities for fibers longer than 5 μm are similar for 0.2 μm pore sized PC filters and all three preparation methods using 0.22 μm porosity MCE filters. In particular, plasma etching apparently has no effect on the reported fiber densities of fibers longer than 5 μm. At the 1% significance level, there were no statistically-significant differences between the mean fiber densities for any of the preparations illustrated in Figure 3.

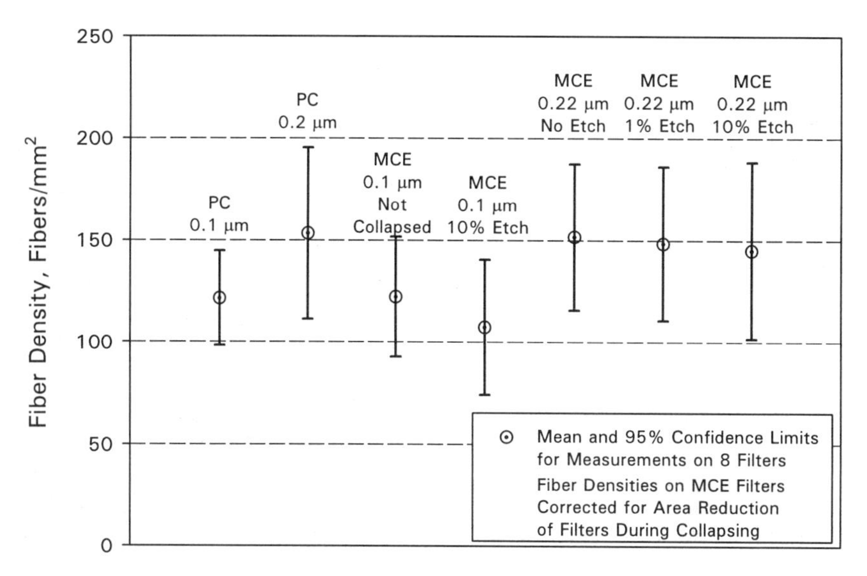

Figure 3--*Fiber densities (fibers longer than 5 μm) observed on different types of filter*

Filter Challenge Experiments Using Chrysotile in Distilled Water

Experimental Procedure

Testing of membrane filters to determine whether chrysotile fibers can pass through them, and if so, what proportion, cannot be performed reliably using simple laboratory vacuum filtration systems in which an open reservoir is clamped or screwed to a filtration base. The equipment must be designed so that filter integrity can be tested prior to the experiment, and so that upstream/downstream concentration ratios as high as 10^8 can be measured. If a simple laboratory vacuum filtration system is used, the test filter is not sufficiently well sealed, and minor leaks can result in downstream fiber concentrations greatly exceeding those which result from passage of fibers through an intact filter. The required integrity test involves application of a differential pressure of up to 700 kPa across the wet filter, and this test cannot be applied to a filter installed in a vacuum filtration unit.

In these filter challenge experiments, 0.1 μm, 0.2 μm and 0.4 μm pore size PC filters, and 0.1 μm, 0.22 μm and 0.45 μm porosity MCE filters were tested. Using a procedure similar to that used by the pharmaceutical industry for testing the retention of bacteria by membrane filters, each of the filters was challenged with a constant volume aliquot of a high concentration chrysotile suspension (approximately 3×10^{11} fibers/liter) in distilled water, and the chrysotile concentration in the filtrate was measured. Prior to each experiment, the test filter was installed in a stainless steel pressure filtration holder, and the integrity of the filter was confirmed using the bubble point test conditions specified by the manufacturer. Then the filter was washed by pumping 200 mL of 0.1 μm MCE filtered water through it, and the last 50 mL of filtrate was collected and analyzed to provide a system background measurement. A volume of 110 mL of the challenge suspension was then pumped through the filter. Half of the filtrate was filtered through a cleaned 0.1 μm pore size PC filter, and the other half was filtered through a 0.1 μm porosity MCE filter; TEM specimens were prepared from the PC filter. Each TEM examination for determination of the concentration of asbestos fibers in the filtrate from a test filter was terminated after either approximately 300 fibers had been counted, for those filters which allowed significant numbers of fibers to pass through, or, for filtrates which contained very low concentrations of fibers, the examination was terminated after approximately 0.2 mm^2 of the TEM specimens had been examined.

Results

It was found that 0.1 μm and 0.2 μm pore size PC filters and 0.1 μm, 0.22 μm and 0.45 μm porosity MCE filters could yield reduction factors (the ratio of the upstream to downstream fiber concentrations) higher than 10^7, corresponding to no fibers found in the filtrate. Where no fibers were found, the downstream fiber concentration was assumed to be equal to the analytical sensitivity, which depends on the area of the TEM specimens examined, and variations in the calculated reduction factors are a result of differences in the analytical sensitivity of the downstream fiber concentration measurements. For 0.4 μm pore size PC filters, the reduction factors were much lower, since significant numbers of fibers had passed through this pore size of filter. Figure 4 is a TEM micrograph showing fibers which passed through a 0.4 μm pore size PC filter when it was challenged with chrysotile suspended in distilled water.

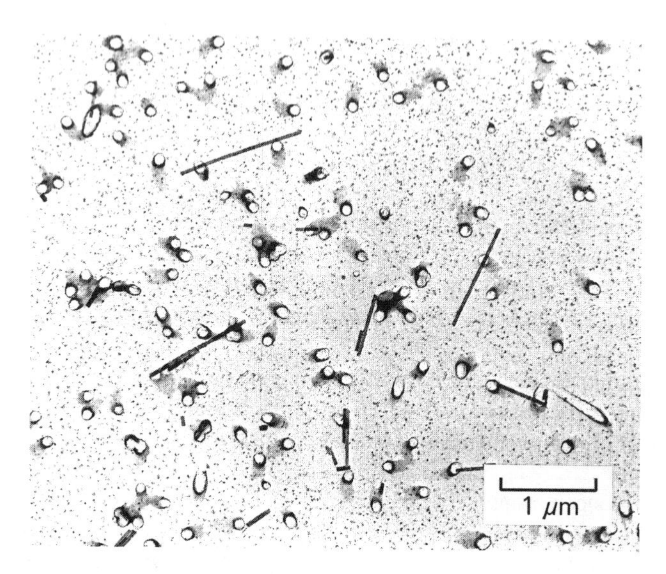

Figure 4--*TEM micrograph of chrysotile fibers which passed through a 0.4 μm pore size PC filter*

For the experiments using chrysotile in distilled water, the results for PC filters are shown in Figure 5, and the results for MCE filters are shown in Figure 6.

It was found that 0.22 μm porosity MCE filters yielded sporadic results. Retention efficiencies higher than 10^7 were often obtained; however, in one of the experiments on a 0.22 μm porosity MCE filter, a reduction factor of approximately 10^2 was obtained. The lengths of the chrysotile fibers which passed through the filter were all less than approximately 2 μm. Sporadic behavior in the retention of polystyrene latex particles by MCE filters has been observed by Pall et al. [*11*], and in order to obtain consistent particle retention characteristics, these authors elected to use a low concentration of surfactant as a standard component of their testing procedures, even though this resulted in lower retention values. On the basis of these observations by Pall et al., an additional study was conducted to examine the effects of surfactant in the challenge chrysotile suspension on filtration efficiency.

Effects of Surfactant on Filtration Efficiency

Filter Challenge Experiments Using Chrysotile in 0.1% Aerosol OT® Solution

A series of filter challenge experiments was conducted in which 0.1 μm, 0.2 μm and 0.4 μm pore size PC filters, and 0.1 μm, 0.22 μm and 0.45 μm porosity MCE filters

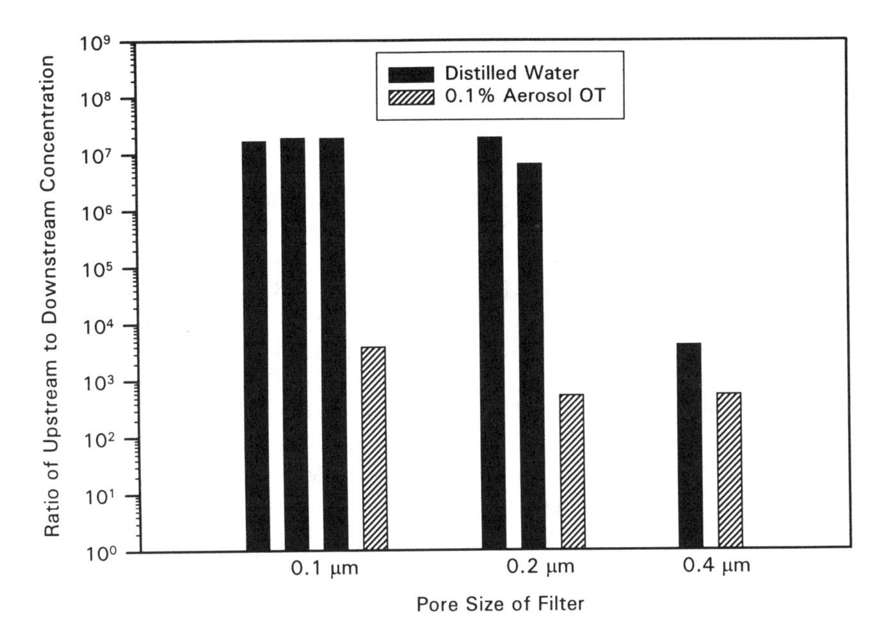

Figure 5--*Transmission of chrysotile fibers through polycarbonate filters*

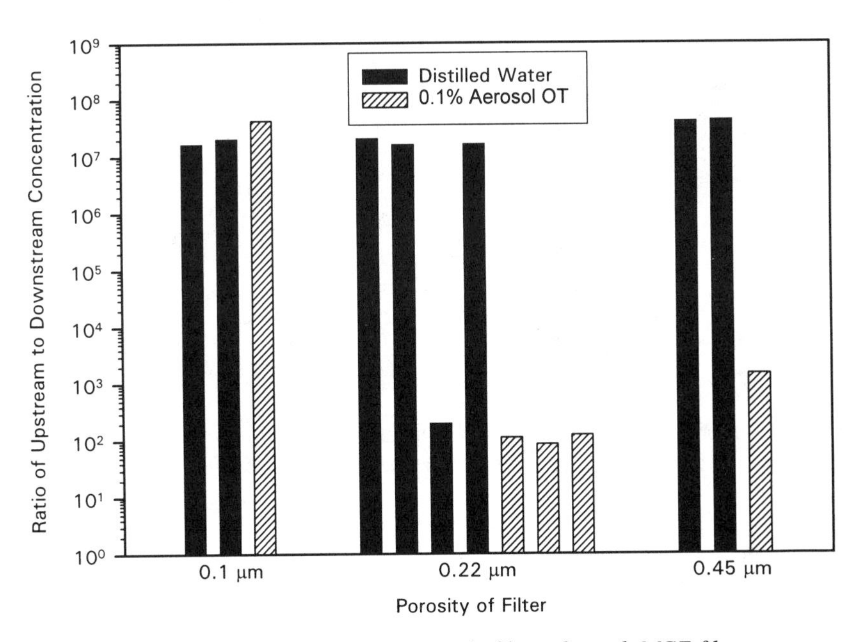

Figure 6--*Transmission of chrysotile fibers through MCE filters*

were tested following the same procedure as previously described for the filter challenge experiments using chrysotile in distilled water. For these experiments, each of the filters was challenged with a constant volume aliquot of a high concentration chrysotile suspension (approximately 3×10^{11} fibers/liter) in 0.1% Aerosol OT® solution.

It was found that addition of Aerosol OT® to the chrysotile suspension prior to filtration caused transmission of fibers to occur through all of the pore sizes of PC filter tested. The results are shown in Figure 5, where it can be seen that the reduction factors were approximately 10^3.

The results for MCE filters of 0.1 μm, 0.22 μm and 0.45 μm porosities are shown in Figure 6. MCE filters of 0.1 μm porosity were the only type of filter for which no transmission of chrysotile fibers was observed when challenged by chrysotile suspended in Aerosol OT® solution. However, these filters became blocked very rapidly when suspensions of chrysotile in Aerosol OT® were filtered, and it was generally not possible to complete filtration of the planned volume of suspension. The reasons for this filtration behavior have not been determined. For 0.22 μm porosity MCE filters, reduction factors of approximately 10^2 were consistently observed when the filters were challenged with chrysotile fibers suspended in Aerosol OT® solution. Figure 7 is a TEM

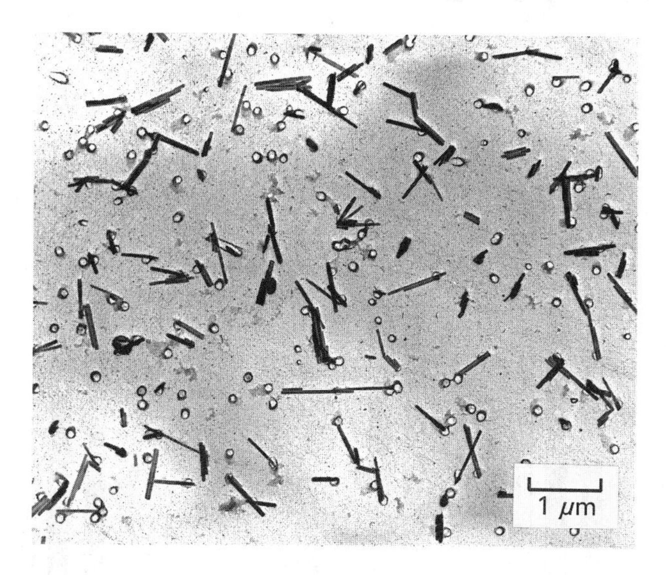

Figure 7--*TEM micrograph of chrysotile fibers which passed through a 0.22 μm porosity MCE filter when surfactant was present in the water*

micrograph which shows chrysotile fibers detected in the filtrate. The lengths of the chrysotile fibers which passed through 0.22 μm porosity MCE filters did not exceed approximately 2 μm. The proportion of chrysotile fibers observed to pass through 0.22 μm porosity MCE filters, as a function of fiber length, is given in Figure 8.

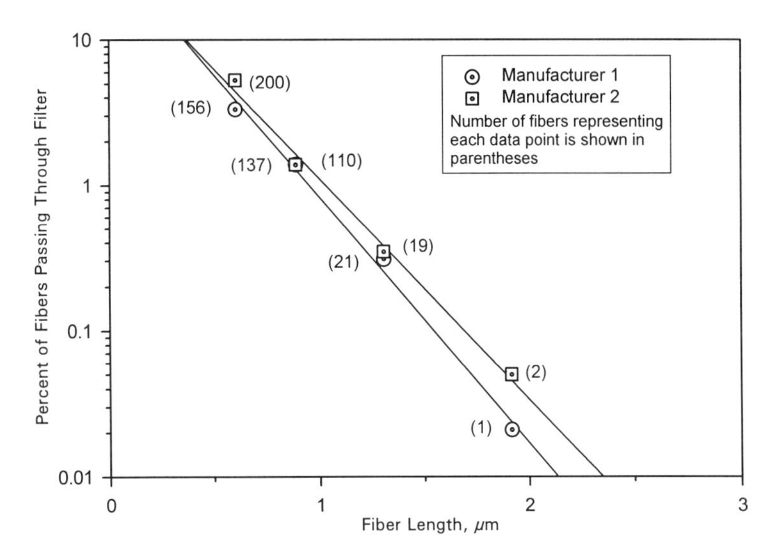

Figure 8--*Percentage of chrysotile fibers passing through 0.22 μm porosity MCE filters as a function of fiber length*

Figure 6 shows that a reduction factor of approximately 10^3 was obtained for 0.45 μm porosity MCE filters when challenged by chrysotile in Aerosol OT® solution.

Fiber Density Comparisons Using Chrysotile in 0.1% Aerosol OT® Solution

After it was discovered that the presence of surfactant could result in measurable transmission of chrysotile fibers through both PC and MCE filters, a second series of filters for measurement of fiber densities was prepared, half of which were prepared using a chrysotile suspension in distilled water, and half of which were prepared using a chrysotile suspension in 0.1% Aerosol OT®. PC filters of 0.1 μm, 0.2 μm and 0.4 μm pore size and MCE filters of 0.1 μm, 0.22 μm and 0.45 μm porosity were used. For each suspension, TEM grids were prepared from one PC filter of each pore size using the method described [*9*], and TEM grids were prepared from one MCE filter of each porosity using the method of Burdett and Rood [*10*], incorporating removal of 10% of the filter weights by plasma etching. Figure 9 shows the initial results which have been obtained for fibers longer than 0.5 μm. Except for 0.4 μm pore size PC filters, the reported fiber densities found on filters prepared by filtration of chrysotile in distilled water were closely similar. The 0.4 μm PC filters behaved differently, and it is clear that as many as 40% of the fibers were not retained on the surface of these filters. The fiber densities obtained from filters prepared by filtration of chrysotile suspended in 0.1% Aerosol OT® solution were invariably lower than those obtained from the distilled water suspension.

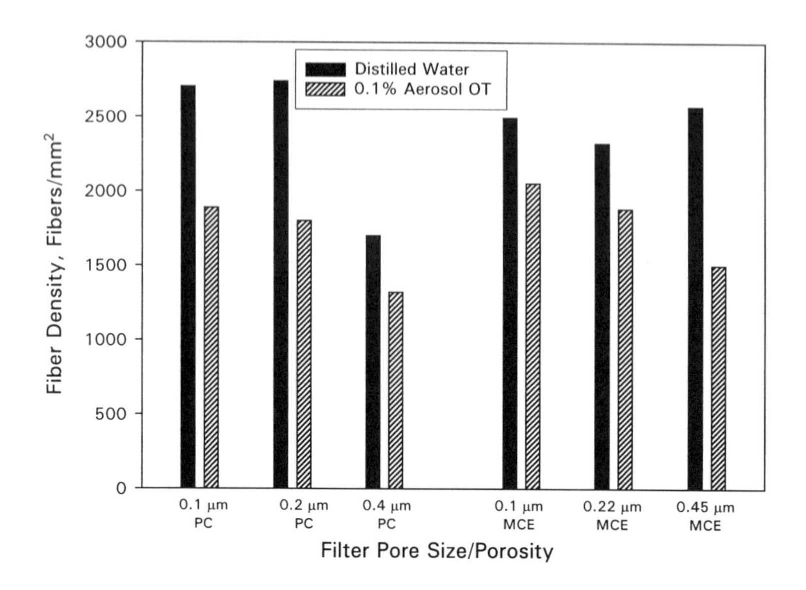

Figure 9--*Effect of Aerosol OT® on chrysotile fiber densities (fibers longer than 0.5 μm) observed on PC and MCE filters*

Discussion

Reduction in Area During MCE Filter Collapsing Procedure

The question arises as to whether the area reduction observed when using the collapsing procedure of Burdett and Rood [*10*], for the preparation of TEM specimens from MCE filters, would occur if the acetone vapor filter collapsing method [*12*] were used, in view of the fact that the edges of the filter are held in position by an adhesive page reinforcement during acetone vapor collapsing. When using the acetone vapor collapsing method, the filter expands considerably when it begins to absorb acetone vapor, resulting in buckling because the edges are restrained. Eventually, one or more of the depressions in the buckled filter make contact with the surface of the microscope slide, and the filter adheres to the glass slide at these points. The filter then settles completely on to the slide, and the final overall area is the same as that of the original filter. However, settling of the buckled filter requires the increased filter area to be accommodated within the fixed area of the hole in the adhesive page reinforcement, and the extent to which this introduces distortion of the filter surface and therefore variability in the fiber densities, on the size scale of a TEM grid opening, is not known.

Unpredictable Retention Efficiencies of 0.22 μm Porosity MCE Filters

Although it is clear that the presence of surfactant resulted in consistent transmission of chrysotile fibers through 0.22 μm porosity MCE filters, transmission of chrysotile through this type of filter was also sometimes observed when no deliberate action was taken to reduce the surface tension of the water suspension being filtered. The reason for the unpredictable retention efficiency of 0.22 μm porosity MCE filters for chrysotile fibers in distilled water suspension has not currently been determined, although it may be related to the surfactant incorporated into these filters by the manufacturers.

Comparison of Fiber Density Experiments and Filter Challenge Experiments

Quantitative comparison of the results of the fiber density measurements with those of the filter challenge experiments is problematic. In anticipation of very high retention efficiencies for the filters, the filter challenge experiments were designed and conducted using very high chrysotile fiber concentrations in the challenge suspensions, so that upstream/downstream concentration ratios of the order of 10^7 would be measurable. The fiber concentrations, used for the filter challenge experiments, corresponded to approximately 70-250 fibers per filter pore in the case of the PC filters, depending on the pore size of filter. It is therefore possible that a log-jam of fibers could have occurred in the passages through all of the filters, leading to a retention efficiency which progressively increased during filtration of the challenge suspension. In effect, the fibers in the challenge suspension could have partially filled the openings of the test filters, thereby resulting in modification of the filtration characteristics. If an experiment were designed using more dilute challenge suspensions of chrysotile fibers, such that, in the same filtration experiment, the concentrations of both the fibers retained on the filter and those passing through the filter could be determined, it is possible that different results would be obtained.

Although a quantitative comparison of the two sets of experiments cannot be made, the addition of surfactant in the suspension being filtered invariably reduces the fiber densities observed on the surfaces of all of the test filters, and, with the exception of 0.1 μm porosity MCE filters, it results in passage of short chrysotile fibers through the filters. The reduction in the fiber densities observed on the surfaces of the filters was considerably larger than can be accounted for by the transmission of fibers through the corresponding types and pore sizes of filters in the fiber challenge experiments. The reason for the discrepancy has not been determined, although it may be related to increased penetration of fibers into the filter structure when surfactant is present.

Implications For Sterilizing Filtration of Parenteral Drugs

As a consequence of publicity in which detection of asbestos fibers in injectable (parenteral) drugs was reported, in 1993 the German Federal Health Agency issued notices, requiring measures to reduce contamination of parenteral medicinal products by asbestos [*13,14*]. In 1994, a method for determination of asbestos in such products was published [*15*]. However, there appeared to be no convincing explanation as to how asbestos fibers could have occurred in these products, given that the products concerned had all been filtered through sterilizing (0.22 μm porosity) filters. The results of the current filter challenge experiments, made on 0.22 μm filters, go some way to explaining how asbestos could have occurred in these products, in that, although reduction factors of better than 10^7 were generally obtained, some filters exhibited sporadic and much lower reduction factors of approximately 10^2. Moreover, since many of the parenteral products, by their nature, have surfactant properties, the filtration reduction factors for asbestos when these products are filtered through sterilizing filters of 0.22 μm porosity may well be of the order of 10^2, rather than a value higher than 10^7 which would frequently be determined using suspensions of chrysotile fibers in distilled water. Accordingly, testing of filters for their efficiency in retaining asbestos fibers, using suspensions of chrysotile fibers in distilled water, would probably not be representative of the filtration efficiency when actual products are being filtered.

Conclusions

For determination of chrysotile fibers longer than 0.5 μm in drinking water samples, either PC filters of 0.1 μm or 0.2 μm pore size, or MCE filters of 0.1 μm porosity should give comparable results. MCE filters of 0.22 μm porosity also should give comparable results provided that plasma etching is used to remove 10% of the filter weight and corrections are made for the filter shrinkage during the collapsing step. At the 1% significance level, differences between the mean fiber densities for these combinations of filters and preparation methods were not statistically significant. The measurements of fiber densities using suspensions of chrysotile in distilled water do not confirm the conclusions of Brackett et al. [*3*], that water filtration using 0.1 μm pore size PC filters or 0.22 μm porosity MCE filters incurs significant losses of fibers relative to the results obtained using 0.1 μm porosity MCE filters.

A single measurement on a 0.45 μm porosity MCE filter appears to indicate that filters of this porosity may also be suitable for analysis of fibers longer than 0.5 μm in water samples, but more work is required to confirm this. A single measurement on a 0.4 μm pore size PC filter appears to indicate that these filters are not suitable for analysis of fibers longer than 0.5 μm in water samples, because significant losses into and through the filter appeared to occur regardless of the presence or absence of surfactant.

For determination of chrysotile fibers longer than 5 μm in chrysotile suspensions in distilled water, the fiber densities observed on all of the filters tested were similar, although there was a trend for the fiber densities on 0.1 μm pore size PC and 0.1 μm porosity MCE filters to be lower. However, at the 1% significance level, differences between the mean fiber densities for the combinations of filters and preparation methods tested were not statistically significant. Plasma etching of collapsed 0.22 μm porosity MCE filters had no effect on the reported fiber densities of fibers longer than 5 μm, and therefore it is concluded that, for measurement of fibers longer than 5 μm, plasma etching is unnecessary.

For determination of fibers longer than 10 μm in drinking water, the size range on which the EPA water quality regulation is based, all of the filters tested should give comparable results. It is unlikely that plasma etching would have any effect on this measurement, since it did not affect the determination of fibers longer than 5 μm in distilled water.

Regardless of the type or pore size/porosity of filter in use, the presence of surfactant in the chrysotile suspension being filtered reduces the fiber densities observed on the TEM specimens prepared from the filters. For all of the filters tested, other than the 0.1 μm porosity MCE filters, the presence of surfactant consistently caused significant numbers of chrysotile fibers to pass through the test filter, corresponding to reduction factors of approximately 10^2 to 10^3.

The filter challenge experiments show that some of the chrysotile fibers in distilled water suspensions can pass through PC filters of 0.4 μm pore size. No fibers were detected in the filtrates when 0.2 μm and 0.1 μm pore size PC filters were challenged with chrysotile suspensions in distilled water with concentrations of the order of 3×10^{11} fibers/liter. The upstream/downstream ratios generally exceeded 10^7. These data do not support the conclusion of Brackett et al. [*3*] that fibers as long as 4 μm can pass through 0.1 μm pore size PC filters.

The reduction factor observed for 0.22 μm porosity MCE filters was generally higher than 10^7, but with sporadic values as low as approximately 10^2. The maximum fiber length for fibers passing through the 0.22 μm porosity MCE filters was approximately 2 μm.

For reasons not currently understood, the flow-rates observed during filtration of fiber suspensions in Aerosol OT® solutions through 0.1 μm porosity MCE filters rapidly decreased.

The observation that sporadic transmission of chrysotile fibers through 0.22 μm porosity filters can occur, even when there is no intentional addition of surfactant to the liquid being filtered, means that filtration of a liquid through a 0.22 μm porosity MCE filter cannot provide assurance that the filtrate contains no asbestos fibers. Of the filters tested, the only filter that can provide this assurance, regardless of the presence or absence of surfactant, is a 0.1 μm porosity MCE filter.

The observed sporadic transmission of short chrysotile fibers through 0.22 μm porosity MCE filters, and the consistent transmission of such fibers through these filters in the presence of surfactant, possibly explain the occurrence of chrysotile fibers found in some parenteral products after filtration through sterilizing filters of 0.22 μm porosity. It is concluded that measurements of the efficiencies with which sterilizing filters retain chrysotile fibers, using suspensions of chrysotile in distilled water, may not represent the performance of these filters when filtering the actual products. The reduction factors for chrysotile during sterilizing filtration of these products could be as low as 10^2.

References

[*1*] Federal Register, Environmental Protection Agency 40 CFR Parts 141 and 143; "Analytical Methods for Regulated Drinking Water Contaminants; Final Rule", Federal Register / Vol. 59, No. 232 / Monday, December 5, 1994 / Rules and Regulations, pp. 62456 - 62465.

[*2*] Chatfield, E. J. and Dillon, M. J., "Analytical Method for Determination of Asbestos Fibers in Water", U.S. Environmental Protection Agency Report EPA-600/4-83-043, National Technical Information Service, 5285 Port Royal Road, Springfield, VA 22161, U.S.A., Order No. PB 83-260471, 1983.

[3] Brackett, K. A., Clark, P. J. and Millette, J. R., "Further Comparisons of the Efficiency of Polycarbonate and Mixed Cellulose Ester Filters For Use in the Filtration of Water Samples", Environmental Information Association Journal, Summer 1994, pp. 11-13.

[4] Federal Register, Environmental Protection Agency 40 CFR Part 141; "*National Primary Drinking Water Regulations*", Federal Register / Vol. 54, No. 20 / Wednesday, January 30, 1991 / Rules and Regulations, pp. 3578 - 3597.

[5] American Water Works Association, "2570 Asbestos", *Standard Methods For the Examination of Water and Wastewater, 18th Edition Supplement.* American Public Health Association, 1015 Fifteenth Street, NW, Washington, DC 20005, 1994.

[6] Chatfield, E. J., Dillon, M. J., and Stott, W. R., "Development of Improved Analytical Techniques For Determination of Asbestos in Water Samples", U.S. Environmental Protection Agency Report EPA-600/4-83-042, National Technical Information Service, 5285 Port Royal Road, Springfield, VA 22161, U.S.A., Order No. PB 83-261651, 1983.

[7] Chopra, K. S., "Inter-Laboratory Measurements of Amphibole and Chrysotile Fiber Concentration in Water", National Bureau of Standards Special Publication 506, Proceedings of the Workshop on Asbestos: Definitions and Measurement Methods, Gaithersburg, Md., 1978, pp. 377-380.

[8] Simonetti, J. A., Schroeder, H. G., and Meltzer, T. H., "A Review of Latex Sphere Retention Work: Its Application to Membrane Pore-Size Rating", Ultrapure Water, July/August 1986.

[9] Chatfield, E. J., "A Rapid Procedure for Preparation of Transmission Electron Microscopy Specimens from Polycarbonate Filters", *Advances in Environmental Measurement Methods for Asbestos, ASTM STP 1342*, M. E. Beard and H. L. Rook, Eds., American Society for Testing and Materials, West Conshohocken, PA, 1999 (These Proceedings).

[10] Burdett, G. J. and Rood, A. P. "Membrane-Filter, Direct Transfer Technique for the Analysis of Asbestos Fibres or Other Inorganic Particles by Transmission Electron Microscopy", *Environmental Science and Technology*, 1982, 17:643-648.

[11] Pall, D. B., Kirnbauer, E. A., and Allen, B. T., "Particulate Retention by Bacteria Retention Membrane Filters", Colloids and Surfaces, 1, pp. 235-256, 1980.

[*12*] National Institute for Occupational Safety and Health. "Asbestos Fibers, Method 7402", NIOSH Manual of Analytical Methods, Issued: 8/15/87, Revision #1: 5/15/89, U.S. Department of Health and Human Services, Public Health Service, Centers for Disease Control and Prevention, 4676 Columbia Parkway, Cincinnati, Ohio 45226.

[*13*] Bundesgesundheitsamt, "Notice On Measures to Reduce Asbestos Contamination in Parenteral Pharmaceuticals", Bundesgesundheitsamt, Berlin, Germany, 21 July 1993.

[*14*] Bundesgesundheitsamt, "Notice Concerning Measures to Reduce Contamination of Parenteral Medicinal Products by Asbestos", Bundesgesundheitsamt, Berlin, Germany, 19 November 1993.

[*15*] Der Verband der Chemischen Industrie, "Analytical Method for the Determination of Asbestos Minerals in Parenteral Medicines - Direct-transfer transmission electron microscopy procedure", Verband der Chemischen Industrie, e.V., Karlstraβe 21, D-60329 Frankfurt, Germany, 1994.

R. Mark Bailey[1] and Meisheng Hu[2]

SLUDGE, CRUD AND FISH GUTS: CREATIVE APPROACHES TO THE ANALYSIS OF NON-STANDARD WATER SAMPLES FOR ASBESTOS

REFERENCE: Bailey, R.M. and Hu, M., **"Sludge, Crud and Fish Guts: Creative Approaches to the Analysis of Non-Standard Water Samples for Asbestos,"** *Advances in Environmental Measurement Methods for Asbestos, ASTM STP 4149,* M.E. Beard, H.L. Rook, Eds., American Society for Testing and Materials, 2000.

ABSTRACT: The analysis of asbestos in non-standard water samples (i.e. surface water, waste water, liquid toxic waste, fish entrails & others) by transmission electron microscopy is an on-going challenge. Samples are highly variable, and commonly containing multiple interfering components. The choice of sample preparation technique is critical in order to obtain meaningful results. Techniques discussed include UV/ozonation, dissolution, dilution, ashing and solvent washing. A variety of case histories are presented, and the approaches taken discussed.

KEYWORDS: asbestos, waste water analysis, drinking water analysis, UV/Ozone, electron microscopy

---

With EPA's promulgation of drinking water regulations (Federal Register, December 5, 1994) requiring the nation's drinking water sources be tested for asbestos, asbestos analysis laboratories have seen a large increase in requests for water sample analysis. The increased concern with respect to asbestos in water generated by the EPA regulations has also caused an increase in requests for asbestos analysis of waters other than drinking water, i.e. effluent, sludge, waste, hazardous waste liquids and others. At present, there are no approved test methods for non-standard asbestos water samples. Analytical laboratories, generally staffed by individuals who welcome the opportunity to exercise their creativity, perform analysis on these materials by modifying the established

[1]President and Laboratory Director, Asbestos TEM Laboratories, Inc., 1409 Fifth St., Berkeley, CA

[2]Senior Electron Microscopist, Asbestos TEM Laboratories, Inc., 1409 Fifth St., Berkeley, CA

drinking water methods [1,2,3] with varying degrees of success. This paper discusses the approaches the staff of Asbestos TEM Laboratories uses to deal with these intractable materials including UV/ozonation, dissolution, dilution, ashing and solvent washing.

Analysis of Asbestos in Water Samples

Asbestos fibers suspended in water samples are well documented as being of a size requiring transmission electron microscopy (TEM) for accurate identification [1,2,3 & 4]. As a result, all water sample methods described here used TEM as the analytical method. The issue for non-standard water sample analysis is not which analytical method to use, but one of what sample preparation technique to apply to get the contained asbestos fibers onto the filter substrate with a minimum of interfering particulate. The current EPA methods [2,3] allow for three basic preparation techniques as given below:

- Direct transfer of 10ml or more of water sample onto a mixed cellulose ester (MCE) or polycarbonate (PC) filter by vacuum filtration.
- Direct transfer of 10ml or more of water sample diluted with particle free water onto a MCE or PC filter by vacuum filtration.
- Ultrasonication and UV/Ozone treatment prior to emplacement on a filter as above.

These techniques assume, as would be logical for drinking water samples, that the water will be relatively pure with a minor amount of contained interfering substances. This is not the case for most effluent, waste and hazardous waste liquids.

Effluent, waste and hazardous waste water samples come from a much wider range of sources than drinking water. We have received samples of pond water, groundwater in well borings, standing water in open pits, liquids extracted from 55 gallon drums, watery extracts from sewage treatment plants, and samples from numerous other unidentified locations. These samples commonly contain a wide range of inorganic and organic particulate, and can also contain a variety of miscible and immiscible organic liquids. Sometimes the high sediment load or concentration of other liquids present in these samples makes it questionable as to whether these samples should even be referred to as "water" samples, but rather as sludges or high liquid content bulk samples. And in some cases, non-liquid materials representing water samples are received for asbestos analysis, such as samples of pureed fish entrails. In all, the wide range of sample types makes it such that no one sample preparation technique can possibly address the full range of materials that the analytical laboratory must contend with.

In order to perform TEM analysis of liquid, semi-liquid or liquid-representing solid samples like those discussed above it is necessary to emplace some fraction of the solid components found in the samples onto a filter. For such a preparation to be analyzed and provide meaningful data, a variety of problems must be overcome. These problems include:

- Particulate overloading of filters
- Risk of explosion, fire or toxic exposure to organic-based hazardous liquids
- Immiscible, often viscous, organic liquids that plug filters and/or leave a residue
- Partial or total dissolution of filters by contained organic liquid solvents
- Solid or semi-solid samples that are marginally water samples, or that represent a type of water sample (i.e. sludge or fish entrails) that must be liquefied in some fashion before emplacement onto a filter.

Real World Problems of Non-Standard Water Sample Analysis

A combination of standard and non-standard water sample preparation techniques, some more effective than others, have been attempted and used by Asbestos TEM Laboratories to deal with the problems associated with preparing non-standard water samples. In some few cases, no special preparation is needed and the samples can be analyzed as regular water samples. For most non-standard water samples, this is not the case. Before analysis of these materials begin, it is important to think through the type of problem presented by the materials to be sure that the sample preparation method or methods to be used are the most appropriate.

Particulate Overloading of Filters

Overloading of water samples containing suspended particulate is commonplace for non-drinking water samples, and occasionally occurs for standard drinking water samples as well. These samples can often be identified before sample preparation by their murky color, often with observable settled particulate on the bottom of the container. Preparation of water samples using a standard 25 mm diameter filter requires the filtration of a minimum of 10 ml to assure uniform distribution of particulate on the filter surface. If less liquid is used, the risk of uneven distribution due to currents in the water becomes high. Water samples containing a heavy particulate concentration (i.e. clay, silt, algae) commonly either prevent a TEM specimen from being made at all, or the resulting TEM specimen has too much particulate present to allow for identification of the contained fibers. Usually the particulate loading is not so heavy as to be a barrier to complete filtration of the 10ml or greater solution. On occasion, however, the filter will become entirely plugged with particulate, effectively halting liquid filtration.

The general solution to the problem of overloaded samples is to dilute aliquots of the water sample with various amounts of pure, particle free water. Aliquots of decreasing volume, each diluted to a minimum volume of 10ml, are filtered until a filter with appropriate particulate loading is obtained. Proper loading is typically signified by filters showing only the slightest hint of coloring as seen by the naked eye. Numerous attempts may have to be made before the right concentration is obtained. It has been found that multiple sample preparations using dilutions that range by powers of ten is often the best way to get a filter with the appropriate loading.

The main problem with dilution is that it necessitates running additional grid openings to attain something close to a reasonable analytical sensitivity. Achieving a sensitivity of 0.2 MFL (Millions of Fibers per Liter) as required for drinking water samples, is generally not attainable. However, sensitivities in the range of the MCL (Maximum Contaminant Level) of 7.0 MFL or better are usually achievable.

In cases where the offending particulate is inorganic, such as mud or silt found in many murky water samples and sludges, dilution is the only effective technique. However, if the interfering material is organic (i.e. algae) we have found UV/ozone treatment prior to filtering very effective for samples from surface waters, lakes and canals. Following the method given by Chatfield [3], the sample is ultra-sonicated for 15 minutes before UV/ozone treatment. UV/ozone treatment consists of producing a gas containing 4% ozone by passing pure oxygen at a rate of 2 liters/minute through a Union Carbide Model 4060 ozone generator operating a 45 watts. The resulting gas stream is split and the two gas streams are bubbled through two different samples, each in their original containers. Each sample receives 1 liter/minute of ozonated oxygen. At the same time an intense UV source (3600 $\mu W/cm^2$ @ 254nm and 105 $\mu W/cm^2$ @ 365nm) is generated from a UVP Pen-Ray UV Lamp immersed in the sample. Each sample is treated for 3 hours so that each sample receives approximately 10 grams of ozone. The resulting treated water sample is biologically sterile with virtually all of its algae and contained organic material oxidized. This removes sample preparation problems including the obscuring of asbestos fibers by organic particulate during analysis.

If the algae content is extreme, such as occurs in stagnant waters, an extended period of ozone treatment may be required. Alternatively, a plasma ashing and sample remobilization technique may be performed as suggested by Anderson & Long [1]. Filters with very heavy loading of algae are first allowed to dry, then put into 50ml Pyrex beakers and ashed in a plasma etcher until only a fine white ash remains. The ash residue is then mixed with particle free water and emplaced onto a new filter.

Explosion Hazard and Toxic Exposure Risk of Samples Containing Organic Liquids

A wide range of organic-based liquids can be found in hazardous waste liquids and samples of sludge. It is extremely important to be aware of and to address the potential hazards these materials represent which include both explosion and toxic hazards. Asbestos TEM Laboratories commonly obtains this type of sample from large environmental testing labs where asbestos is being run as part of a larger suite of organic and other hazardous material tests. If at all possible, it important to ask the client whether they have identified any organic-based compounds as being present in the liquids before performing sample preparation so that appropriate precautions may be taken. If no information is available, assume the worst and prepare in a properly vented hood using appropriate personal protective equipment. (Note of Caution: DO NOT treat any water sample with ozone if there is even the slightest suspicion that an organic solvent material is present in the liquid sample as a major explosion hazard would be created).

Plugging of Filters w/ Immiscible or Viscous Liquids

Samples containing oily sludge and/or immiscible organic liquids can be extremely problematic to prepare. These materials are often identifiable prior to sample preparation by the presence of oily product floating on the surface of the water sample. Sample preparation of the materials usually begins in a straightforward fashion of pouring an aliquot of the sample into the filtration apparatus after ultrasonication. Filtration usually starts off without any problem, but at some point problems occur where either liquid ceases to pass through the filter, and/or a residue, often irregular in outline, is left on the filter.

The first case given above is the biggest problem - if a known volume of liquid can not be completely run through a filter, then the sample can not be analyzed. The solution is to find a solvent that can be added to the liquid sample that will dissolve the offending organic liquid, but not dissolve the filter. This may require trying various combinations of both solvents and filters. A good place to start is the chemical compatibility guide that the filter manufacturer provides (See Table 1). After a period of trial and error, it is usually possible to find the right combination to filter the liquid. Running of method blanks is particularly important with this type of sample because the solvents used to dissolve the immiscible liquids are usually poorly characterized with regard to their asbestos concentration.

TABLE 1 -- *Microfiltration Membrane Characteristics - Partial List* [5]

	POLYCARBONATE (PC)	POLYESTER (PE)	MIXED CELLULOSE ESTER (MCE)
FUELS			
Gasoline	+	+	+
Jet Fuel	+	+	+
Kerosene	+	+	+
HYDROCARBONS			
Benzene	0	+	+
Toluene	0	+	+
Xylene	+	+	+
OILS			
Silicones	+	+	+
Petroleum Oils	+	+	+
Skydroil	+	0	0
MISCELLANEOUS			
DMSO	-	+	-
Mineral Spirits	+	+	+
Turpentine	+	+	+

(+ = no change in material, 0 = testing recommended, - = material is not compatible)

After filtration of the liquid has been achieved, an additional difficulty is commonly encountered with the presence an organic liquid residue on the filter. This happens in virtually all cases where an oily immiscible liquid is present, even though a good solvent may have been found that allowed the liquid to pass through the filter. Also, the residue can often be unevenly distributed on the filter surface, making the filter completely unacceptable for a direct filter preparation. If the attempt is made to continue with plasma etching and/or carbon coating of the residue covered filter, in virtually every case the carbon film will not adhere to the filter. When an attempt is made to dissolve the filter in an appropriate solvent, the sample disintegrates.

To analyze such samples, further preparation must be performed. Unfortunately, the extended plasma etching method suggested by Anderson and Long [1] as discussed above for algae covered samples, has not been found to work well. Plasma etching of heavy oils has not been found to be of great success; they take a very long time to etch away. The method we have found to work best is to place the oily residue covered filters into a muffle furnace and to ash them at 480^0 C for 4 to 8 hours. The ash residue is then remobilized in pure water and refiltered onto a new filter. The resulting sample is generally quite readable.

Partial or Total Dissolution of Filter by Solvents Present in the Liquid

If organic solvents that are incompatible with filter substrate materials are present in a water sample, drastic problems will occur during sample filtration. The solution, similar to the approach above, is to consult the chemical compatibility guide that the filter manufacturer provides (Table 1) and try using a different filter type. Usually a filter type can be found that will resist attack by the solvents present in most hazardous liquid samples. PC and MCE filters are resistant to dissolution by most chemicals, and if one is, the other almost always is not. Unfortunately, the samples also commonly leave an oily residue requiring a muffle furnace ashing step as above. Since ashing is almost inevitable for this type of sample, using nylon filters as listed in Table 1 is also possible.

Solid or Semi-Solid Samples

Sludge samples fall into a gray area in the regulatory framework. Are they water/liquid samples or are they bulk samples? Assuming they are considered as water samples, the treatment depends upon the concentration of organic compounds. If it is a material containing organic-free mud, then dilution with particle free water can work well. In many cases, however, a high organic content is present if the sample is a hazardous waste containing sludge, a sewage effluent or biogenic, organic-laden mud. In these cases, the sample must be dried, weighed, ashed in a muffle furnace, and then remobilized in particle free water and emplaced onto a filter.

A rather unusual form of water sample analysis is the analysis of the body tissues of marine animals, such as fish, in an attempt to determine the effect of asbestos present in the host water. Typically some portion of the creature expected to hold the asbestos is to be analyzed (i.e. fish entrails). Chatfield [3] describes a technique using high pressure oxidation of organic material using a stirred reactor vessel to break down and digest a ground beef slurry. Using this technique he was able to obtain a filterable liquid sample. This type of analysis is so rarely called for, and the reaction vessel equipment needed is so expensive, it is not a viable technique for the average lab. We approached the problem by ashing the material at 480^0 C overnight similar to the methods discussed above. The ash residue was finely ground in a mortar and pestle, suspended in particle free water and emplaced onto a filter.

Conclusion

The analysis of non-standard water samples is a difficult and often problem-filled venture. For overloaded samples, dilution is a good approach, though the analytical sensitivity of the technique can be severely affected by the amount of other particulate matter in the sample. When algae is the primary problem, UV/ozone treatment works very well to clean up the sample. Extreme concentrations of algae may require a plasma etching and remobilization approach. Immiscible and often highly viscous organic liquids present in water samples can plug filters requiring the addition of solvents to allow filtration. Oily, and often unevenly distributed, residues require muffle furnace ashing and sample remobilization. Solvent-containing samples may partially or totally dissolve the filter material, requiring the use of other, non-reactive filter substrates. Solid organic materials (marine plant or animal products) representing water samples require high pressure oxidation or muffle furnace ashing. In all, non-standard water sample preparation techniques can be summarized as follows: dilute, dissolve, try a different filter, oxidize, or some combination of the above.

References

[1] Anderson, C.H. and Long, J.M.(1980), “Interim Method for Determining Asbestos Concentration in Water”, EPA-600-4-80-005.

[2] Brackett, K.A., Clark, P.J. and Millette, J.R.(1994), “Determination of Asbestos Structures Over 10μm in Length in Drinking Water”, EPA Method 100.2 (EPA-600-R-94-134) Available from NTIS, Order Number PB94-201902.

[3] Chatfield, E.J., and Dillon, M.J.(1983), “Development of Improved Analytical Techniques For Determination of Asbestos in Water Samples” EPA 600/4-83-043

[4] Chatfield, E.J., and Dillon, M.J.(1983), “Analytical Method For the Determination of Asbestos Fibers in Water” EPA Method 100.1 (EPA 600/4-83-042) Available from NTIS, Order Number PB83-260471.

[5] Corning, Inc. - Science Products Division (1997), ”Products for Laboratory and Process Filtration”, Catalog #SLF-CAT97.

Gary B. Collins[1], Paul W. Britton[1], Patrick J. Clark[1], Kim A. Brackett[2], and Eric J. Chatfield[3]

ASBESTOS IN DRINKING WATER PERFORMANCE EVALUATION STUDIES

REFERENCE: Collins, G. B., Britton, P. W., Clark, P. J., Brackett, K. A., and Chatfield, E. J., "**Asbestos in Drinking Water Performance Evaluation Studies,**" *Advances in Environmental Measurement Methods for Asbestos, ASTM STP 1342*, M. E. Beard, H. L. Rook, Eds., American Society for Testing and Materials, 2000.

ABSTRACT: Performance evaluations of laboratories testing for asbestos in drinking water according to USEPA Test Method 100.1 or 100.2 are complicated by the difficulty of providing stable sample dispersions of asbestos in water. Reference samples of a graduated series of chrysotile asbestos concentrations dispersed in glass-distilled water were prepared in December of 1989 to address this concern.

Sealed glass ampuls, containing the reference dispersions, were sent to volunteer asbestos testing laboratories. The number of participating labs varied from 33 to 50. The data reported here was compiled from the four most recent test rounds performed.

Statistical analysis of data from the volunteer laboratories indicate that further work is needed to develop asbestos reference samples that produce data with less variability.

Details of the preparation of the samples, laboratory analytical procedures and statistical analysis of the results are presented, along with a discussion of issues suggested by the data.

KEYWORDS: Asbestos, drinking water, chrysotile dispersions

The United States Environmental Protection Agency (US EPA) announced the Proposed Federal Drinking Water Standards in May of 1989 [1]. These standards mandated a limit for the concentration of asbestos at 7 million fibers, longer than 10 μm, per liter. Transmission electron microscopy (TEM) was designated as the method of analysis to determine compliance. A program was started at that time to develop stable

[1]US Environmental Protection Agency, Cincinnati, Ohio 45268.

[2]TN & Associates, Milwaukee, WI 53226

[3]Chatfield Consulting, Ltd., Mississauga, Ontario, Canada.

reference standards consisting of known concentrations of asbestos dispersed in water, which could subsequently be used for performance evaluations of laboratories involved in compliance testing.

The specific requirements for the drinking water performance evaluation samples were dictated by the nature of the regulation, i.e., the determination of the number of fibers which fell above a minimum length of 10 μm. Therefore, it was decided to prepare a series of concentrations of asbestos centered around the 7 MFL limit, using chrysotile asbestos from sources known to have two different fiber size distributions. Equipment for filling and sealing 20 mL ampuls was available, which necessitated the production of a concentrated suspension that could be diluted to a volume suitable for filtration by the participating laboratories. Approximately 19 mL of suspension could be introduced into each ampul while allowing sufficient air space for shaking and possible expansion of volume from heat absorbed during sterilization or sonication. Therefore, it was decided to make 15 mL the working volume for dilution by the participating laboratories, which would allow for any possible variation in the 19 mL volume dispensed into the ampul by the filling equipment. A target number of 600 ampuls dictated production of 12 liters of each concentration of asbestos.

Previous studies [2] have indicated that the presence of small amounts of high molecular weight organic compounds, such as are synthesized by bacteria, can result in the aggregation of chrysotile fibers dispersed in water. In severe cases, the fiber aggregates adhere to the interior surface of the container and are effectively removed from suspension. Preparation of suspensions using carefully cleaned glassware and freshly distilled water combined with storage in flame-sealed ampuls, which are autoclaved after filling, prevents the introduction of these bacterial contaminants and should produce suspensions which are stable over a period of years. Previous studies dealt with solutions which were prepared at lower concentrations and used within several weeks of their production. It was also known that there could be some difficulty in obtaining complete dispersions of the chrysotile at higher concentrations, but the practical limit for successful dispersions was not known [3]. Thus, test dispersions of concentrated chrysotile in water were necessary to determine if such samples could be obtained.

Determination of mass concentrations of asbestos for test dispersions

Test dispersions of asbestos in water were prepared at the US EPA facility in Cincinnati, Ohio, under the guidance of Dr. Eric Chatfield of Chatfield Technical Consulting, Ltd., Mississauga, Ontario. Two types of asbestos were chosen for the dispersions, purified Calidria COF-25 chrysotile and U.I.C.C. chrysotile B. Data available for COF-25 indicated that a mass concentration of 1 μg of COF-25 suspended in 1 liter of water would produce a numerical fiber concentration of approximately 2 million fibers per liter (MFL) in the required size range of over 10 μm in length. Similar data were not available for U.I.C.C. chrysotile, so experimental dispersions of each type of chrysotile were prepared to determine the mass which would be needed to produce the desired suspensions.

A weight of 19.95 mg of COF-25 chrysotile was placed in a small agate mortar with a few milliliters of water and ground with an agate pestle to suspend the fibers. This suspension was made up to 2 liters with freshly distilled water in a glass bottle, and the pH was adjusted to between 3 and 4 by addition of glacial acetic acid. The suspension was then sonicated for 30 minutes with vigorous shaking at 5 minute intervals. A volume of 10 mL of the resulting suspension was removed by pipette, diluted to 500 mL with freshly distilled water, and sonicated for 15 minutes. Then, 10 mL of the diluted suspension was pipetted and diluted to 500 mL with freshly distilled water. A volume of 100 mL of the final suspension was removed and filtered through a 0.22 μm pore mixed cellulose ester (MCE) filter, using a 5 μm pore MCE filter as a backing filter. Samples from this filter were prepared for observation in the transmission electron microscope by the direct transfer method of Burdett and Rood [4]. TEM counts indicated that the concentration of fibers over 10 μm in length was 2.68 MF per μg of reference solid. A duplicate experiment yielded a value of 3.08 MFL. Similar tests using 20.2 mg U.I.C.C. chrysotile, but with final filtration of 200 mL of the serial dilution, yielded a result of 0.588 MF/μg over 10 μm in length.

Preparation of sealed ampuls of chrysotile dispersions

The concentrated stock suspension used for the final COF-25 standards were prepared by placing 20.34 mg of Calidria chrysotile in the agate mortar and grinding it in a small volume of water with an agate pestle. The suspension was then made up to 500 mL with fresh glass-distilled water in a 1 liter glass bottle. The pH of this suspension was adjusted to between 3 and 4 with glacial acetic acid, as indicated by pH indicator strips. The suspension was sonicated for 30 minutes, with the bottle being removed from the ultrasonic bath every five minutes and shaken vigorously. The stock suspension was then made up to 2000 mL with fresh glass-distilled water in a 4 liter glass bottle. This dilution underwent the same sonication/shaking procedure previously described. This entire sequence was repeated using 39.2 mg of the U.I.C.C. chrysotile B sample to prepare the stock suspension for the second set of reference concentrations.

The highest concentration in the series of standards was intended to be in the range of 28 MFL when diluted to 500 mL. Calculations (summarized in Table 1) indicated that 364 mL of the COF-25 stock suspension, or 953 mL of the U.I.C.C.-B stock suspension, diluted to 12 liters should produce ampuls containing enough fibers over 10 μm to closely approach this concentration when diluted to 500 mL by the laboratories. A series of lower concentrations were prepared by diluting progressively smaller volumes of the stock suspensions into 12 liters of water. Water with no stock solution added was used to prepare "blank" ampuls.

Each of the 12 liter solutions was prepared in a clean polyethylene bottle using freshly distilled water. The bottle was shaken vigorously and then placed on a magnetic stirrer. Using constant stirring, the bottle was attached to the ampul filling machine and approximately 19 mL of suspension was injected into each ampul. The ampul was flame-sealed immediately after filling. The blank solution was packaged first, followed

TABLE 1--Asbestos Content of Maximum Concentrations*

Reference Chrysotile	Stock Solution	12 Liter Solution	15 mL of Ampul Contents	MFL in Laboratory Dilutions
COF-25	10.17 μg/mL	0.308 μg/mL	4.635 μg	
	2.75×10^7 f/mL	8.27×10^5 f/mL	1.24×10^7 f/mL	24.8
U.I.C.C.-B.	19.6 μg/mL	1.557 μg/mL	23.335 μg	
	1.15×10^7 f/mL	9.13×10^7 f/mL	1.37×10^7 f/mL	27.4

* Values are based on the results of the preliminary experimental determination of mass concentrations for COF-25 and U.I.C.C.-B. All numerical fiber data refers to fibers over 10 μm in length.

by increasing concentrations of the COF-25 chrysotile, which eliminated the need for cleaning of the equipment between filling runs. The filling equipment was then thoroughly cleaned before processing the increasing concentrations of U.I.C.C.-B chrysotile. Each lot of ampuls was autoclaved for 30 minutes at 121° C to sterilize the contents.

Quality Assurance Testing of Ampuls

A few ampuls were randomly selected from each lot, filtered and prepared for TEM observation by the electron microscopy laboratory, US EPA National Risk Management Research Laboratory, Cincinnati, Ohio and at Chatfield Technical consulting, Mississauga, Ontario. At least one ampul from each concentration was prepared for TEM observation following the procedure outlined in Method 100.1 with the exception that a 0.22 μm pore-size MCE filter membrane was used in place of the specified polycarbonate membrane. Direct-transfer TEM samples were prepared from the filters using the Burdett and Rood method [4]. The results of the calculation of concentrations when made up to 500 mL appear in Table 2, along with the calculated values obtained using the formula in method 100.1. The concentrations from both Chatfield and the EPA are based on the results from a single ampul prepared using the protocol in EPA Method 100.1.

Preliminary Study

The first Water Supply Laboratory Performance Evaluation Study that unofficially included asbestos ampuls was WS032. The resulting asbestos data were highly variable, which caused the viability of using these ampuls for testing performance to be questioned. As a result, asbestos ampuls were not available during WS033 and WS034, while the cause of the variability was investigated. It was subsequently determined that the WS032 instructions caused the excessive variability. These instructions did not make it clear to the study participants whether the sample ready for analysis should be 200 mL or 500mL, or that the final result should be in terms of the sample ready for analysis rather than the

15 mL of concentrate used to prepare the study sample. Since the high study variability was judged to have been caused by correctable problems in the study instructions rather than by problems involving the ampuls, asbestos ampuls were officially available for the first time in WS035. Since it was not feasible to correct the WS032 data to make it consistent with data from the subsequent studies, it is not considered further in this paper. The instructions used in these official studies are given in the next section.

TABLE 2--Results of Initial Quality Assurance Testing of Ampuls

	Percentage of Calculated Concentration (measured by Chatfield) (%)	Percentage of Calculated Concentration (measured by EPA) (%)
Lot #1	Blank	Blank
Lot #2	42.3	53.7
Lot #3	50	66.6
Lot #4	34.7	67.4
Lot #5	15.4	53.5
Lot #6	31.8	56.6*
Lot #7	71.4	73.1
Lot #8	58	38.3
Lot #9	91.9	67.1
Lot #10	126.9	66.8*
Lot #11	128.6	72.8

* Values are from counts made one year after production of ampules (initial QA samples did not meet minimum criteria for acceptable preparations.)

Instructions for Preparation and Analysis of Asbestos Suspensions

The following set of instructions, condensed from EPA Method 100.2, was sent with each ampul to the participating laboratories.

Approximately one hour before beginning analysis:

1. Vigorously shake the ampul for approximately 15 seconds.

2. Place the ampul in an ultrasonic bath for at least 30 minutes

3. Shake the ampul again for 15 seconds.

4. Place the ampul in the ultrasonic bath for another 30 minutes and continue to sonicate the ampul until it is opened.

5. Place 100 mL of freshly-distilled water into a very clean 500 mL glass bottle with a screw cap. Precautions should be taken to ensure that no organic materials of bacterial origin remain in the bottle from any previous use or from washing with unsterile water.

6. Break the ampul at the pre-scored neck and complete sample preparation as quickly as possible.

7. Pipet 15 mL from the ampul into the glass bottle containing the 100 mL distilled water, and dilute to 500 mL with freshly distilled water.

8. Shake the bottle vigorously for 15 seconds.

9. Adjust the pH of the suspension to between 3 and 4 with glacial acetic acid. The acid is most conveniently added using a microliter pipet and non-bleeding pH indicator strips to determine the pH. After each addition of acetic acid, shake the suspension vigorously before checking the pH.

10. Place the bottle in an ultrasonic bath for at least 30 minutes.

11. Shake the bottle vigorously before removing any aliquot from the suspension.

12. To avoid loss of fibers settling on tapered funnel sides, use a filtration apparatus with straight vertical sides.

13. Only 0.1 μm pore-size polycarbonate filter membranes or 0.1 to 0.22 μm pore mixed cellulose ester (MCE) membranes should be used. Filters should be taken from a lot which has been pre-screened for background contamination. This is particularly important if fibers <10 μm are to be counted. The filter should be backed by a ≤ 5 μm pore-size MCE membrane filter to diffuse the vacuum across the membrane.

Notes:

- Analyze the sample with "Analytical Methods for Determination of Asbestos Fibers in Water", 40 CFR Parts 141, 142 and 143 p. 3582, Wednesday, January 30, 1991.

- An analytical sensitivity of 200,000 fibers per liter (0.2 MFL) is required subject to the following analysis termination rules:

  - Analysis may be terminated either at completion of the grid opening during which an analytical sensitivity of 0.2 MFL is achieved or the grid opening that contains the 100th asbestos fiber over 10 μm in length, whichever occurs first.

  - A minimum of four grid openings must be counted, even if this results in counting more than 100 asbestos fibers over 10 μm in length.

  - The grid openings examined should be drawn about equally from a minimum of three specimen grids.

- Counting rules:

  - Count fibers with an aspect ratio of ≥3:1.

  - Count a fiber bundle as a single fiber with width equal to an estimate of the mean bundle width and length equal to maximum length.

  - Count individual asbestos fibers and bundles within clusters and matrices, as long as they meet the above definitions of fibers and bundles.
  - Fibers that intersect the top and left sides of the grid opening are counted and recorded as twice their visible length. Fibers intersecting the bottom and right sides of the grid are not recorded.

  - Count only one end of each fiber to avoid possibly counting a fiber more than once.

- Fiber identification criteria:

  - Each fiber suspected to be chrysotile must first be examined by electron diffraction (ED) following the procedure in Figure 15 of the method. If the characteristic ED pattern is observed, the fiber shall be classified as CD. If no pattern is observed or the pattern is not distinctive, the fiber shall be examined by EDXA (energy dispersive x-ray analysis) and classified according to the method. Only chrysotile fibers classified as CD, CMQ or CDQ should be included in the calculation of the concentration.

  - Each fiber suspected to be amphibole must first be examined by electron diffraction following the procedure in Figure 18 of the method. Each fiber must then be examined by EDXA. If a random orientation ED pattern showing a 0.53 nm layer spacing is obtained and the elements and peak areas of the EDXA spectrum correspond to those of a known amphibole asbestos, the fiber shall be classified as ADQ. If the random orientation ED pattern cannot be obtained, is incomplete, or is not recognizable as a non-amphibole pattern but the elements and peak areas of the EDXA spectrum correspond to those of a known amphibole asbestos, the fiber shall be classified as AQ. Only amphibole fibers classified as ADQ, AQ, or identified by zone axis ED pattern(s) should be included in the calculation of the concentration.

- A calibrated magnification of at least 10,000X ± 5% is adequate for counting fibers over 10 μm in length. A minimum 250 nm spot size is adequate for this analysis.

- The mass concentration of asbestos is not needed.

- Calculation of results. The concentration of asbestos in a given sample is calculated using the following formula:

$$\frac{\text{no str} \times \text{efa}}{\text{GO} \times \text{GOA} \times \text{V} \times 1000}$$

where

No str = number of asbestos fibers counted

efa = effective filter area of the samp:ing filter in square millimeters

GO = number of grid openings counted

GOA = area of grid openings in square millimeters

V = volume of sample filtered (not the 15 mL of ampul solution) in mL

- Report results as millions of fibers >10 μm long per liter (MFL) with three significant figures.

Results of Laboratory Performance Test Rounds

Data from the first four studies are given in Table 3 and statistics from these data are presented in Table 4. The biweight mean and standard deviation are calculated as defined by Kafadar [5], using C=4. The objective of using biweight statistics is to minimize the effect of outliers that are clearly present in each set of study data.

Use of Lot 10 in two of the studies provides an opportunity to estimate the within-laboratory component of the standard deviation at that concentration, pooled over both studies. Table 5 shows the results of an Analysis of Variance (ANOVA) that was performed on the data reported by the 27 laboratories that reported data in both of the studies in which Lot 10 was used.

TABLE 3. Asbestos Data From USEPA Water Supply (WS) Studies

	Test Round Number			
Laboratory Number	**WS035 (MFL)**	**WS036 (MFL)**	**WS037 (MFL)**	**WS038 (MFL)**
1	0.01	1.42	0.087	0.714
2	1.8	0.828	3.646	0.9
3	2.4	1.75	2.62	2.1
4	3.51	4.8	5.24	0.9
5	3.77	2.4	2.1	0.59
6	3.85	12.6	--	--
7	4.2	1.69	5.88	0.82
8	4.22	3.59	5	0.668
9	4.3	7.3	12.5	--
10	4.64	2.69	9.24	1.24
11	5.5	6.83	5.92	0.56
12	5.61	3.79	2.07	0.507
13	5.67	8.55	--	--
14	6.33	1.04	4.53	1.49
15	6.91	--	--	--
16	6.98	--	--	--
17	7	37.5	0.878	--
18	7.37	--	--	--
19	7.43	4.29	9.54	--
20	7.9	1.6	6.3	0.995
21	7.92	5.66	5.15	0.46
22	8.16	--	--	--
23	8.189	2.6	2.039	0.441
24	9	4.25	2.6	1.5

	Test Round Number			
Laboratory Number	WS035 (MFL)	WS036 (MFL)	WS037 (MFL)	WS038 (MFL)
25	9.46	6.076	12.8	--
26	9.86	7.951	18.47	--
27	10.06	4.586	6.973	0.769
28	10.6	7.66	7.87	1.6
29	10.6	1.62	8.8	1.75
30	10.7	4.75	5.43	1.14
31	11.5	6	12.4	--
32	13	4.75	5.89	0.162
33	15	4.22	13.81	2.6
34	--	4.66	--	--
35	--	2.4	9.19	0.699
36	--	2.81	--	--
37	--	3.13	4.26	1.88
38	--	1.64	7.14	2.26
39	--	1.399	--	0.419
40	--	5.249	13.68	--
41	--	8.61	--	--
42	--	4.132	5.538	--
43	--	6.764	9.369	1.433
44	--	7.96	20.9	1.92
45	--	8.64	87.3	--
46	--	7.49	17.1	1.72
47	--	4.7	12.5	1.59
48	--	47.31	--	--
49	--	1.067	7.644	0.913

	Test Round Number			
Laboratory Number	**WS035 (MFL)**	**WS036 (MFL)**	**WS037 (MFL)**	**WS038 (MFL)**
50	--	1.71	3.37	0.32
51	--	4.661	--	35.19
52	--	11.88	13.51	0.571
53	--	10.4	6.09	--
54	--	3.96	5.97	--
55	--	--	10.2	1.43
56	--	--	9.8	9.93
57	--	--	4.73	--
58	--	--	4.896	3.107
59	--	--	--	1.03
60	--	--	--	2.42
61	--	--	--	2.46
62	--	--	--	0.79
63	--	--	--	2.42
64	--	--	--	2.957

Discussion

Production Problems

1. The initial TEM testing of the ampuls immediately after production indicated that there were some problems with the solutions. The first problem encountered was that all of the suspensions prepared from the Calidria COF-25 chrysotile were lower in concentration than expected by a value of approximately 2.5. In essence, the >10 μm fiber yield was only 1.15 MF/μg of solid, rather than the 2.68 to 3.08 obtained in the preliminary trials. Additionally, the TEM preps indicated that there were still a significant number of bundles and clusters present in the dispersions of both types of chrysotile. The original instructions for preparation of the laboratory dilutions were modified by increasing the minimum recommended sonication periods from 15 minutes to 30 minutes each in an attempt to further reduce these complex structures to individual

TABLE 4--Statistical Summary of Data From the First Four Studies

Study	Lot Number Used	Conc. as measured by Chatfield (MFL)*	Number of Labs Reporting	Biweight Mean (MFL)	Biweight Standard Deviation (MFL)	Relative Standard Deviation (% of Mean)
WS035	10	17.76	33	7.008	3.474	49.6
WS036	6	8.89	50	4.499	3.035	67.5
WS037	10	17.76	46	7.097	4.644	65.4
WS038	3	1.76	42	1.248	0.8540	68.4

*Measured at time of production.

TABLE 5--Results of an ANOVA on the Available Data From Analysis of Lot #10

Effect of :	Sums of Squares	Degrees of Freedom	Mean Squares
Difference in Studies	5.181	1	5.181
Difference in Laboratories	643.495	26	24.750
Error	211.834	26	8.147
Totals	860.510	53	

fibers. These problems were not obvious when the preliminary fiber per unit mass determinations were made, which suggests that complete dispersion of chrysotile at this level of concentration is difficult to achieve consistently.

2. The results from Lot #5 indicated that a mistake had been made in the formulation of this lot, which apparently was made up to the same concentration as Lot #4.

3. Particles of polymer debris were observed in some of the TEM preparations from the ampuls. Presumably this debris originated from the polyethylene bottles used for collection of the glass-distilled water and preparation of the 12 liter suspensions. In particular, the particles may have been introduced by the action of the magnetic stirrer used to agitate the suspension during filling of the ampuls. The effect of this material on the stability of the suspensions is unknown, but it would be desirable to avoid the use of plastic bottles in any future production of ampuls.

Changes in suspensions with storage

The appearance of the contents of some of the ampuls has changed over time. Initially, close inspection of randomly selected ampuls revealed a clear suspension with no visible material discernible in the water. After a couple of years, close inspection revealed small visible "dust" floating in some of the ampuls. Assuming that all of the ampuls were clear initially, this change in appearance suggests that the chrysotile fibers re-aggregated into larger, more complex structures. Unless these aggregates can be broken down by sonication, subsequent dilutions prepared from these ampuls will produce lower fiber concentrations than expected. A small test of this concern was performed in the EPA TEM laboratory two years after the ampuls were prepared. A series of filtrations of multiple aliquots prepared from the same 500 mL dilution of the U.I.C.C.-B. Chrysotile suspension made from an ampul in Lot #9 produced filter membranes with concentrations ranging from 4.81 to 13.10 MFL.

Earlier interlaboratory comparisons involving the analysis of chrysotile dispersions from sealed ampuls also produced unacceptably high variability of results [6]. That study was the first to determine the role that bacterial contamination played in the instability of dispersions over even a short time period. Tests of ampuls prepared under sterile conditions indicated that dispersions were stable over a period of at least 60 days. Given the results in the current series, the long term stability of concentrated reference dispersions prepared in the manner employed in this study remains in doubt.

Laboratory Variables

An obvious variable in the preparation of the ampul contents for dilution is that of the power level to which the sample is subjected during the sonication steps prior to filtration. It would be expected that increasing levels of power could increase the number of fibers released from complex structures. It is not likely that any power level attainable with a tabletop ultrasonic bath would be sufficient to cause fracture of longer fibers into shorter lengths. However, the authors do not know of any systematic study on the effects of different power levels or types (probe versus bath) of ultrasonic treatment on chrysotile in suspension.

Analysis of Statistics

One observation that can be made from Table 4 is that the means of the study data are universally lower than the concentration as measured by Dr. Chatfield. This could reflect the development of persistent clumps of the asbestos fibers after production and initial analyses, i.e., clumps that could not be disrupted by the sonication procedures followed during the studies.

If persistent clumps have developed in the solutions since their production, did this clump formation continue in Lot #10 during the approximate 17 month period between WS035 and WS037? This question may be answered from the unbalanced data in Table

4, where the biweight means for WS035 and WS037 are quite obviously not different, or from the statistics in Table 5 by the following F test:

$$F = MS_{studies} / MS_{error} = 5.181 / 8.147 < 1$$

An F value less than 1 indicates there is truly no difference between the study means, so there is no clear evidence to suggest continuing significant and persistent clump formation in Lot #10 during the period between these two studies.

Another interesting question is whether the overall study data variability is significantly greater than the data variability within the average laboratory? In other words, were the differences among laboratories a significant source of addition variability in the study data? To answer this question, conduct the following F test using the statistics from Table 5:

$$F = MS_{laboratories} / MS_{error} = 24.750 / 8.147 = 3.04$$

The probability of getting an F value this larger or larger when there is no difference among laboratories is less than 1 percent. Such a small probability suggests that differences among the laboratories does add significant variability to asbestos data. If the differences among laboratories could be reduced or eliminated, much of this added variability could be avoided, thereby reducing the variability of asbestos results among laboratories that analyze the same sample.

Since the within-laboratory standard deviation (s_{wl}), which can be estimated from Table 5, is quite large,

$$S_{wl} = (MS_{error})^{1/2} = (8.147)^{1/2} = 2.854,$$

some benefit may also be derived from improving the consistency of analytical procedure and analyst interpretation within each laboratory over time, i.e., from one analysis to the next.

Conclusions

1) Results from analysis of the ampuls studied, provide a viable basis for evaluating the analytical performance of laboratories measuring asbestos in drinking waters.

2) To avoid contamination of the sample concentrates with non-asbestos particles, plastic bottles should not be used during sample production.

3) Although the sterile techniques used for ampul production during this study were apparently not completely successful in avoiding the subsequent formation of clumps/bundles, the change in study results over the 17 months between WS035 and WS037 was not significant, suggesting that clump/bundle formation does not continue

indefinitely. Further work should be done to investigate ways to maximize the avoidance of clump/bundle formation.

4) Further work should be done to determine whether there is a sonication power level that maximizes the dispersion of individual asbestos fibers in samples produced from ampuls like those studied, without fracturing the fibers themselves. The results of such an investigation might also have application to the sonication of routine asbestos samples.

5) Modifications of the current TEM method should be studied to see if there are changes which will make the method more rugged in routine application, thereby reducing the variability within and among laboratories.

REFERENCES

[1] 40 CFR Parts 141, 142 and 143. National Primary and Secondary Drinking Water Regulations: Proposed Rule. Federal Register, Vol. 54, No. P7, May 22, 1989, Pages 22062 to 22160.

[2] Chatfield, E.J., Dillon, M.J. and Stott W.R., "Development of Improved Analytical Techniques for Determination of Asbestos in Water Samples" National Technical Information Service, Publication PB 83-260-471. 1983.

[3] Chatfield, E.J., "Preparation and Characterization of Reference Chrysotile Dispersions for Use in Quality Assurance Testing" Report No. 89K009 1990-04-20 International Technology, Inc. 1990.

[4] Burdett, G. J. and Rood A. P., " Membrane-Filter Direct-Transfer Technique for the Analysis of Asbestos Fibers or Other Inorganic Particles by Transmission Electron Microscopy," *Environ. Sci. Technol.* 17:643-648, 1983.

[5] Kafadar, K., "A Biweight Approach to the One-Sample Problem." *Journal of the American Statistical Association*, Vol. 77, No. 38, June, 1982. Pages 416 to 424.

[6] Chatfield, E.J., "Analytical Procedures and Standardization for Asbestos Fiber Counting in Air, Water and Solid samples." Proceedings of the NBS/EPA Asbestos Standards Workshop, Gaithersburg, Maryland. NBS Special Publication 619 1982. Pages 91 to 107.

James S. Webber,[1] Laurie J. Carhart,[2] and Alex G. Czuhanich[2]

PROFICIENCY TESTING FOR ALL FIBER SIZES IN DRINKING WATER: THE LONG AND THE SHORT OF IT

REFERENCE: Webber, J. S., Carhart, L. J., and Czuhanich, A.G. **"Proficiency Testing for All Fiber Sizes in Drinking Water: The Long and the Short of It ,"** *Advances in Environmental Measurement Methods for Asbestos, ASTM STP 1342,* M. E. Beard, H. L. Rook, Eds., American Society for Testing and Materials, 2000.

ABSTRACT: The Environmental Laboratory Approval Program (ELAP) of the New York State Department of Health has administered a proficiency-testing (PT) program for asbestos in drinking water since 1992. While the United States Environmental Protection Agency requires that only fibers longer than 10 μm be counted, ELAP continues to require that all fibers longer than 0.5 μm be analyzed. ELAP mandates the use of polycarbonate filters because short fibers are apparently lost in mixed-cellulose ester filters. Thirteen different waterborne asbestos samples have been distributed to approximately two to three dozen laboratories analyzing asbestos by transmission electron microscopy. Mean total fiber concentrations have ranged from 5.49 to 3340 million fibers per liter (MFL) while mean long (>10μm) fiber concentrations have ranged from 0.377 to 76.7 MFL. Failure rates have ranged from 3.5% to 21% of participating laboratories, with equivalent failure rates for total and long-fiber analyses. As laboratories gained experience in long-fiber analysis, they appeared to report results closer to consensus means.

KEYWORDS: asbestos, proficiency testing, transmission electron microscopy, drinking water

Introduction

A New York State legislative mandate of 1984 [*1*] requires that all laboratories analyzing environmental samples be certified by the New York State Department of Health's Environmental Laboratory Approval Program (ELAP). As part of this certification, a laboratory must successfully participate in ELAP's proficiency-testing

[1]Research scientist and [2]laboratory technicians, respectively, Wadsworth Center, New York State Department of Health, P.O. Box 509, Albany, NY 12201-0509

(PT) program. This paper presents an overview of the PT for asbestos in drinking water from its initiation in 1992 through the end of 1996.

Analytical Requirements in New York State

New York State's requirements for analysis of waterborne asbestos differ from federal requirements in three areas: fiber-counting protocol, filter-medium acceptability, and filtration apparatus.

Short Fibers

The United States Environmental Protection Agency (EPA) in the early 1990s revised the protocol for counting asbestos fibers in drinking water. On the basis of apparent promotion of intestinal polyps in laboratory animals, the EPA limited its Maximum Contaminant Limit (MCL) to include only those fibers longer than 10 μm [2]. This led to an alternative method [*3*] to the original method [*4*], which required counting of all fibers longer than 0.5 μm. The original method could still be used, but laboratories would be required to count only those fibers longer than 10 μm.

New York State was concerned that this exclusion of short fibers overlooked several significant issues. First, numerous laboratory and field studies in the past two decades had revealed that the shortest fibers were the most likely candidates for transmigration from the small intestine to the blood stream [*5,6*]. The effects of these fibers in the body fluids and tissues of humans remains a topic of debate. Second, in one instance in New York State, a water sample with *billions* of fibers per liter did not contain a sufficient number of long fibers to exceed the EPA MCL of 7 million fibers per liter (MFL) [*7*]. Other samples collected from the same site but at a different time exceeded the long-fiber MCL. Hence a single water sample that contains high concentrations of short fibers but very few long fibers can still be an indicator that the drinking water system may be carrying a long-fiber load that exceeds the MCL. Finally, if water contaminated with fibers shorter than 10 μm evaporates on indoor surfaces, these fibers could become air-entrained and pose a much more significant health threat via the respiratory route versus the digestive route [*8*]. For these reasons, water samples originating in New York State must be analyzed for both long and total fiber concentrations. The required analytical sensitivity for long fibers is identical to the EPA sensitivity of 0.2 MFL but New York State requires that total (longer than 0.5 μm) fibers be analyzed, though at a less time-consuming analytical sensitivity of 10 MFL.

Polycarbonate Filters

Commercial laboratories using transmission electron microscopy to analyze airborne asbestos almost universally employ mixed-cellulose ester (MCE) rather than polycarbonate (PC) filters for sample collection and preparation. Because of this familiarity with MCE filters, the EPA in 1993 allowed the use of MCE filters for water sampling as an alternative to the PC filters that had been the required medium for a decade.

Chatfield [9] reports that fibers can penetrate MCE filters to the point where they cannot be easily retrieved. This is consistent with an investigation within our laboratory that showed that short fibers were recovered from some MCE filters only with extensive etching. Because of a lack of definitive guidance on the amount of etching needed for water samples filtered through MCE filters and because of the possible loss of short fibers (still a concern for the New York State Department of Health), samples originating from New York State must be prepared only on PC filters.

Filtration Apparatus

EPA Methods 100.1 and 100.2 simply require "analytical filter holder" [4] or "filter funnel assembly" [3]. Investigations of various filter assemblies in our laboratory have shown a tendency for assemblies with non-vertical walls to produce non-uniform distribution of particles on filters. Using diatoms as the particle suspension, water was filtered through a half dozen different commercially available filter assemblies. Assemblies with a taper that decreased the diameter of the assembly from top to bottom tended to leave a heavier deposit of particles at the edges of the filter, as determined with a densitometer (Figure 1). Hence ELAP mandates the use of filtration assemblies that have a uniform diameter from top to bottom, i.e., perfectly vertical walls.

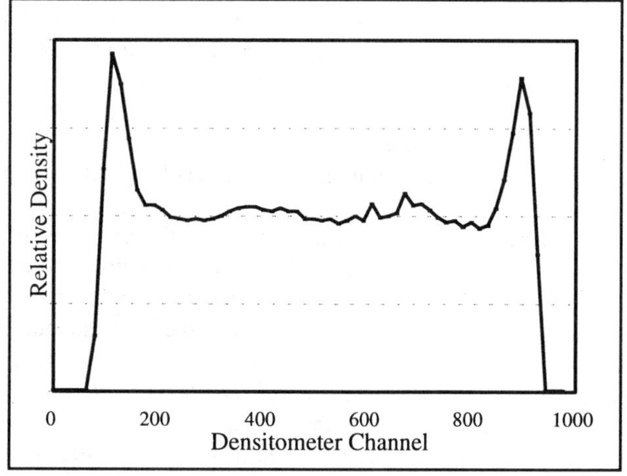

Figure 1 Density of diatom loading on PC filtration apparatus with tapered sides. Edges of filter are near channels 100 and 900 while center of filter is near channel 500.

Administration of Proficiency Testing

Preparation of Ampuls

To minimize contamination, all preparation steps were carried out using fresh double-distilled deionized water. PT samples were prepared with a variety of asbestos types or other mineral fibers, some of which had been cryo-milled. These fibers were often dispersed in water with a sonic probe. These fibers were usually run through a series of filtrations before use, utilizing nylon-mesh netting with mesh openings ranging from 5 to 211 μm. To vary the proportion of long fibers within a single sample, fibers were sometimes collected from the surface of the mesh (predominantly long fiber) or from the filtrate (predominantly short fibers). For example, for the most recent amosite-containing sample, 0.22 g amosite was placed in 350 mL 0.1-μm-filtered water and sonicated at 60% power with a medium probe. This suspension was run through a 210-μm nylon mesh and the trapped fibers discarded. The smaller fibers in the remaining suspension were filtered through a 20-μm nylon mesh and the even smaller fibers remaining in this suspension were filtered through a 10-μm nylon mesh. The fibers trapped on this nylon mesh were transferred to filtered water in a 4-L beaker, where a magnetic stir bar maintained the suspension during dispensing of 20-mL aliquots into 25-mL ampuls. Ampuls were flame-sealed.

Chatfield warns that micro-organisms in water may produce high molecular weight organic materials that irretrievably attach asbestos fibers to container walls [*4*]. ELAP has tried several methods of preparing PT samples to minimize this possibility.

During the initial rounds, ampuls were prepared the day before shipping to minimize time for bacterial growth. A major drawback of this method quickly came to light when laboratories requested replacement ampuls owing to problems such as breakage in transit. The quality of duplicate samples could not be ensured in such cases for two reasons. First, passage of even a few days could have allowed sufficient microbial growth to compromise results. Furthermore, because of the heterogeneous nature of asbestos, it was impossible to guarantee that creating a new suspension under identical conditions would yield the same concentrations.

An attempt was then made to generate samples in a medium that would retain the asbestos fibers at a fixed, retrievable concentration. Fibers were suspended in fresh double-distilled deionized water, gelatin was added, the solution was heated and then dispensed into compartmentalized trays, each with about 1cm^3, for setting. The fibers in the gel remained suspended so that they would not coalesce with time. The gelatin cubes were stored at 4° C until use. Unfortunately, even when the gelatin cubes were dissolved in 1 L of water, the gelatin concentration was still too great to allow easy passage through filter pores.

Finally, samples were frozen immediately after dispensing into the ampuls. Freezing was initially accomplished by plunging a rack of 16 ampuls into liquid nitrogen. Because of substantial (~10%) loss of ampuls due to cracking by this flash freezing, subsequent racks of ampuls were simply placed in a large freezer where all ampuls froze within 90 minutes. Freezing prevented microbial growth and also maintained fibers in rigid suspension so that coalescing would not be a problem. Even with a 90-minute freezing time, fibers as thick as 10 μm would not settle from the top to the bottom of the ampul. Thus samples could be prepared ahead of time and stored in the freezer until needed. This also allowed ELAP to thaw and send a few samples to reference laboratories for evaluation before thawing and sending a larger number to ELAP-participating laboratories. Reference laboratories provided data to evaluate the homogeneity of the samples and also provided helpful feedback on the suitability of the overall test.

Distribution of Samples and Evaluation of Performance

Typically, two ampuls containing a suspension of asbestos or other mineral fiber were sent twice a year to laboratories seeking to maintain certification in analysis of asbestos in drinking water. Samples were shipped to laboratories using next-day delivery service. Laboratories had been notified ahead of time of the samples' arrival date and were instructed to filter the samples within 24 hours of receipt. Filtration was achieved by diluting the contents of the ampul to 1 liter. Laboratories were required to achieve analytical sensitivities of 0.2 MFL for long fibers at 10,000 magnification and 10 MFL for total fibers at 15,000 magnification. Laboratories were allowed seven weeks to submit results to ELAP; ELAP reported scores to the laboratories six weeks after that.

Laboratory results were scored on the basis of consensus statistics for both long and total fiber concentrations. Distribution of data points in a test were evaluated by several statistical tests and then, if needed, an appropriate transformation was applied. Outliers were excluded from the consensus mean. Because of EPA scoring criteria for drinking water PT [2], only those laboratories whose results fell within 95% confidence limits of the mean passed the test.

Results and Discussion

During this PT's brief history, laboratory participation has ranged from a maximum of 30 laboratories in 1994 to 23 laboratories during the last round in 1996. Thirteen different asbestos-containing samples have been distributed during this period. Two of these samples have been sent out in different rounds, constituting a total of 15 asbestos-containing samples distributed since 1992.

Mean long-fiber concentrations have ranged from 0.377 to 76.7 MFL while total fiber concentrations have ranged from 5.49 to 3340 MFL (Table 1). Ratios of long fibers to total fibers have ranged from 0.18 to 24%. Three types of asbestos have been utilized. Amosite (fibrous grunerite) was distributed in six separate samples, with total fiber concentrations ranging from 5.49 to 50.7 MFL and long-fiber concentrations ranging from 0.377 to 4.13 MFL. Table 1 lists five crocidolite samples but only four separate samples were actually prepared. The second and third crocidolite samples in Table 1 are actually from the same batch, with the second set of samples re-labeled as new samples and sent out a year after the batch's original use. Agreement between the two submissions for long fiber concentrations (1.28 versus 1.25 MFL) and for total fiber concentrations (682 versus 702 MFL) is excellent. The two chrysotile samples listed in Table 1 are likewise from a single sample batch, the first half sent in 1994 and the second half sent in 1995 under a different sample number. Agreement between the two samples for total (1470 versus 1180 MFL) and long (6.28 versus 4.76 MFL) fibers is good.

TABLE 1 — *Summary of asbestos types, consensus-mean concentrations (MFL), and standard deviations (parentheses).*

	Amosite Samples					
Total Fiber	5.49 (2.56)	28.2 (13.2)	16.9 (5.91)	19.7 (14.8)	50.7 (29.2)	46.6 (18.1)
Long Fiber	0.642 (0.397)	0.86 (0.723)	4.13 (1.39)	0.377 (0.193)	0.451 (0.14)	5.27 (1.07)
Long/Total Ratio (%)	12.	3.0	24.	1.9	0.89	11.
	Crocidolite Samples					
Total Fiber	368 (187)	682 (438)	702 (312)	20.6 (7.55)	9.52 (2.75)	
Long Fiber	3.08 (1.37)	1.28 (0.684)	1.25 (0.543)	0.769 (0.302)	1.06 (0.393)	
Long/Total Ratio (%)	0.84	0.19	0.18	3.7	11.	
	Chrysotile Samples			Crocidolite/Chrysotile Samples		
Total Fiber	1470 (827)	1180 (451)		651 (352)	3340 (1840)	
Long Fiber	6.28 (3.02)	4.76 (2.35)		18.6 (12.7)	76.7 (53.7)	
Long/Total Ratio (%)	0.43	0.40		2.9	2.3	

A single sample containing a suspension of sepiolite was distributed in 1996. Three laboratories failed this sample when they reported above-background concentrations of anthophyllite; while two other laboratories failed when they reported above-background concentrations of chrysotile.

Results generally followed a Poisson or negative binomial distribution and sometimes required transformations to fit a normal curve (Figure 2). Interlaboratory precision was fairly consistent throughout the PT. Relative standard deviations (RSD) ranged from 0.35 to 0.84, with no significant correlation between MFL and RSD. Nor was there any correlation between total and long-fiber RSD (Figure 3). This indicated a lack of systematic non-uniformity in fiber-length category within specific sample batches. Furthermore, the long-fiber vs total-fiber RSD relationship

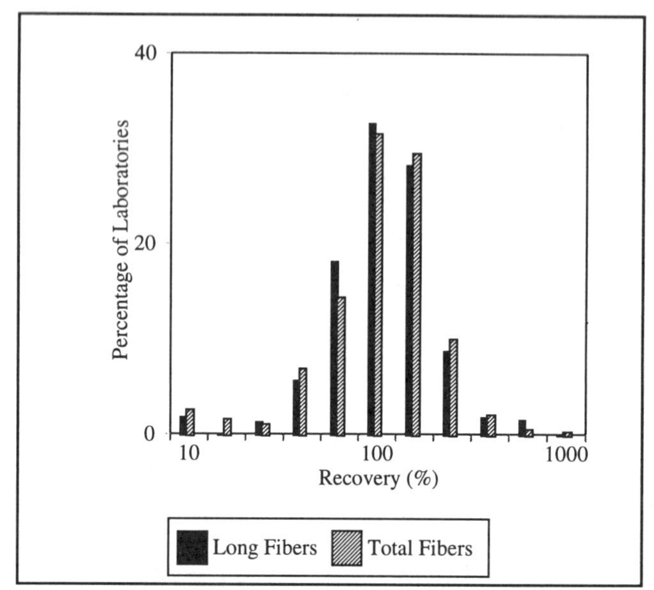

Figure 2 Distribution of normalized laboratory recoveries (100% = consensus mean) for asbestos-containing PT samples.

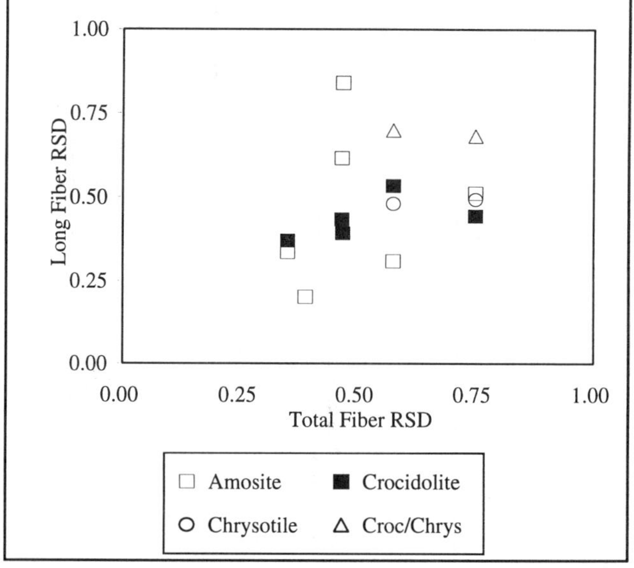

Figure 3 Relative standard deviation (RSD) of total-fiber analyses versus long-fiber analyses for each type of asbestos.

for each type appeared to be random and similar between species.

Failure rates ranged from 3.5% to 21% of participating laboratories for asbestos-containing samples (Figure 4). Laboratory performance did not appear to improve during the PT period, i.e., failure rates did not decline. No appreciable difference in failure rates for long-fiber analysis versus total-fiber analysis was observed. The failure rates for long-fiber analysis and for total-fiber analysis both averaged 11% of participating laboratories. Long-fiber failure rates exceeded total-fiber failure rates in six samples while total-fiber failure rates exceeded long-fiber failure rates in four samples.

Laboratory performance as a function of experience was statistically evaluated. For the 11 laboratories that had analyzed all 15 asbestos-containing

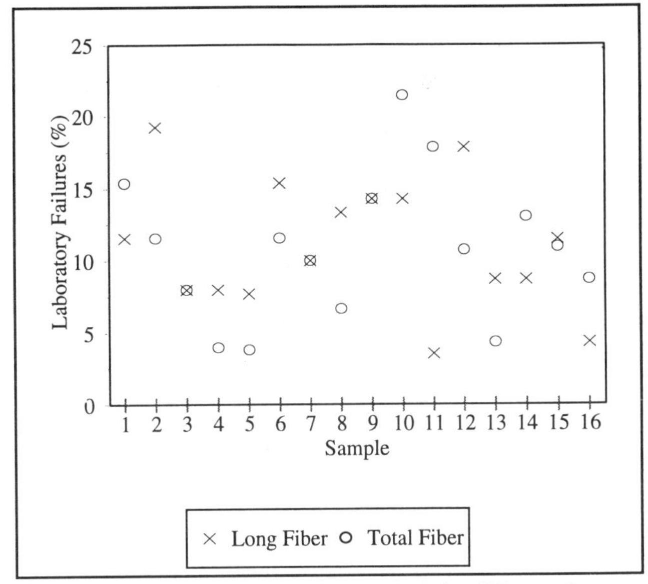

Figure 4 Failure rate for asbestos-containing PT samples.

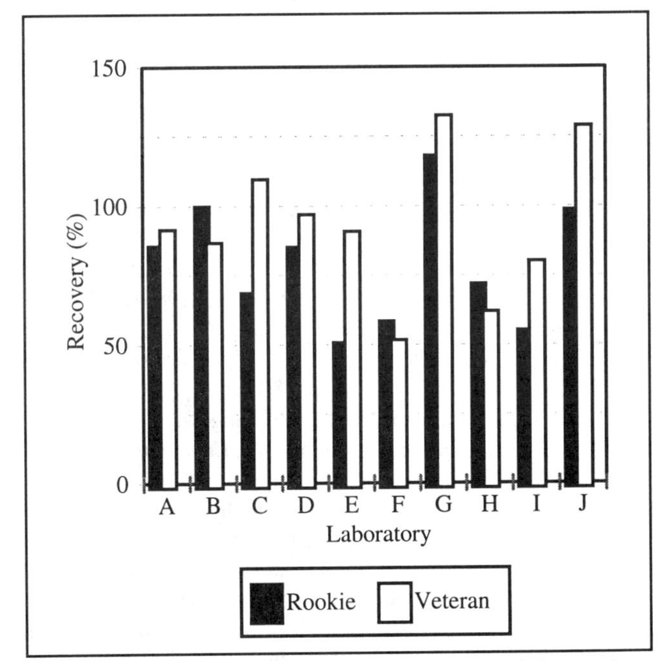

Figure 5 Recovery of asbestos as reported by ten laboratories during their "rookie" and "veteran" periods.

samples, their results were divided into two groups: rookie (first 7 samples analyzed), and veteran (last 8 samples analyzed). Each laboratory's "rookie" results versus "veteran" results were compared for significant differences as determined by the Wilcoxon paired-rank sum test [*10, 11*]. This test evaluated both the number of times one category exceeded the other as well as the magnitude of each difference.

For long-fiber analysis, these laboratories averaged 79% recovery as "rookies" versus 100% recovery as "veterans" (Figure 5), a significant difference (p=0.019). For total fiber analysis, rookie laboratories were marginally closer to the consensus mean (99% recovery versus 107% recovery), but this difference was not significant. The apparent lack of improvement for total-fiber analysis might be due to the similarity of this analysis to the analysis of airborne asbestos for abatement projects [*12*]. Laboratories have been performing such analyses for all airborne asbestos fiber sizes for about a decade, whereas analysis of exclusively long fibers began only in 1993. The lack of significance for total-fiber analyses might also be due to the smaller number of fibers typically counted for this part of the analysis. Experience of laboratories is unquestionably important in producing accurate analyses of asbestos in drinking water.

References

[*1*] Section 502 Amended, New York State Public Health Law, 1 April 1984.

[2] 40 CFR Parts 141, 142, and 143, "National Primary Drinking Water Regulations; Final Rule", *Federal Register*, Vol. 56, No. 20, 30 January 1991, pp. 3526-3597.

[*3*] Brackett, K. A., Clark, P. J., and Millette, J.R., "Method 100.2: Determination of Asbestos Structures Over 10 μm in Length," *EPA/600/R-94/134*, 1994, 21 pp.

[*4*] Chatfield, E. J., and Dillon, M. J. "Analytical Method for Determination of Asbestos Fibers in Water," *EPA-600/4-84-043*. 1983, 277 pp.

[*5*] Patel-Mandlik, K., and Millette, J., "Accumulation of Ingested Asbestos Fibers in Rat Tissues over Time," *Environmental Health Perspectives,* Vol. 53, 1983, pp. 197-200.

[*6*] Cook, P. M., and Olson, G. F., "Ingested Mineral Fibers: Elimination in Human Urine," *Science*, Vol. 204, 1979, pp.195-198.

[*7*] Webber, J. S., Covey, J. R., and King, M. V., "Asbestos in Drinking Water Supplied through Grossly Deteriorated A-C Pipe," *Journal of the American Water Works Association*, Vol. 81, 1989, pp. 80-85.

[*8*] Webber, J. S., Syrotynski, S., and King, M. V., "Asbestos-Contaminated Drinking Water: Its Impact on Household Air, *Environmental Research,* Vol. 46, 1988, pp. 153-167.

[*9*] Chatfield, E. J., "Measurements of Chrysotile Fiber Retention Efficiencies for Polycarbonate and Mixed Cellulose Ester Filters," *Advances in Environmental Measurement Methods for Asbestos, ASTM STP 1342,* M. E. Beard, H. L. Rook, Eds., American Society for Testing and Materials, 1998.

[*10*] Lehmann, E.L.: *Nonparametrics: Statistical Methods Based on Ranks.* San Francisco, Calif.: Holden-Day, Inc., 1975. Table H.

[*11*] Beyer, W.H. (ed.): *Handbook of Tables for Probability and Statistics.* 2nd ed.

Cleveland OH, The Chemical Rubber Company, p. 400, 1968.
[*12*] 40 CFR 763, "Asbestos Hazard Emergency Response Act - Detailed Procedure for Asbestos Sampling and Analysis - Non-Mandatory" *Federal Register*, Vol. 52, No. 210, 30 October 1987, pp. 41845-41905.

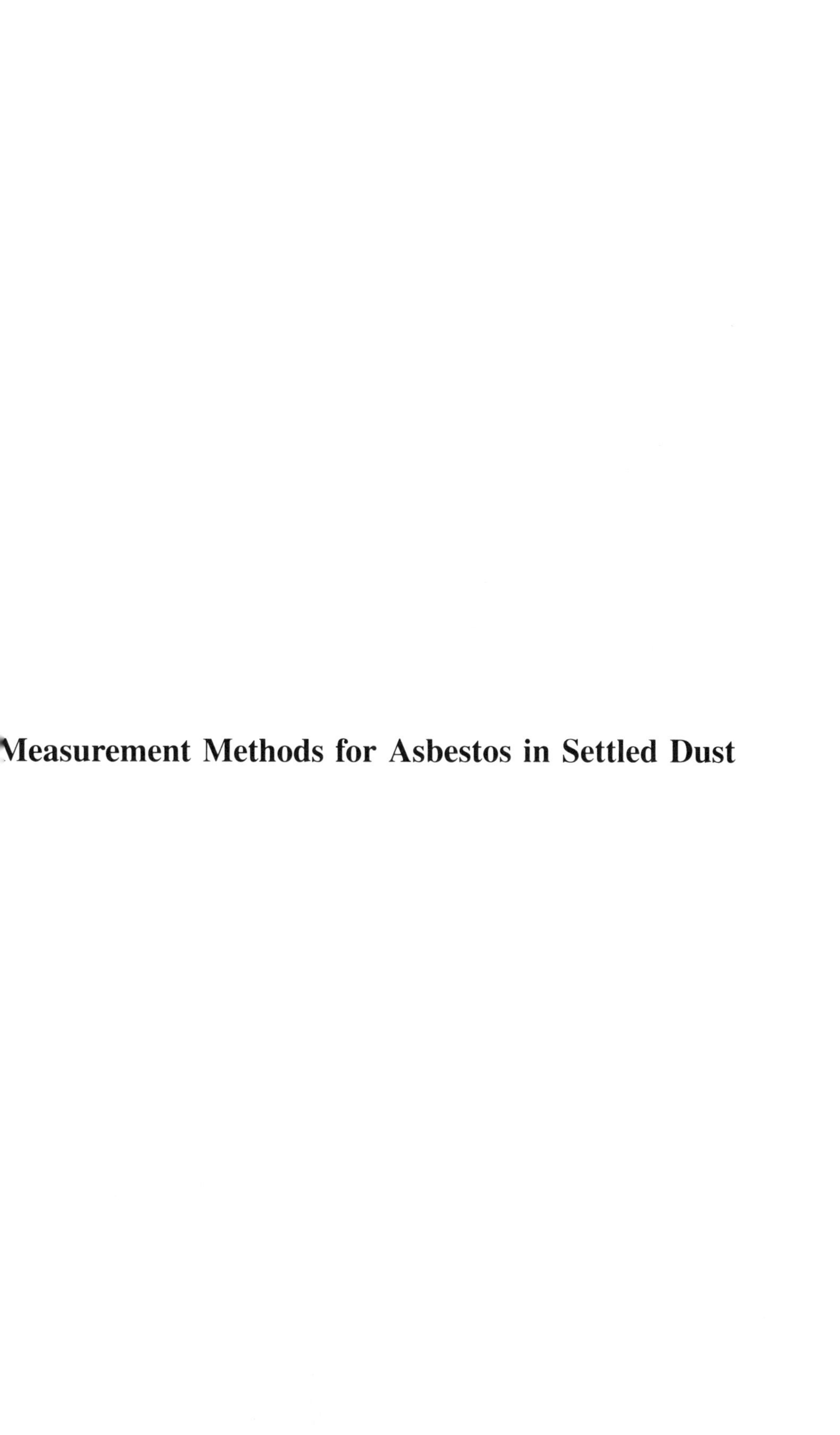

Measurement Methods for Asbestos in Settled Dust

R. L. Hatfield,[1] J. A. Krewer[2] & W.E. Longo[1]

A Study of the Reproducibility of the Micro-Vac Technique As A Tool for the Assessment of Surface Contamination in Buildings with Asbestos-Containing Materials

---

REFERENCE: R. L. Hatfield, J. A. Krewer & W. E. Longo, **"A Study of the Reproducibility of the Micro-Vac Technique As A Tool for the Assessment of Surface Contamination in Buildings with Asbestos-Containing Materials"** *Advances in Environmental Measurement Methods for Asbestos, ASTM STP 1342,* M. E. Beard, H. L. Rook, Eds., American Society for Testing and Materials, 2000.

ABSTRACT: The Standard Test Method for Microvacuum Sampling and Indirect Analysis of Dust by Transmission Electron Microscopy for Asbestos Structure Number Concentrations (D-5755-95) was balloted and passed in 1995. Estimates of the precision of this method was determined by examining historical data from a round robin laboratory study and as actual dust samples taken from buildings with asbestos fireproofing. The analysis followed the draft ASTM dust method that was in use at that time. Like most ASTM methods, the development of the D-5755-95 method involved a series of draft methods, each differing somewhat from its predecessor. While some changes were made to each draft, these changes were primarily made to the sample preparation and analysis sections of the method. No changes were made to the sections addressing sample collection. Since the dust sample data was generated using several draft methods and the final method, the differences between methods may have contributed to the data's variation. The first study consisted of dust samples that were sent to nine independent laboratories. The laboratories involved in the round robin study were provided with known weight amounts of asbestos contaminated dust that was collected from a building with in-place vermiculitic asbestos-containing fireproofing. Good correlation was found between laboratories when the number concentrations were normalized to the weight amount of the dust provided to the labs. The round robin study demonstrated good precision with an overall coefficient of variation (CV) of 0.71. Several laboratories demonstrated a CV in the range of 0.30 to 0.40. These low CV values showed a high degree of reproducibility for the analytical

---

[1] Materials Analytical Services, Inc., Suwanee, Georgia 30024
[2] Environmental Protection Division, State of Georgia, Atlanta, Georgia 30303

method and indicated little if any numerical contribution from large particles (≤ 2mm) that may have been in some samples.

A second study conducted from 1991 to 1996 examined a large database of dust samples that were collected from 38 buildings located throughout the United States. These buildings also contained a vermiculitic asbestos-containing fireproofing product. The samples were collected from surfaces beneath the fireproofing. No physical barriers or obstructions were present between the fireproofing and the surface sampled. Only buildings with at least three or more collected samples were included in the study. These sample sets were selected to control variation due to sample location, collection technique and type of asbestos-containing material (ACM) present. While samples collected within buildings did vary by more than one order of magnitude, statistical comparisons show a greater degree of variability between buildings than between samples collected within individual buildings. Additionally, this study showed good correlation with other similar data.

The two studies showed that asbestos surface dust measurements determined by the micro-vac technique are reproducible and may be used to determine asbestos surface concentrations in buildings during asbestos evaluations.

KEYWORDS: asbestos, analysis, dust, quantitative, round robin, buildings

---

Introduction

Analyzing surface dust for asbestos contamination has been of interest to consultants and industrial hygienists for some time. Early studies date back to at least 1935 when Huribut and Williams collected dust that had settled on rafters in a manufacturing plant *[1]*. They evaluated the asbestos content in the dust by polarized light microscopy (PLM) to help determine the industrial hygiene conditions in asbestos plants. Dr. Irving J. Selikoff suggested that the analysis of settled dust would help to evaluate secondary environmental exposures in homes from asbestos which was released from nearby factories *[2]*. Carter documented an early use of the microvac collection technique for asbestos surface contamination in 1970 *[3]*. Dust and fibers were vacuumed from work clothing using an air-sampling pump equipped with a filter cassette. The filters were analyzed by phase contrast microscopy (PCM) and reported as asbestos fibers per cm^2 of cloth. Asbestos consultants in the early 1980's began to collect information about the asbestos content of surface dust and debris. Various techniques were employed to collect and analyze these materials. Usually samples were gathered from surfaces by scraping or wiping up the dust and/or debris and analyzed by PLM *[4]*. This method worked well on the larger debris type samples. However, fine particulate dust samples typically would result in false negatives because PLM could not resolve most of the respirable-size asbestos structures found in dust. During this time frame the microvacuum method was beginning to be utilized to collect dust from building surfaces and most of these samples were analyzed by transmission electron microscopy (TEM) *[5]*. The TEM detected and identified even the smallest asbestos fibers *[6,7]*. This more sophisticated method was slow to be utilized during the mid 1980's because of the low availability of TEM analysis and limited understanding of how to interpret the data that was generated. Greater interest in dust analysis was generated in 1989 after the USEPA developed a draft method for evaluating surface dust to assist them in their research. This method used the micro-vac technique for sample collection and TEM for sample analysis. This method was published after further refinement in October 1995 by the American Society for Testing and Materials (ASTM) as their D5755-95 Standard Method entitled, "Microvacuum Sampling and Indirect Analysis of Dust of Transmission Electron Microscopy for Asbestos Number Concentrations". The ASTM method employed the microvacuum sampling technique and the analysis of the collected dust by TEM and this has become the method of choice for most evaluations of asbestos contaminated dust. Most commercial TEM laboratories in this country currently provide this analysis for their clients.

A few individuals still question the method's value. They doubt the method's reproducibility and accuracy as well as its usefulness in determining what asbestos concentrations in dust may pose a risk to building occupants and workers *[8]*. These criticisms have been based primarily on the logic that since this method relies on an indirect preparation procedure and ultrasound for particle disbursement that complex asbestos structures must be disaggregated causing artificially high number counts *[9]*. To address some of these questions a study of the analytical method was performed and a review of a large database of dust sample results was conducted.

A round robin study was conducted in 1991 by Lee et al. and published in 1996 that reported that ASTM D 5755-95 method may have uncontrolled positive bias and large variability *[10]*. The study used three types of samples: (1) asbestos particles (0.5 to 1mm in size), (2) single large bundles of 7M chrysotile and (3) a laboratory mixture of asbestos and Arizona road dirt. However, both the selected asbestos particles and the 7M chrysotile bundles were only measured in two dimensions. The third dimension (height) of either the single particles or the chrysotile bundles was never measured and therefore it could not be determined if each of the laboratories in the study was receiving the same amount of material. Different size particles of asbestos would have a major impact on the numerical amount of asbestos present in the samples sent to those laboratories. Because of these problems each laboratory in the Lee round robin study could have received a different concentration of asbestos structures that would lead to significant variability for the test. Additionally, the Lee study used simulated building dust manufactured with asbestos and Arizona road dirt to expand their round robin study. Using this simulated building dust was more appealing than individual uncharacterized asbestos debris particles. However, this introduced variability into the study in the form of unknown collection efficiency of each sample and unknown starting concentration of asbestos structures. These unknowns could have factored large variabilities into the final results. Round robin studies have always required that the participating laboratories receive well-characterized samples to test laboratory differences. Crankshaw [11] documented that the micro-vac method provided good variability (<15%), precision and reproducible results when the starting asbestos dust material was well characterized with known mass concentrations of asbestos and non-asbestos materials.

The ASTM dust method went through five years of balloting before it was finally published in 1995. However, from 1991 to 1995, only minor changes were made to the actual analytical protocol. These changes may have contributed to the data's variation.

This paper will describe two studies that include inter-laboratory testing of a known weight of dust collected from vermiculite based fireproofing and a summarization of nearly 200 dust samples collected from buildings across the country that were analyzed by a draft of the ASTM D5755-95 method or by the final published method.

Experimental Methods

Study 1

A large volume of settled dust was collected during an asbestos evaluation of a high-rise office building located in the southeast. The dust was collected on top of an air distribution duct located directly below vermiculitic asbestos-containing fireproofing. The round robin study was performed using subsets of this dust sample. Before the dust sample was subdivided for the participating laboratories, it was sieved through a 2x2 mm fiberglass screen to remove any large particles. This study was initiated prior to the introduction of the 1 mm particulate screening into the sample preparation section of ASTM D5755-95 method. Therefore, the introduction of some particles greater than 1 mm, but smaller than 2 mm may have contributed to the data's variation. The sample was then mixed to assure homogeneity. Small portions of the sieved dust were placed in pre-weighed glass vials and adjusted to be in a similar weight range (8 to 12 mg each). The weighed dust portions were placed into new pre-weighed, clear plastic 37 mm air cassettes with a 0.8 μm pore size MCE filter. Each cassette was re-weighed to determine the final weight of the added dust. Laboratories were provided sets of two samples during each of the two rounds. Three laboratories participated in both rounds while the host laboratory analyzed seven of these prepared samples. The laboratories were requested to perform a dust analysis according to the proposed 1991 ASTM draft method using a 1/100 dilution and AHERA TEM counting rules *[12]*. Laboratory B reported only 3 of their 4 samples. Results were normalized to the number of asbestos structures per mg of dust. Results were evaluated to determine relative differences between the laboratories as well as differences within the individual laboratories.

Results

Study 1

The results obtained from the round robin, inter-laboratory study are summarized in Table 1. The results are reported in structures per mg of dust x 10^7 in Figure 1. The pooled arithmetic mean was 3.41 x 10^7 per mg with a standard deviation of 2.43 x 10^7 and a CV of 0.71. Results for the intra-laboratory comparison are shown in Table 2. The mean results ranged from 0.92 to 5.97 x 10^7 structures per mg with a standard deviation range of 0.19 to 3.21 x 10^7, and a CV range of 0.12 to 0.76.

FIGURE 1

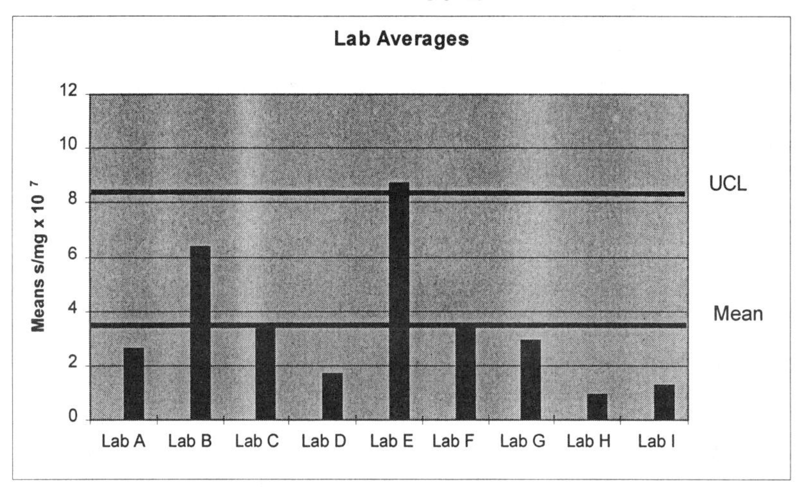

Table 1 - Inter-laboratory dust sampling analyses (asbestos structures / mg x 10^7)

Sample #	Lab A	Lab B	Lab C	Lab D	Lab E	Lab F	Lab G	Lab H	Lab I
1	2.58	3.54	2.39	1.54	11	4.17	2.34	0.73	0.59
2	3.49	8.7	4.25	1.82	6.46	2.72	2.97	1.12	1.96
3	2.69	6.9					3.75		
4	1.67						1.88		
5							2.17		
6							3.11		
7							4.22		
Arithmetic Mean	2.61	6.38	3.32	1.68	8.73	3.45	2.92	0.92	1.28
Geometric Mean	2.52	5.97	3.19	1.67	8.43	3.37	2.81	0.90	1.08
Standard Deviation	0.75	2.62	1.32	0.20	3.21	1.03	0.86	0.28	0.97
CV	0.29	0.41	0.40	0.12	0.37	0.30	0.29	0.30	0.76

Pooled by Laboratory	
Arithmetic Mean	3.41
Geometric Mean	2.75
Standard Deviation	2.44
CV	0.71

These results show when laboratories receive a known amount of well homogenized asbestos-containing dust, the results show low variability for this type of analytical method as shown by the overall CV of 0.71 for all laboratories. This study is in contrast to the Lee inter-laboratory round robin study that used somewhat uncharacterized asbestos particles and Arizona road dust samples materials, but are in better agreement with the Crankshaw work at Research Triangle Park. However, if in Lee's study, individual laboratory CV's were calculated, most would have been between 0.30 and 1.00. The major variability in Lee's study is between laboratories with one laboratory clearly an outlier.

Discussion

Study One

One of the factors that may affect the variability of an inter-laboratory round robin study is the experience a particular laboratory may have performing the micro-vac dust analysis at that time. This inter-lab study was performed in 1991. During this time many laboratories may not have had much experience with the analysis of dust samples. This lack of experience is most likely the cause of the significant differences found between laboratories in this study as shown in Table 3. The significant differences detected between the laboratories is, however, primarily driven by high variability of measurements made by Lab B and particularly Lab E.

The two laboratories believed to have the most experience in analyzing dust samples by this method in 1991 were laboratories A and G. It was estimated that each of these two laboratories had analyzed in excess of 1000 dust samples by the time of this study. The results for laboratories A and G are nearly identical and each lab had a CV of 0.29.

Table 3 – *Anova: Single Factor*

SUMMARY

Groups	*Count*	*Sum*	*Average*	*Variance*
Lab A	4	10.43	2.61	0.555
Lab B	3	19.14	6.38	6.859
Lab C	2	6.64	3.32	1.730
Lab D	2	3.36	1.68	0.039
Lab E	2	17.46	8.73	10.306
Lab F	2	6.89	3.45	1.051
Lab G	7	20.44	2.92	0.733
Lab H	2	1.85	0.92	0.077
Lab I	2	2.55	1.28	0.937

ANOVA

Source of Variation	*SS*	*df*	*MS*	*F*	*P-value*	*F crit*
Between Groups	114.801	8	14.350	7.192	0.000338	2.548
Within Groups	33.922	17	1.995			
Total	148.723	25				

Additionally, there have been some studies suggesting that the use of the ultrasonic water bath in the sample preparation would cause the uncontrolled breakdown of large (0.5 to 2.0 mm diameter) asbestos-containing particles in the dust sample, thereby producing anomalistic concentrations of asbestos structures [8,10]. Since the preparation of the test samples for this study included some particles as large as 2mm in diameter, following the theories proposed in the above cited papers, the expected inter-sample variations would be high. Each of the laboratories in this study used their own ultrasonic water bath for sample preparation. The make, model, and power setting of the bath was left up to the laboratory. If the ultrasonicator caused uncontrolled asbestos structure disruption, the coefficient of variation should have been much higher than the CV of 0.71 determined in this study. Also, the asbestos structure concentration range should have been found to be several orders of magnitude instead of the one order of magnitude observed. This study demonstrated low intra-lab variations and low overall variations. Therefore, there does not appear to be any appreciable contribution from randomly occurring particles as large as 2mm in diameter. The findings of this study are in agreement with an Environmental Protection Agency Study [13] which found that the indirect method does not cause asbestos structure breakup. Its main affect is to homogenize the particle suspension and provide a better particle distribution on the filter. This in turn provides for a more precise measurement of asbestos air and dust concentrations. To further improve the precision of the method, the current ASTM D5755-95 method instructs the sampler not to include particles greater than 1mm in diameter in the sampled area. The sample preparation section provides for the screening out of particles larger than 1 mm. Therefore, little concern should be made of this issue in the future.

Experimental Methods

Study 2

From 1991 to 1996, a large number of surface dust samples have been collected and evaluated. From this database of sample results, a subset of samples was selected from buildings which contained spray applied vermiculite based asbestos-containing fireproofing materials on structural components. Only buildings in which at least three or more samples were collected were included in this study. These samples were collected from areas either above the ceiling or in rooms with no ceilings or other obstructions which might interfere with the dust fallout from the fireproofing. The selected samples were collected primarily by one individual and were analyzed by the same laboratory in general accordance with a draft or the final version of ASTM D5755-95 method.

Results

Study 2

The results of 195 dust samples collected in 38 different buildings are summarized in Table 5. Table 6 displays the results of dust samples collected from six buildings that contain non-asbestos spray-applied fireproofing. Table 5 summarizes the 195 sample results collected from 38 buildings with spray-applied vermiculitic asbestos-containing fireproofing. The fireproofing contains approximately 10 to 12 percent chrysotile asbestos. These samples were collected from horizontal surfaces located in the ceiling plenum space or from mechanical or storage spaces that did not have ceilings or major obstructions between the fireproofing and the sample locations. Typically, these areas were dusty and frequently fireproofing debris was visible in nearby locations. While debris pieces (< 1 mm in diameter) were not collected, reasonable quantities of settled dust were easily collected from the surfaces. Since dust sample concentration data, like air concentration data, is most likely distributed in a log-normal manner, the geometric means along with the arithmetic means are presented.

A single factor Analysis of Variance (ANOVA) was performed on the data to determine if there were significant differences between buildings. The analysis of variance resulted in a significant F test indicating differences in asbestos dust concentrations between buildings.

While the ANOVA indicated that the data should not be combined, the summary in Table 5 was made to compare to other studies' results. It is also useful to determine an overall coefficient of variation (CV) for the method. The weighted geometric means for all 195 samples is 10.3 million s/cm^2 and the corresponding arithmetic mean is 15.8 million s/cm^2. The pooled standard deviation is 14.8 million s/cm^2 and the weighted CV is 0.97. This 0.97 CV included the recognized variations between buildings. The CV of 97% appears to be as good as if not better than the 100 to 150% EPA [14] estimated for direct analysis of air samples. This lower CV for dust analysis may be partly due to the indirect preparation of the samples, which provides for a more homogeneous distribution of the dust particles on the filter resulting in a more precise measurement.

TABLE 5- Summary of Dust Sample Results from 38 Buildings (structures/cm^2 x 10^6)

Building Number	No. of Samples	Geometric Mean	Arithmetic Mean	Standard Deviation	CV	Range
1	11	4.091	20.089	42.904	2.136	1.6 - 114.9
2	8	5.031	11.436	8.407	0.735	0.143 - 21.9
3	6	1.590	3.456	3.937	1.139	0.198 - 10.1
4	4	17.794	35.240	44.992	1.277	4.58 - 101
5	6	0.083	0.209	0.316	1.513	0.006 - 0.2
6	4	6.905	13.225	15.468	1.170	1.2 – 35.5

TABLE 5- Summary of Dust Sample Results from 38 Buildings (structures/cm^2 x 10^6) (Continued)

Building Number	No. of Samples	Geometric Mean	Arithmetic Mean	Standard Deviation	CV	Range
7	7	0.495	1.347	2.223	1.651	0.19 - 6.3
8	4	3.465	6.500	9.135	1.405	1.7 – 20.2
9	4	4.371	4.575	1.717	0.375	3.4 - 7.1
10	9	13.078	21.283	15.706	0.738	0.947 - 49
11	6	1.470	1.705	1.085	0.636	0.073 - 3.7
12	4	2.046	3.881	3.924	1.011	0.222 - 9.4
13	4	9.732	20.825	26.892	1.291	1.3 - 60.6
14	6	5.486	28.950	47.969	1.657	0.313 - 37.7
15	3	2.714	2.833	0.961	0.339	1.8 - 3.7
16	6	14.464	19.583	16.847	0.860	7.2 - 41.9
17	6	3.135	5.392	4.609	0.855	0.35 - 11.1
18	6	8.895	12.433	9.063	0.729	1.6 - 25.8
19	8	12.509	14.538	5.958	0.410	2.4 - 20.5
20	5	3.500	4.640	3.993	0.861	1.1 - 11.4
21	6	8.497	17.905	19.454	1.087	0.929 - 49.4
22	3	15.901	32.400	43.923	1.356	5.9 - 83.1
23	5	8.886	12.100	11.384	0.941	3.7 - 31.6
24	5	15.523	16.940	7.611	0.449	7.7 - 28.1
25	4	14.746	19.675	14.301	0.727	4.4 - 32.7
26	3	6.861	8.600	7.299	0.849	3.8 - 17
27	5	65.282	67.880	20.544	0.303	43.9 - 93.9
28	6	5.135	7.617	6.909	0.907	1.9 - 17.4
29	3	67.405	67.633	6.801	0.101	60.9 - 74.5
30	5	11.492	19.840	23.353	1.177	2.3 - 60.4
31	5	20.181	23.360	16.665	0.713	13.9 - 53
32	4	6.718	9.750	8.713	0.894	1.6 - 22
33	3	18.164	51.500	77.529	1.505	5 - 141
34	5	8.953	11.240	7.384	0.657	2.7 - 20.7
35	4	9.863	9.975	1.750	0.175	8.3 - 12.2
36	3	9.135	13.567	12.646	0.932	2.6 - 27.4
37	3	2.860	4.844	5.713	1.179	0.933 - 11.4
38	6	9.574	13.300	10.488	0.789	1.8 - 31.6
Buildings Total	Sample Total	Weighted Geo. Mean	Weighted Mean	Pooled Std. Dev.	Weighted CV	Overall Range
38	195	10.271	15.826	14.845	0.972	0.006 -114.9

Table 6 summarizes dust sample data collected from 4 buildings that did not contain any major applications of asbestos-containing materials such as fireproofing. In only 3 of 15 samples were any asbestos structures counted. In each case, only a single asbestos structure was counted. This is an insufficient count to conclude asbestos was present in the samples. These results demonstrate that the source of the asbestos contamination found in the dust from buildings with asbestos fireproofing is from the in-place ACM and not from some source outside the building.

TABLE 6 - *Dust Sample Results From Non-ACM Buildings*
Structures/cubic centimeter

Sample #	Building 1	Building 2	Building 3	Building 4
1	ND*	ND	ND	ND
2	ND	ND	ND	ND
3	ND		ND	ND
4	557		1300	
5	1300			
6	ND			

* Not Detected

Of the 35 field blank samples associated with 195 dust samples collected in all of the buildings, only 3 were found with 1 fiber counted and in 1 sample, two asbestos fibers were counted during the analysis.

Discussion and Conclusions

Study Two

The geometric mean of asbestos concentrations in the 38 buildings ranged from 83 thousand s/cm^2 to 67.4 million s/cm^2. The arithmetic mean ranged from 209 thousand s/cm^2 to 67.8 million s/cm^2. The lowest concentration on any of the surfaces was 6 thousand s/cm^2 and the highest concentration was 141 million s/cm^2.

In 1996 William Ewing and others [15] published similar data for surface dust samples collected in buildings with fireproofing. This data contained buildings with either vermiculite or mineral fiber based asbestos-containing fireproofing. The geometric mean of asbestos concentrations from 47 different buildings ranged from 17 thousand s/cm^2 to 52 million s/cm^2 . The arithmetic mean ranging from 19 thousand s/cm^2 to 76 million s/cm^2. The lowest measured asbestos concentration was 1 thousand s/cm^2 and the highest was 220 million s/cm^2.

It appears that in both studies a similar range of asbestos concentrations and building average concentrations was found. All of the upper and lower ranges for sample concentration and averages were in the same order of magnitude. The exception was the lower limit of the arithmetic mean that differed by one order of magnitude.

The range of asbestos concentrations found between buildings and, to a lesser degree within buildings, is due to a variety of factors. Certainly among these factors are the condition of the fireproofing material, past activities which have affected the fireproofing and any past cleaning of the surfaces which were sampled. Generally, dust collected in areas where asbestos-containing materials are in poor condition or where activities have impacted the materials will contain higher asbestos concentrations than dust collected in areas where materials appear in good condition and have little evidence of contact.

Maybe more important to an investigator than the absolute quantity of asbestos present in the dust is the location of the asbestos contaminated dust and its likelihood of disturbance. In many past studies the reentrainment of asbestos-laden dust from housekeeping, maintenance and renovation activities has been demonstrated. The resulting asbestos air concentrations from these reentrainment activities give rise for concern.

References

[1] Huribut, C.S., and Williams, C.R., "The Mineralogy of Asbestos Dust", *Journal of Industrial Hygiene*, Vol. 17, p. 289, 1935.

[2] Selikoff, I.J., and Nicholson, W.J., "Mortality Experience of Residents in the Neighborhood of an Asbestos Factory," Health Hazards of Asbestos Exposure, Annals of the New York Academy of Science, Vol. 330, p. 417, 1979.

[3] Carter, R.F., The Measurement of Asbestos Dust Levels in a Workshop Environment, Technical Report 028-70, United Kingdom Atomic Energy Authority, Aldermaston, United Kingdom, 1970.

[4] Nichols, G., "Scotch Magic Tape – An Aid to the Microscopist for Dust Examination," Microscope, Vol. 33, p. 247, 1985.

[5] Millette, J.R., Kremer, T., and Wheels, R.K., "Settled Dust Analysis Used in Assessment of Buildings Containing Asbestos," Vol. 38, p. 215, 1990.

[6] Small, J.A., Steel, E.B., and P.J. Sheridan, "Analytical Standards for Analysis of Chrysotile Asbestos in Ambient Environments," Analytical Chemistry, Vol. 57, p. 204, 1985.

[7] Steel, E.B., and Small, J.A., "Accuracy of Transmission Electron Microscopy for the Analysis of Asbestos in Ambient Environments," Analytical Chemistry, Vol. 57, p. 209, 1985.

[8] Lee, R.J., Dagenhart, T.V., Dunmyre, G.R., Stewart, L.M., and Van Orden, D.R., "Effect of Indirect Sample Preparation Procedures on the Apparent Concentration of Asbestos Settled Dusts," Environmental Science and Technology, Vol. 29, p. 1728, 1995.

[9] Sahle, W., and Laszlo, I., "Airborne Inorganic Fiber Level Monitoring by Transmission Electron Microscopy (TEM)" Comparison of Direct and Indirect Sample Transfer Method," Annals of Occupational Hygiene, Vol. 40, p. 29, 1996.

[10] Lee, R.J., Van Orden, D.R., and Dunmyre, G.R., "Interlaboratory Evaluation of the Breakup of Asbestos-Containing Dust Particles by Ultrasonic Agitation," Vol. 30, p. 3010, 1996.

[11] Crankshaw, O.S., Perkins, L.R., and Beard, M.E., "Quantitative Evaluation of the Relative Effectiveness of Various Methods for the Analysis of Asbestos in Settled Dust," Environmental Information Association Technical Journal,Vol. 4, p. 6, 1996.

[12] USEPA, "Asbestos-Containing Materials in School: Final Rule and Notice," Federal Register, 40CFR Part 763, Appendix A to Sub-part E, October 30, 1987.

[13] USEPA, "Comparison of Airborne Asbestos Levels Determined by Transmission Electron Microscopy (TEM) Using Direct and Indirect Transfer Techniques," EPA 560/5-89-004, Washington, DC, 1990.

[14] USEPA, "Measuring Airborne Asbestos Following an Abatement Action," EPA 600/4-85-049, Washington, DC, 1985.

[15] Ewing, W.M., Dawson, T.A., and Alber, G.P., "Observations of Settled Asbestos Dust in Buildings," Environmental Information Association Technical Journal, Vol. 4, p. 13, 1996.

Richard J. Lee[1], Drew R. Van Orden[2], Ian M. Stewart[3]

DUST AND AIRBORNE CONCENTRATIONS - IS THERE A CORRELATION?

Lee, R. J., Van Orden, D. R., and Stewart, I. M., " **Dust And Airborne Concentrations - Is There A Correlation?**", *Advances in Environmental Measurement Methods for Asbestos, ASTM STP 1342,* M. E. Beard, H. L. Rook, Eds., American Society for Testing and Materials, West Conshohocken, PA, 2000.

ABSTRACT: An extensive data set of air and dust samples collected in residential properties has been evaluated to determine whether any correlation exists between airborne concentrations of asbestos and asbestos concentrations measured in settled dust. Interior airborne concentrations were higher than outdoor concentrations, but there was no significant difference between levels detected in the presence of damaged ACM and those detected where no ACM was present. Airborne levels did not correlate with asbestos concentrations in the dust as measured by the indirect TEM methods. The presence of asbestos in the surface dust is shown to be independent of the presence of asbestos in bulk samples collected from the residences. The presence of airborne asbestos is independent of the presence of asbestos in surface dust. Thus, observing asbestos in a dust sample does not imply that asbestos will be found in an air sample nor does it imply that asbestos will be found at some predictable concentration.

KEYWORDS: airborne asbestos, surface dust, correlation, residence

Introduction

The possible risk to human health resulting from the presence of asbestos-containing materials (ACM) in buildings has prompted widespread public and governmental concern. Coupled with the US Environmental Protection Agency's Asbestos Hazard Emergency Response Act (AHERA) [*1*], National Emission Standards for Hazardous Air Pollutants (NESHAP) [2] regulations and the ban on the

[1] President, RJ Lee Group, Inc., 350 Hochberg Road, Monroeville, PA 15146
[2] Manager, Quality Assurance, RJ Lee Group, Inc., 350 Hochberg Road, Monroeville, PA 15146
[3] Vice-President, Analytical Services, RJ Lee Group, Inc., 350 Hochberg Road, Monroeville, PA 15146

manufacturing of ACM, this concern has spawned a large industry involved in removing asbestos from buildings. Recently there have been suggestions that occupants of the buildings are exposed to hazardous airborne levels of asbestos resulting from spontaneous emission of asbestos fibers from the in-place building product or disturbance of the product or surface dust containing fibers and particulate produced during routine housekeeping, maintenance or renovation.

Extensive testing of the air in buildings has been performed as a way of establishing the airborne concentrations of asbestos in buildings under conditions of normal occupancy. These studies [*3, 4*] have shown buildings have average concentrations of 0.00013 f/ml for asbestos fibers ≥ 5 μm. Even in buildings sampled immediately after an earthquake, concentrations averaged only 0.00043 asbestos f/ml [*5*], compared with the OSHA Permissible Exposure Level (PEL) of 0.1 f/ml 8-hour TWA, predicated on exposure over an entire working life.

More recently, although air concentrations in buildings have been found to be very low, studies of the surface dust and debris found in a building have been used to support evidence of building "contamination". The numbers generated by these analyses are alleged to represent a history of exposure and a potential for future release of asbestos from the product as well as the dust and debris into the air.

Part of the problem with the use of dust to evaluate the potential for airborne exposures relates to the use of an indirect preparation technique that is used on the microvacuum dust samples. The large, positive bias resulting from indirect preparation is widely recognized but poorly understood. It has been shown [*6*] that increasing sonication times result in higher apparent asbestos fiber concentrations as a result of physical release of asbestos from matrices and from the comminution of the fibers into more numerous, shorter and thinner fibrils. Additional studies [*7, 8, 9*] confirmed this result on samples of airborne asbestos. A secondary issue is that the smaller the particle, the harder it is to reentrain it into the air from surface dust.

Recently, RJ Lee Group participated in the recovery effort following the 1995 Northridge earthquake. As part of our work, air and dust samples were collected from a number of locations throughout residences. The results of this large body of data offer an opportunity to examine whether there is any correlation between airborne asbestos and surface dust concentrations. This report contains a summary of these data.

Source of Samples

Samples of air and surface dust were collected from 142 homes in the Los Angeles area. These residences all claimed damage to the structure resulting from the Northridge earthquake. All samples were collected by California Certified Asbestos Consultants under the supervision of RJ Lee Group personnel, and in accordance with normal industrial hygiene practices. The residents of the homes were allowed to follow their normal daily patterns with the request that they do not disturb the air pumps. In general, between three and eight people were present during the inspection and sampling.

Bulk Samples

Bulk samples were taken only of building materials that either were damaged or were expected to be damaged during repair work. The samples were collected to determine if the material meets the definition of asbestos-containing material (ACM, >1% for Federal requirements, >0.1% under California regulations). These homes have been divided into 3 groups based on the results of these analyses using the following definitions: 1) no ACM detected in any bulk sample (No ACM); 2) asbestos was detected in at least 1 bulk sample in the building, but not in ceiling material (Other ACM) ; and 3) asbestos was detected in samples of ceiling material (Ceiling ACM). Included in the second group are homes that may have ACM ceilings, but where the ceiling was not sampled because the material was either not damaged or expected to become damaged. Groups 2 and 3 were combined into a fourth category of ACM homes. There were 28 homes in group 1, 50 in group 2, and 64 in group 3.

Air Samples

Over 1300 air samples were collected in the homes to use in the evaluation of the potential risk to the occupants during periods of normal activity. Area air samples were collected in several locations throughout the home and outside of the home. At each location, the air was sampled at two elevations - one roughly at the breathing height for adults (5 ft), the other at breathing height for small children (1.5 ft). In addition to the area samples, personal samples were collected from the survey crew. The samples were analyzed using PCM (NIOSH 7400) and TEM (NIOSH 7402) methods.

Surface Dust

Surface dust was collected from the various homes using two different sampling methods - adhesive lift sampling and microvacuum sampling. There were 2000 adhesive lift and 900 microvacuum samples collected. Adhesive lift samplers[4] collect the dust in a manner that allows the spatial distribution and the form of the dust to be examined [*10*]. Microvacuum samplers collect the dust into a cassette and in a form that requires the dust to undergo an intermediate preparation step prior to examination. The sampling protocol required both samples to be collected in proximity to each other.

The adhesive lift samples (Lifts) were examined using phase contrast microscopy (PCM, NIOSH 7400) and scanning electron microscopy (SEM) following the procedure described in VDI 3492 (Verein Deutscher Ingenieure; messen anorganischer faserfömiger Partikel in der Außenluft: Rasterelektronenmikroskopisches Verfahren). PCM and SEM were used to determine the fiber content of the samples.

The microvacuum samples were prepared for analysis following either ASTM Standard Test Method for Microvacuum Sampling and Indirect Analysis of Dust by Transmission Electron Microscopy for Asbestos Structure Number Concentrations (D 5755) or ASTM Standard Test Method for Microvacuum Sampling and Indirect Analysis of Dust by Transmission Electron Microscopy for Asbestos Mass Concentration (D

[4] A standard test method is being developed by ASTM Committee D22.07 as item Z3505Z.

5756). The samples were analyzed to determine the concentration of asbestos structures ≥ 0.5 μm long, asbestos fibers ≥ 5 μm long, and the weight percent of asbestos. For some samples, because of the very light loading of dust in the cassette, the weight percent of asbestos could not be determined.

Statistical Analysis

The data were analyzed to determine the percentiles, median, and mean of the as-reported data. Nonparametric statistical procedures (Mann-Whitney U, Wilcoxon signed rank, Spearman rank correlation, Kruskal-Wallis, and Chi-square tests) were used to compare the data. The Spearman rank correlation coefficient is interpreted in a manner similar to the more familiar Pearson product-moment correlation coefficient. It ranges from -1 to 1 and is an estimate of the strength of the relationship between two variables in the sampled population. The Chi-square test is another test used to determine whether two variables are associated, in this report, to test whether the variables are independent. For this test, the variables are cross-classified according to two criteria, so that each observation belongs to one and only one level of each criterion. The tables were created to compare the presence of asbestos in air and dust samples. In any room, if either air sample contained asbestos, then the air data was classified as positive for the presence of asbestos. Similarly, if any dust sample contained asbestos, then the dust data was classified as positive for asbestos. Nonparametric tests were chosen because the true distribution of the data was unknown.

Results

Summaries of the air sample concentrations are contained in Table 1. A comparison of the interior and exterior samples showed that, although virtually all samples had very low concentrations, in general, the interior samples had statistically higher concentrations than the exterior samples (p=0.0003). When considering the homes for which the bulk samples were negative for asbestos, the interior and exterior samples analyzed by TEM were not statistically different (p=0.0569). This appears to be an effect of the number of samples in this set, since there was also no statistical difference in the airborne concentrations in homes containing damaged ACM and that in homes containing no ACM. There was also no statistical difference in the airborne concentrations in homes with damaged ACM in the ceilings and homes with no ACM in the ceilings. There was also no statistical difference between the upper and lower elevation samples (p=0.9).

A summary of the adhesive lift dust samples is shown in Table 2. There were statistically significant differences in the total fiber concentrations (as measured by either PCM or SEM) for any comparison of buildings classified on basis of the bulk PLM results (for example, ACM vs. Non-ACM homes, p=0.0095 for PCM concentrations). There were no statistically significant differences between ACM and Non-ACM homes when considering the SEM asbestos concentration (p=0.2560). A Chi-square test (p=0.3028) found the presence of asbestos in the dust to be independent of the presence of ACM in the building materials.

TABLE 1 — Summary of the concentration of survey air samples, f/ml

Location	Analysis	# Samples	Elevation	10 %-ile	Median	Mean	90 %-ile
Personal	PCM	221		0.0063	0.0300	0.0455	0.0955
	TEM	221		0	0	0.0001	0
Interior	PCM	747	All	0.0027	0.0078	0.0116	0.0213
		382	Upper	0.0025	0.0078	0.0129	0.0224
		365	Lower	0.0029	0.0079	0.0102	0.0204
	TEM	747	All	0	0	0.0001	0
		382	Upper	0	0	0.0001	0
		365	Lower	0	0	0.0001	0
Exterior	PCM	295	All	0.0005	0.0015	0.0019	0.0032
		151	Upper	0.0004	0.0016	0.0018	0.0034
		144	Lower	0.0006	0.0015	0.0019	0.0032
	TEM	295	All	0	0	<0.0001	0
		151	Upper	0	0	<0.0001	0
		144	Lower	0	0	<0.0001	0

TABLE 2 — Summary of fiber concentration from adhesive lift dust samples, f/cm^2

Location	Method	# Samples	10 %-ile	Median	Mean	90 %-ile
Interior	PCM	832	191	637	1,090	2,390
	SEM (all fibers)	1837	76.6	1150	1,670	4,100
	SEM (asbestos)	1837	0	0	10.7	0
Exterior	PCM	84	153	383	530	1,090
	SEM (all fibers)	181	45.9	958	1,560	3,510
	SEM (asbestos)	181	0	0	6.8	0

The results of the microvacuum dust samples are summarized in Table 3. There were no statistically significant differences between interior and exterior samples (lowest p-value was p=0.2434 for D 5755 fibers ≥ 5 μm). The presence of asbestos in the dust was found to be statistically independent of the presence of asbestos in the bulk building materials (p>0.05 for all analyses).

Tables 4 and 5 contain the results of paired correlations of dust concentrations with air concentrations for the adhesive lifts and microvacuum samples, respectively. The largest correlation (0.306) in the adhesive lift samples occurs in homes with no ACM in the building materials. No correlation in the microvacuum data exceeds 0.3 in magnitude.

Chi-square tests were conducted to determine if the presence of asbestos in the air is independent of the presence of asbestos in the surface dust. The p-values from these tests are summarized in Table 6. A p-value of < 0.05 indicates the variables are not independent. Only with the adhesive lifts does the test show some dependence, due primarily to samples collected from homes with ceiling ACM.

TABLE 3 — Summary of TEM analysis of microvacuum samples

Samples	Preparation	Units	# Samples	10 %-ile	Median	Mean	90 %-ile
Interior	D 5755	s/cm^2	489	0	0	1,130	1,340
		f/cm^2	485	0	0	69.6	0
		Wt %	479	0	0	0.0440	0.0006
	D 5756	s/cm^2	199	0	1,770	17,800	41,600
		f/cm^2	146	0	0	3,100	4,910
		Wt %	196	0	0.0002	0.0078	0.0120
Exterior	D 5755	s/cm^2	68	0	0	2,360	1,300
		f/cm^2	66	0	0	16.4	0
		Wt %	68	0	0	0.0112	0.0031
	D 5756	s/cm^2	88	0	2,410	10,600	18,600
		f/cm^2	67	0	0	1,920	2,430
		Wt %	88	0	0.0002	0.0026	0.0042

s/cm^2 - asbestos concentration for structures ≥ 0.5 μm in length
f/cm^2 - asbestos concentration for fibers ≥ 5 μm in length
One Exterior sample, analyzed by D 5756 and not included in this table, contained 8,342,000 s/cm^2, 802,000 f/cm^2, and 0.073 wt.%.

Further statistical tests were conducted using the average of all dust samples in each home. Based on the average dust results, homes having damaged ACM in ceilings had higher dust concentrations than homes without ACM in the ceilings, and both were statistically higher than homes with no ACM.

Discussion

The subject of asbestos in dust has received a great deal of attention in recent years. This study is the first to evaluate a large set of data where simultaneous measurements of airborne dust concentrations and surface dust concentrations are compared. It is also the first to make the comparison in circumstances where damaged ACM was in the active living space, thus creating a potential for on-going release of fibers and ACM particulate. In addition, the study compared airborne levels of asbestos in the breathing zone of adults with that in the breathing zone of children, since the potential exists that a significant gradient in airborne concentration could exist if surface dust was indeed being entrained. The study demonstrates that the presence and/or the amount of asbestos in dust, as measured by indirect sample preparation techniques, is not a predictor of airborne concentrations.

TABLE 4 — Spearman rank correlations for air samples and adhesive lift dust samples

Building Class	Air Analysis Level	Air Analysis Measure	Adhesive Lift Measurement PCM	SEM (all fibers)	SEM (asbestos)
All Buildings	Upper	PCM	0.270	0.043	0.056
		TEM	-0.092	0.047	0.098
	Lower	PCM	0.221	0.046	0.078
		TEM	0.060	0.037	0.189
No ACM	Upper	PCM	0.235	-0.057	0.041
		TEM	-0.021	0.010	-0.045
	Lower	PCM	0.159	-0.058	-0.034
		TEM	0.236	0.034	0.306
ACM	Upper	PCM	0.264	0.058	0.053
		TEM	-0.098	0.048	0.111
	Lower	PCM	0.225	0.061	0.083
		TEM	-0.004	0.029	0.176
Ceiling ACM	Upper	PCM	0.244	0.104	0.027
		TEM	-0.025	0.100	0.188
	Lower	PCM	0.200	0.124	0.031
		TEM	0.037	0.120	0.244
Other ACM	Upper	PCM	0.309	-0.006	0.086
		TEM	-0.175	-0.021	-0.020
	Lower	PCM	0.294	-0.034	0.152
		TEM	-0.046	-0.147	0.010

The presence of damaged ACM in occupied living space is not correlated with airborne asbestos concentrations in either adult or child breathing zones. Thus, the combined effects of entrainment of surface dust and shedding from damaged ACM are negligible. While there is some evidence that, for adhesive lift samples, the presence of asbestos in the air is dependent of the presence of asbestos in the dust, this is a casual relationship at best.

It had been suggested that airborne concentrations of asbestos may not be homogeneously distributed throughout a room, but may exist as a density gradient extending from the source of the asbestos. Air samples collected at two different heights were found to statistically be the same. There were no differences in the correlation coefficients of air and dust concentrations for the two sample elevations. Except for the data from the adhesive lift samples, the presence of asbestos in the air is independent of the presence of surface dust, regardless of the elevation of the air sample.

TABLE 5 — Spearman rank correlations for air samples and microvacuum dust samples

Building Class	Air Analysis Level	Air Analysis Measure	D 5755 s/cm^2	D 5755 f/cm^2	D 5755 Wt%	D 5756 s/cm^2	D 5756 f/cm^2	D 5756 Wt%
All Buildings	Upper	PCM	-0.021	-0.007	-0.021	0.01	-0.045	0.048
		TEM	-0.021	0.044	0.007	0.099	0.007	0.048
	Lower	PCM	-0.038	-0.023	-0.020	0.006	-0.029	0.043
		TEM	0.034	0.006	0.047	0.072	-0.021	0.047
No ACM	Upper	PCM	0.083	0.019	0.026	0.189	0.042	0.134
		TEM	-0.037	0.096	-0.094	0.120	-0.009	0.002
	Lower	PCM	0.081	0.001	0.173	0.287	0.194	0.147
		TEM	-0.089	-0.057	-0.082	0.082	0.000	0.082
ACM	Upper	PCM	-0.045	-0.015	-0.029	-0.029	-0.056	0.026
		TEM	-0.262	0.023	0.140	0.099	0.009	0.053
	Lower	PCM	-0.067	-0.029	-0.030	-0.038	-0.048	0.024
		TEM	0.061	0.040	0.081	0.071	-0.031	0.038
Ceiling ACM	Upper	PCM	-0.002	0.027	-0.001	-0.064	-0.106	0.016
		TEM	0.121	0.094	0.157	0.145	-0.004	0.113
	Lower	PCM	0.015	0.044	0.031	-0.055	-0.065	0.020
		TEM	0.083	0.071	0.103	0.080	0.039	0.047
Other ACM	Upper	PCM	-0.107	-0.105	-0.075	0.004	-0.039	0.016
		TEM	-0.152	-0.050	-0.144	-0.012	0.064	-0.084
	Lower	PCM	-0.197	-0.195	-0.132	-0.035	-0.081	-0.023
		TEM	0.025	-0.037	0.039	0.008	-0.142	-0.032

Of the 161 data pairs used for this correlating D 5756 (f/cm^2) and air data, only five contained asbestos in both the air and dust analyses. For these five samples, the correlation was 0.872 ($p=0.0811$), which is a high correlation (though not statistically significant). However, this result can only be obtained by ignoring the other 97% of the data where there is no correlation, obviously an improper statistical procedure [*11*].

The study used two different techniques to collect the dust samples, adhesive lifts and microvacuuming. No statistical analyses were performed to compare the results of the two collection techniques, or to compare the results from the two analytical techniques for the microvacuum samples D 5755 and D 5756. Because of the differing preparation techniques, it is not surprising that the results vary between sampling techniques and between analytical techniques. The relative merits of the indirect preparation methods have been discussed elsewhere [*6, 9, 12*].

In a subset of homes, in addition to the collection of air and dust samples, aggressive simulations of household activities; cleaning and vacuuming which would occur for shorter periods of time, were performed. The results of these studies, which will be described in detail elsewhere, showed that airborne levels are slightly increased during these activities but remain far below the OSHA PEL. The airborne levels during such activity did not correlate with the results of indirect dust measurements.

TABLE 6 —P-values from the contingency testing of the presence of asbestos in dust with the presence of asbestos in air

Building Class	Air Samples	Lifts	D 5755			D 5756		
		f/cm^2	s/cm^2	f/cm^2	Wt%	s/cm^2	f/cm^2	Wt%
All	Upper	0.0679	>0.9999	0.3432	0.7651	0.5715	>0.9999	0.5733
	Lower	0.0009	0.7449	>0.9999	0.4910	0.7733	0.3575	0.7774
Non-ACM	Upper	>0.9999	>0.9999	0.3880	>0.9999	>0.9999	>0.9999	>0.9999
	Lower	0.1138	>0.9999	>0.9999	>0.9999	>0.9999	0.5800	>0.9999
ACM	Upper	0.0646	>0.9999	0.5226	0.7400	0.7642	>0.9999	0.7648
	Lower	0.0043	0.4819	0.4731	0.2632	0.7529	0.4750	0.7598
Ceiling ACM	Upper	0.0148	0.1461	0.3332	0.0938	0.4516	>0.9999	0.4539
	Lower	0.0024	0.3967	0.4229	0.2011	0.7320	0.6872	>0.9999
Other ACM	Upper	>0.9999	0.3517	>0.9999	0.3442	0.6246	0.5868	0.6287
	Lower	>0.9999	>0.9999	>0.9999	0.5842	0.5089	>0.9999	0.5156

References

[1] U.S. Environmental Protection Agency (1987). *Asbestos Hazard Emergency Response Act*, 40 CFR Part 763, Appendix A to Subpart E.

[2] U.S. Environmental Protection Agency (1973). National Emission Standards for Hazardous Air Pollutants, 40 CFR Part 61. (Revised 1975, 1978, repromulgated 1984, revised 1990)

[3] Corn, M., Crump, K., Farrar, D. B., Lee, R. J., and McFee, D. R. (1991). Airborne Concentrations of Asbestos in 71 School Buildings. *Regulatory Toxicology and Pharmacology*, **13**, 99-114.

[4] Lee, R. J., Van Orden, D. R., Corn, M., and Crump, K. S. (1992). Exposure to Airborne Asbestos in Buildings. *Regulatory Toxicology and Pharmacology*, **16**, 93-107.

[5] Van Orden, D. R., Lee, R. J., Bishop, K. M., Kahane, D., and Morse, R. (1995). Evaluation of Ambient Asbestos Concentrations in Buildings Following the Loma Prieta Earthquake. *Regulatory Toxicology and Pharmacology*, **21**, 117-122.

[6] Lee, R. J, Dagenahrt, T. V., Dunmyre, G. R., Stewart, I. M., and Van Orden, D. R. (1995). Effect of Indirect Sample Preparation Procedures on the Apparent Concentration of Asbestos on Settled Dusts. *Environmental Science & Technology*, **29**, 1728-1736.

[7] Kauffer, E., Billon-Galland, M. A., Vigneron, J. C., Veissiere, S., and Brochard, P. (1996). The Effect of Preparation Methods on the Assessment of Airborne Concentration of Asbestos Fibers by Transmission Electron Microscopy. *Annals of Occupational Hygiene*, **40**, 321-330.

[8] Sahle, W., and Lazlo, I. (1996). Airborne Inorganic Fibre Level Monitoring by Transmission Electron Microscopy (TEM) : Comparison of Direct and Indirect Sample Transfer Methods. *Annals of Occupational Hygiene*, **40**, 29-44.

[9] Fowler, D. P., and Chatfield, E. J., (1996). Surface Sampling for Asbestos Risk Assessment, *Inhaled Particles VIII*, 26-30.

[10] Lange, J. H., Lee, R. J., and Dunmyre, G. R. (1992). Monitoring asbestos in an industrial facility using surface dust and passive samplers. Emerging Technologies for Hazardous Waste Management. American Chemical Society, Atlanta, GA, September 21-23, 1992.

[11] Oehlert, G. W., Lee, R. J., and Van Orden, D. R. (1995). Statistical Analysis of Fibre Counts. *Envirometrics*, **6**, 115-126.

[12] Lee, R. J., Van Orden, D. R., and Dunmyre, G. R. (1996). Interlaboratory Evaluation of the Breakup of Asbestos-Containing Dust Particles by Ultrasonic Agitation. *Environmental Science & Technology*, **30**, 3010-3015.

William M. Ewing[1]

FURTHER OBSERVATIONS OF SETTLED ASBESTOS DUST IN BUILDINGS

REFERENCE: Ewing, W. M., **"Further Observations of Settled Asbestos Dust in Buildings,"** *Advances in Environmental Measurement Methods for Asbestos,* ASTM STP 1342, M. E. Beard and H. L. Rook, Eds., American Society for Testing and Materials, 2000.

ABSTRACT: Surface sampling for asbestos is a tool used to determine the potential for asbestos exposure. It is also employed to estimate when a surface is sufficiently clean of asbestos structures.

The sampling and analytical procedure employed has been described in ASTM standard method D 5755-95. Sampling was conducted in 66 buildings and from out-of-doors. The geometric mean surface concentration was 3.7 million asbestos structures per square centimeter (s/cm^2) in areas with asbestos-containing fireproofing. Samples collected from areas having asbestos-containing acoustical plaster had a geometric mean of 160,000 s/cm^2. Samples collected in six buildings without friable asbestos-containing surfacing materials indicated a geometric mean of 1000 s/cm^2.

KEYWORDS: asbestos, settled dust, surface sampling, chrysotile

---

INTRODUCTION

Asbestos surface dust sampling has gained wide acceptance as a tool to evaluate exposure from in-place asbestos-containing materials. The primary purpose of conducting such tests is to estimate the concentration of asbestos structures

---

[1] Technical Director, Compass Environmental, Inc., 2231 Robinson Road, Suite B, Marietta, GA 30068

on surfaces. This information, coupled with the investigator's knowledge of building occupancy, activities, ventilation, and surfaces is valuable to reasonably anticipate when exposures to airborne asbestos may occur.

Much of the early surface sampling consisted of "rafter dust" or settled dust sampling for silica and on occasion, asbestos. These samples were usually analyzed by x-ray diffraction with a limit of detection around 2 - 4 percent.[1] This sampling was confined to industrial settings and not employed in buildings. During the 1980's some building inspectors collected wipe samples or bulk samples of surface dust and debris.[2] These samples were usually analyzed by polarized light microscopy (PLM) with results reported in percent asbestos by types, or simply as detected or not detected. A report by R.F. Carter in 1970 represents some early research on asbestos dust, including surface deposition and re-entrainment.[3] This study evaluated asbestos dust in an industrial setting and relied heavily on optical microscopy. It was not until the late 1980's that a standardized method for collecting asbestos surface dust began to evolve. In the past few years the method of choice has been the mircovacuum technique followed by analysis with transmission electron microscopy (TEM).

SAMPLING AND ANALYTICAL METHODS

The samples herein were collected and analyzed as described in ASTM Standard Test Method for Microvacuum Sampling and Indirect Analysis of Dust by Transmission Electron Microscopy for Asbestos Structure Number Concentrations (D 5755-95) or the earlier draft method prepared by the US Environmental Protection Agency (USEPA).[4, 5] The method incorporates the use of a personal air sampling pump calibrated at 2 L/min. A mixed cellulose ester filter membrane housed in a 25 mm sampling cassette is used as the collection media. At the inlet of the cassette is attached a 25 mm long piece of Tygon™ tubing with the end cut at a 45 degree angle.

All samples were collected from a known surface area, usually 100 cm^2. The surface of interest was examined visually and sometimes with the aid of a 10x hand lens for an area of evenly dispersed surface dust with no particles greater than 1 mm diameter. The samples were collected by vacuuming the dust from the surface as the cassette nozzle was passed across it.

For all samples discussed herein the surface would be considered non-porous. Such surfaces included polished stone, metal, polished wood, and plastic. The sample locations usually had visible dust on the surface. The cleaning history of the surface was usually unknown. Field blanks were collected at a rate of one sample per batch or 10 percent, whichever was greater.

Upon completion of sampling each sample was capped and sealed into a plastic bag for transport to the laboratory. In the laboratory, the samples were transferred into an aqueous solution and an aliquot was dispersed onto a new filter. This filter was analyzed by TEM for asbestos structures. Structures were classified as fibers, bundles, clusters, or matrices. Results were reported as asbestos structures per square centimeter of surface sampled (s/cm^2).

PRESENTATION AND DISCUSSION OF RESULTS

The data summarized in Table 1 was presented previously but is reviewed here to provide a basis for comparison with new data presented herein.[6] The data in Table 1 includes 275 surface dust samples collected from various buildings throughout the United States. Additionally, there were 44 field blank samples collected in which one chrysotile asbestos structure was detected in each of six field blank samples , and no fibers detected in 38 blanks. The fibers detected were in field blanks and likely the result of opening and closing the sampling cassette in the field. Blank correction of samples collected in the same lot resulted in no change to the calculated value.

The samples collected from inside buildings with no surfacing ACM, and the samples collected from outside buildings in a large city demonstrate that asbestos structures can be found on such surfaces. Virtually all the asbestos structures found were chrysotile. In two of these buildings, area and personal air sampling was conducted while intentionally disturbing dust laden surfaces. These activities included dry dusting of suspended acoustical ceiling tiles with a 20+ year build-up of settled dust. No significant increase in the airborne asbestos concentration, as measured by TEM, was found in the personal or area samples. These observations suggest a surface concentration of 1000 s/cm^2 may be considered clean.

The geometric mean for 79 samples collected outside of buildings in a large city was 5100 s/cm^2. Higher values were generally found closer to street level. While emissions from break pads and clutch facings are a possible source, this was not confirmed.

TABLE 1 -- Summary of Asbestos Surface Dust Samples by Building Category

Building Category	Number of Buildings	Number of Samples	Range (s/cm^2)	Geometric Mean (s/cm^2)
Outside buildings in a large city	5	79	<400 – 140,000	5100
Inside buildings with no surfacing ACM	6	28	<240 – 210,000	1000

TABLE 1 (cont.)-- Summary of Asbestos Surface Dust Samples by Building Category

Building Category	Number of Buildings	Number of Samples	Range (s/cm^2)	Geometric Mean (s/cm^2)
Areas of buildings with acoustical plaster	12	34	<3500 – 74 million	160,000
Areas of buildings with exposed fireproofing	18	41	7000 – 140 million	3.6 million
Above ceilings tiles with fireproofing	29	93	<3500 – 220 million	3.8 million

In the areas of buildings with acoustical plaster ceilings the geometric mean for 34 samples was 160,000 s/cm^2. The acoustical plaster in these buildings typically contained chrysotile asbestos in a range of 5 - 15 percent. These ceilings were often painted and usually had small areas of visible damage comprising less than 10 percent of the surface area. The range of surface concentrations found within buildings and across buildings is quite large. This is probably due to differences in deposition rates and the cleaning frequency of the surfaces sampled.

The highest concentrations of asbestos surface dust were found on surfaces below spray-applied friable asbestos-containing fireproofing. In 41 samples from surfaces below exposed fireproofing the geometric mean was 3.6 million s/cm^2. This was not significantly different from the geometric mean of 3.8 million s/cm^2 for 93 samples collected from above suspended ceiling tiles. Visually, these surfaces appeared dusty and probably represent a dust accumulation of 20+ years. It is not possible to state a deposition rate based on these data alone. One theory is the deposition of asbestos dust occurs at a steady rate over time. It is perhaps more likely that the deposition rate of asbestos onto surfaces accelerates over time as the source ages. The deposition of dust may also be at least partially attributed to disturbance of the source material which can cause episodic releases of dust.

Various studies have been conducted by this author and others designed to characterize asbestos exposure due to disturbance of ACM, or dust and debris from ACM.[7 – 11] Others have described in some detail the concept of "K factors."[3, 12] The K factor being the ratio of the airborne asbestos level to the asbestos surface dust concentration resulting from a known activity. The results of these studies indicate that routine building maintenance and custodial activities (replacing ceiling tile, sweeping, dusting, vacuuming) can produce significantly elevated airborne asbestos exposures through the reentrainment of surface dust. This statement assumes the surfaces are non-porous and special dust suppression methods are not employed. A significantly elevated airborne asbestos exposure is defined here as a 10-fold rise in concentration.

Asbestos-containing fireproofing was often applied to structural steel and decking of ceiling plenums. These plenums are frequently part of the building ventilation system. Return air from occupied spaces is pulled into the plenum, travels across the suspended ceiling system to return air ducts, and enters air shafts to return to the fan coil unit. At the fan coil unit, the return air is mixed with some fresh air from outdoors, heated or cooled, and distributed back to occupied spaces through supply air ducts and diffusers. To determine the extent to which these ductwork accumulate asbestos dust, surface dust samples were collected inside the return and supply air ducts of three multistory buildings. These buildings had fireproofing with approximately 15% chrysotile in a matrix of gypsum and vermiculite. The results are presented in Tables 2 and 3.

TABLE 2 -- Asbestos Dust Sampling Results from Return Air Ducts in Three Buildings with Structural ACM Fireproofing

Location/Description	s/cm^2
Building A, return air duct, 3rd floor	20 million
Building B, return air duct, 4th floor	3.3 million
Building B, return air duct, 3rd floor	4.4 million
Building B, return air duct, 2nd floor	6.0 million
Building C, return air duct, 1st floor	1.2 million
Building C, return air duct, 3rd floor	3.5 million
Building C, return air duct, 2nd floor	2.2 million
Building C, return air duct, 3rd floor	1.1 million

TABLE 3 -- Asbestos Dust Sampling Results from Supply Air Ducts/Diffusers in Three Buildings with Structural ACM Fireproofing

Location/Description	s/cm^2
Building A, supply air duct, 2nd floor	3.8 million
Building B, supply air duct, 4th floor	27 million
Building B, supply air duct, 3rd floor	830,000
Building B, supply air duct, 2nd floor	490,000
Building C, supply air duct, 3rd floor	170,000
Building C, "new" supply air duct (6 years old), 1st floor	27,000
Building C, "new" supply air duct (6 years old), 1st floor	110,000
Building C, "new" supply air duct (6 years old), 1st floor	79,000

TABLE 3 (cont.)-- Asbestos Dust Sampling Results from Supply Air Ducts/Diffusers in Three Buildings with Structural ACM Fireproofing

Location/Description	s/cm^2
Building A, supply air diffuser, 1st floor	160,000
Building A, supply air diffuser, 3rd floor	79,000

The return air duct results in Table 2 show that 7 of the 8 samples were in a range of 1.1 million - 6.0 million s/cm^2 with a geometric mean of 2.6 million s/cm^2. If the one sample from building A is included the geometric mean value rises to 3.4 million s/cm^2.

The supply air duct results in Table 3 present findings of samples from original ductwork (about 25 - 30 years old), new ductwork (6 years old), and supply air diffusers. While the data is limited, it does indicate that asbestos dust concentrations in the supply air ducts are generally an order of magnitude lower than that found in return air ducts.

In the late 1980s surface dust samples were sometimes collected without a nozzle using an open face 37 mm diameter sampling cassette. In order to compare the 1980's technique with ASTM D 5755 a set of side by side samples were collected using each technique.

Fifteen pairs of samples (30 total) were collected in five buildings from horizontal metal surfaces having a visual accumulation of dust. The surfaces were all located beneath spray-applied asbestos-containing fireproofing. One sample of each pair was collected following ASTM method D 5755-95 from a surface area of 100 cm^2. The second sample of the pair was collected as follows.

1. An open face 37 mm cassette with a 0.8 μm pore size MCE filter.
2. A flow rate of 2.0 L/min using a battery operated personal sampling pump.
3. A collection area of 929 cm^2 (1 ft^2).

Three pairs of field blanks were also collected and submitted for analysis. All samples were submitted to the same laboratory and analyzed in the same manner with two exceptions. The 37 mm cassette samples were placed into an aqueous suspension of particle free water without a screening step to remove 1 mm or larger particles. The 25 mm cassette samples were placed into an aqueous suspension of 50 percent water and 50 percent alcohol and passed through a 1 mm screen. The classification of asbestos structures and the stopping rules followed were the same for all samples.

The results of the side-by-side samples are presented in Table 4. In every instance, the sample collected using the ASTM method D 5755-95 with the nozzle was greater than the sample collected with the open face cassette.

TABLE 4 -- Results of Side-by-Side Asbestos Surface Dust Samples Using Open Face Versus Nozzle Equipped Cassette

Sample Location	Sample Results (s/cm^2) Open Face Cassette	Nozzle Cassette
1	232,000	4.4 million
2	820,000	14.2 million
3	642,000	9.0 million
4	216,000	1.3 million
5	382,000	541,000
6	142,000	657,000
7	403,000	3.3 million
8	1.1 million	13.9 million
9	539,000	612,000
10	17,000	2.4 million
11	98,000	2.6 million
12	1.2 million	23.5 million
13	3.3 million	18.3 million
14	2.1 million	4.2 million
15	410,000	2.4 million
16B	ND[1]	ND[1]
17B	ND[1]	ND[1]
18B	ND[1]	ND[1]

[1]ND = no asbestos structures detected in blank sample

The one significant difference between the two methods is in the sample collection. The ASTM method employs a nozzle with a known inside diameter of 0.63 cm (0.25 in) and a flowrate of 2.0 L/min. This provides a face velocity of 106 cm/sec. The open face 37 mm cassette has an effective collection area (inside diameter) of 33 mm. Operating at 2.0 L/min, this provides a face velocity at the point of dust collection of 6.4 cm/sec. Accordingly, the ASTM method provides a collection velocity over 16 times the collection velocity of the open face method. The use of the 1 mm screen would not have had an effect since particles greater than 1 mm were not collected in any samples. The alcohol addition to the aqueous suspension for the D 5755-95 samples would not tend to provide higher structure concentrations.

OBSERVATIONS AND CONCLUSIONS

Surface dust sampling for asbestos can be a useful tool in preventing unnecessary exposures to activity generated asbestos aerosols. With knowledge of where high surface loadings of asbestos dust are located, preventive measures can be designed and implemented to reduce or eliminate exposures before they occur. Surface dust sampling may also be employed as a measure of cleanliness at the conclusion of an asbestos abatement project. When used in this manner it augments the information provided by the visual inspection and air sampling.

A surface may be considered clean when the asbestos concentration is below 1000 s/cm^2, and contaminated when above 100,000 s/cm^2. The problem arises when the asbestos concentration is between 1000 s/cm^2 and 100,000 s/cm^2, and requires professional judgment by the investigator. The investigator must consider who will be exposed and what activities can reasonably be expected to impact the surface. The example below illustrates this point.

EXAMPLE: Surface dust sampling in a series of storage rooms for an office building finds an average asbestos concentration of 50,000 s/cm^2 on the shelving and stored boxes. Are the rooms sufficiently contaminated with asbestos to require they be locked and cleaned before access is permitted? The users of the storage rooms are secretaries and clerks of tenants. The activities include placing boxes in storage, retrieving boxes, and occasionally dusting off boxes with paper towels. The activities of moving boxes alone might or might not result in significant exposures. The dry dusting of boxes would almost certainly raise exposures. In this instance, use of the rooms should be discontinued until the surfaces are thoroughly cleaned.

There are other surface dust sampling and analytical methods available for asbestos dust. These include wipe samples, tape samples, and ASTM method D 5756-95. Each have their advantages and disadvantages. These are reviewed briefly below.

Wipe samples probably have the highest collection efficiency from non-porous surfaces and provides a measure of total asbestos surface dust concentration. This compares to the mircovacuum technique which measures that dust easily removed and therefore available for re-entrainment. A significant disadvantage to wipe sampling is the lack of a standard method for both collection and analysis. As a result, data developed to date often cannot be compared.

Tape samples, and other direct sampling techniques have been employed in some instances. Its principal advantage is to characterize particles as they exist on the surface. Disadvantages include a poor collection efficiency and the inability to use the method on very dusty surfaces.

ASTM method D 5756-95 is similar to ASTM method D 5755-95 except the analysis includes additional steps which provide for results to be reported as a mass quantity.[13] To date very little data, if any, has been published to provide a basis for understanding the results obtained. Should this method gain wider use, the mass of asbestos on a surface may become a useful index for investigators.

The ASTM Committee D-22.07 is currently working on a draft standard practice for employing asbestos dust sampling. If this practice is finalized, it may provide guidance on the number of samples, selection of sampling locations, quality control procedures, and interpretation of results. Recognizing that the ASTM practice guide may be years away, outlined below are some suggestions and comments which practitioners may find helpful.

The selection of sampling locations depends on the purpose of the sampling. Target sampling is used to pinpoint specific sites of interest, such as a location where an asbestos spill is suspected. Representative sampling is used to characterize a larger surface area based on a small number of samples. One method of obtaining representative samples is to divide the surface area of interest into a grid. Number each grid opening consecutively with two digit numbers (i.e., 01, 02, 03 . . . NN). Beginning in the middle of a random number table select pairs of numbers until the number of samples desired is obtained. Repeat this procedure to select replacement sample locations should the first choice not be suitable (i.e., not accessible).

The method suggests a minimum of three samples be used to characterize a surface. While this number was chosen rather arbitrarily, it appears to be a good rule of thumb considering the range of values encountered. Additional samples may be appropriate if the results of the three samples are close to the concentration of interest. For example, if the questions is, "Is the surface clean, defined as an average of 1000 s/cm^2?" If three samples are in the range of 500 - 2000 s/cm^2, additional sampling would be appropriate. If the results are 10,000 - 300,000 s/cm^2, additional sampling is not appropriate.

Dust sampling may be used to provide some qualitative information, such as where did the asbestos originate. If the dust analysis also finds vermiculite and calcium sulfate attached to asbestos matrices, the source is likely the product with these other constituents.

Additional research is needed to evaluate the relationship between specific activities and concentrations of asbestos in dust. Of particular interest is the amount of air movement and turbulence necessary to re-entrain dust from a surface.

REFERENCES

[1] Taylor, D.G., *NIOSH Manual of Analytical Methods, Part I, Vol I (2nd Edition)*, [Method P&CAM 259, Free Silica (Quartz, Cristobalite, Tridymite) in Airborne Dust], U.S. Dept. of Health, Education and Welfare (NIOSH) Publication No. 77-157-A, Cincinnati, OH, April 1997

[2] Light, E.N. and J.T. Jankovic, "Assessment of Asbestos Fiber Release in Buildings through Analysis of Settled Dust," Proceedings of the National Asbestos Council Fall Technical Conference, New Orleans, LA, September 25, 1986

[3] Carter, R.F., "The Measurement of Asbestos Dust Levels in a Workshop Environment," AWRE Report No. 0 28/70, United Kingdom Atomic Energy Authority, H.M. Stationary Office, U.K., July 1970.

[4] ASTM Standard Method D 5755-95, "Standard Test Method for Microvacuum Sampling and Indirect Analysis of Dust by Transmission Electron Microscopy for Asbestos Structure Number Concentrations," ASTM, West Conshohocken, PA, 1995.

[5] Clark, P., "Standard Test Method for Sampling and Analysis of Dust for Asbestos Structures by Transmission Electron Microscopy," (Draft), US Environmental Protection Agency, Cincinnati, OH, August 1989.

[6] Ewing, W.M., Dawson, T.A. and G.P. Alber, "Observations of Settled Asbestos Dust in Buildings," *EIA Technical Journal*, Environmental Information Association, Bethesda, MD, August 1996, pp. 13 - 17.

[7] Ewing, W.M., et. al., "An Investigation of Airborne Asbestos Concentrations During Custodial and Maintenance Activities in a Boiler Room," Paper presented at the American Industrial Hygiene Conference, Boston, MA, June 3, 1992.

[8] Keyes, D.L., et. al., "Baseline studies of Asbestos Exposure During Operations and Maintenance Activities," *Applied Occupational and Environmental Hygiene* 9 (11), 1994, pp. 853 - 860.

[9] Ewing, W.M., et. al., "Asbestos Exposure During and Following Cable Installation in the Vicinity of Fireproofing," *Environmental Choices Technical Supplement* 1 (2), 1993, pp. 12 - 18.

[10] Keyes, D.L., et. al., "Re-entrainment of Asbestos From Dust in a Building with Acoustical Plaster," *Environmental Choices Technical Supplement* 1 (1), 1992, pp. 6 - 11.

[11] Keyes, D.L., et. al., "Exposure to Airborne Asbestos Associated with Simulated Cable Installation Above a Suspended Ceiling," *American Industrial Hygiene Association Journal*, 52 (11), 1990, pp. 479 - 484.

[12] Millette, J.R. and S.M. Hays, *Settled Asbestos Dust Sampling and Analysis*, Lewis Publishers, Boca Raton, FL, 1994, pp. 59 - 65.

[13] ASTM Standard Method D5756-95, "Standard Test Method for Microvacuum Sampling and Indirect Analysis of Dust by Transmission Electron Microscopy for Asbestos Mass Concentrations," ASTM, West Conshohocken, PA, 1995.

Douglas P. Fowler[1] and Bertram P. Price[2]

SOME STATISTICAL PRINCIPLES IN ASBESTOS MEASUREMENT AND THEIR APPLICATION TO DUST SAMPLING AND ANALYSIS

REFERENCE: Fowler, D. P. and Price, B. P., "**Some Statistical Principles in Asbestos Measurement and Their Application to Dust Sampling and Analysis,**" *Advances in Environmental Measurement Methods for Asbestos, ASTM STP 1342,* M. E. Beard, H. L. Rook, Eds., American Society for Testing and Materials, 2000.

ABSTRACT: Basic statistical principles of deterministic and stochastic variability are examined for their application to asbestos sampling and analysis methods, especially the approved and draft ASTM sampling and analysis methods for asbestos in dust. The sensitivities of these methods are shown to be inadequate for reliably determining the presence of asbestos structures in dust at 1000 s/cm^2 using currently-required counting protocols. Appropriate sensitivities can only be attained by counting impractically large numbers of grid openings. The current ASTM practice of stating detection limits for asbestos counts based on the confidence intervals for Poisson-distributed variables is shown to be wrong. Basic physical and statistical considerations show that detection limits are almost never appropriate for these methods, except in the presence of irremediable contamination of the sampling/analytical process.

KEYWORDS: asbestos, dust, statistics, detection limit, sensitivity

Every ASTM method for measuring asbestos, whether in air or surface dust, includes a discussion of sensitivity and limit of detection (DL or LOD -- hereafter referred to as DL). These concepts apply to the interpretation of the measurement data. They contain information about the degree of uncertainty in the measurements and, therefore, the degree of uncertainty in conclusions and decisions based on the measurements. However, the presentation of these concepts in ASTM methods typically is confusing and does not lead to an understanding or useful application of these concepts. Our purpose is to provide guidance for understanding and applying sensitivity and DL where appropriate.

Uncertainty is central to our discussion. We differentiate deterministic uncertainty from

[1] President, Fowler Associates, Inc., 643 Bair Island Road, Suite 305, Redwood City, CA 94063

[2] President, Price Associates, Inc., 1800 K Street, N.W. Suite 718, Washington, D.C. 20006

stochastic uncertainty. Sensitivity is principally associated with deterministic uncertainty. The DL is associated with stochastic uncertainty. We take the perspective of one who is designing a measurement system, a monitoring program, or an exposure study where the quantitative data will be asbestos measurements. We use the measurement of airborne concentrations of asbestos as a vehicle for explaining the concepts. Later, we relate these concepts to methods for measuring asbestos in dust.

Sensitivity and DL Applied to Airborne Asbestos Measurements

As an aid in the subsequent discussion, we use a simplified characterization of air sampling and analysis for measuring airborne asbestos concentrations. Although our characterization of the measurement process may lack some important details from a microscopist's perspective, it is adequate for our intended purpose.

Air sampling is accomplished by drawing air through a filter at a specified rate for a specified period of time. Airborne particles consisting of asbestos and other matter are deposited on the filter. After air sampling has been completed, a section of the filter is prepared for inspection under a microscope. The preparation may be "direct" or "indirect." If the preparation is "direct," then the microscopist inspects the air sampling filter directly. If the preparation is "indirect," then the air sampling filter is dissolved in an appropriate liquid, and a portion of the liquid-suspended particles is redeposited on a secondary filter, which is then analyzed.

A specified number of fields of view of known size, (grid openings in electron microscopy, or reticle fields in optical microscopy -- here generalized as "fields") are selected and inspected microscopically. The particles found in each field are classified (e.g., asbestos or not, or fiber or not, etc.) and a count is recorded for the particles meeting the classification criteria for the method. A ratio is applied to the count obtained from the fields inspected to produce an estimated count for the total filter. This count is then adjusted (if an indirect preparation is used) to take into account the "dilution factor," calculated from the proportion of the total suspension redeposited and the size of the secondary filter. This estimate is divided by the volume of air collected. The result, the asbestos or fiber measurement, is interpreted as an estimate of the asbestos concentration in the air, and is reported in units of fibers or structures per unit volume of air, e.g., fibers or structures per milliliter or cubic centimeter of air (f/ml or f/cc or f/cm^3) or structures per milliliter or cc (s/ml or s/cc or s/cm^3).

From a measurement design perspective, this description leads to a number of questions concerning specifications. For air sampling, specifications are needed for the following:

- particle classification criteria;
- filter material, pore size, and total filter size;
- sampling rate, sampling time, and total air volume sampled;
- temporal distribution of samples;
- number of samples; and

- location of samples.

For inspection of the filter under a microscope (analysis), specifications are needed for:

- method;
- dilution;
- field size;
- number of fields inspected; and
- degree of magnification.

Sensitivity and DL are performance characteristics of a measurement system that are used to design the system. Consideration of sensitivity and DL, therefore, should provide answers to some of the specification questions listed above. In the subsequent discussion, we first address sensitivity, then DL. We demonstrate that sensitivity is, in fact, an important measurement design concept. We argue, however, that DL is usually not a meaningful concept for asbestos measurement methods and its consideration in ASTM documents addressing asbestos measurement is ordinarily unnecessary.

Sensitivity

Definition - Sensitivity is a concept that principally applies to a single measurement. In common analytical usage, sensitivity is usually considered to be the amount of analyte that must be added to a sample in order to have a discernable difference in analytical response from the baseline. In asbestos analyses or any other particle-counting procedures, sensitivity is defined as the *concentration corresponding to a count of one qualifying particle (fiber or structure) in the sample.*[3]

Sensitivity, therefore, depends on air volume and the fraction (a proportion) of the filter that is inspected under the microscope. The proportion depends on the size of the total filter collection area, the size of the fields, and the number of fields inspected.

Sensitivity = [FCA/(#Fs*FA)]/V

where:

FCA is filter collection area in square millimeters (mm^2);
#Fs is number of fields;
FA is average field area in mm^2; and
V is air volume in cubic centimeters (cc).

Given any value as a requirement for sensitivity, the air volume and number of fields may be varied (within the constraints imposed by the method) to achieve the required value.

---

[3] "Sample" in this context refers to the portion of the filter that is inspected, and thereby, embodies the "sample" of air and the "sample" of particles collected from the air on the filter.

(The filter collection area and field area typically are fixed - 385 mm^2 for a 25 mm^2 diameter filter and 0.00785 mm^2 for the field area for NIOSH Method 7400, for an example - and other variables may also be fixed.) It follows immediately from the definition of sensitivity that the fiber concentration for a sample is the number of fibers counted multiplied by sensitivity:

f/cc = (# fibers) * sensitivity

Selecting a Target Value for Sensitivity

Controlling Deterministic Uncertainty -- Before discussing how to vary the three measurement parameters in the definition of sensitivity, we discuss the process for establishing a target or design value for sensitivity. We simplify the discussion by addressing a specific example, the sampling and analysis requirements for air measurements intended to determine compliance with OSHA's Short Term Exposure Limit (STEL) of 1.0 f/cc average for 30 minutes. (Forget for the moment that OSHA already has specified sampling and analysis requirements for measurements used to determine compliance with its exposure limits.) The value selected for sensitivity for measurements intended to determine compliance with the STEL clearly should not be larger than 1.0 f/cc. If, for example, sensitivity were set at 2.0 f/cc, measured concentrations could only take on the values 0.0 f/cc if no fibers were found in the sample, 2.0 f/cc if one fiber were found in the sample, and multiples of 2.0 f/cc corresponding to higher fiber counts. Such a measurement system would embody substantial deterministic uncertainty. The information provided by the measurement would make it virtually useless for assessing STEL compliance. If the sample count of fibers were zero, the true concentration could be anywhere between 0.0 f/cc and 2.0 f/cc. The measurement would not provide enough information even to make a reasoned guess as to whether the true concentration is above or below the 1.0 f/cc STEL. In this circumstance, the only rational decision rule would be to decide in favor of "compliance" or "noncompliance" based on the outcome of a coin flip assigning, for example, heads to mean "compliance" and tails to mean "noncompliance."

We note at this point that selecting sampling and analysis parameter values within the ranges required by OSHA for airborne asbestos measurements used for testing compliance with its exposure limits leads to a sensitivity quite adequate for evaluating compliance with the STEL. (Refer to 29 CFR 1910.1001 or 1926.58) For example, sampling at 2.0 liters per minute for 30 minutes and inspecting 100 microscope fields of size 0.00785 mm^2 on a filter with collection area equal to 385 mm^2 results in a sensitivity value of 0.008 f/cc:

$$\text{Sensitivity} = 1 * (385/ (100*0.00785)) / (30*2.0*1000)$$

$$= 0.008 \text{ f/cc.}$$

Considering deterministic uncertainty alone, OSHA's specifications are overly stringent for a measurement intended to test compliance with the STEL. Since the critical measurement value is 1.0 f/cc, a sensitivity equal to 0.10 f/cc may be adequate because it provides sufficient differentiation of the measurement scale. This sensitivity value would be approximated if 10 fields were inspected and all other sampling and analysis parameters in the example above were not changed. (The actual sensitivity value for 10 fields would be 0.08 f/cc.)

We reemphasize that sensitivity addresses deterministic uncertainty. It allows the planner to specify a measurement system that includes enough information in terms of air collected and filter area inspected to assure a measurement scale, sufficiently differentiated, to achieve the intended purpose of the measurement. The measurements, however, also are subject to stochastic uncertainty. Consideration of stochastic uncertainty may be used to select a target value for sensitivity from among a number of values that would be acceptable for controlling deterministic uncertainty.

Controlling Stochastic Uncertainty -- Continuing with the STEL example, we consider the role of stochastic uncertainty in designing the measurement system. We set a high probability of making a correct compliance decision as a design objective and ask how the probability of a correct compliance decision varies with sensitivity. For example, what is the likelihood of a correct compliance test outcome using a measurement with sensitivity equal to 0.008 f/cc as opposed to the likelihood of a correct test outcome using a measurement with sensitivity equal to 0.08 f/cc? Stated differently, how does sensitivity affect the probability that a measurement will be less than 1.0 f/cc when the true concentration is less than 1.0 f/cc, and how does it affect the probability that a measurement will be greater than 1.0 f/cc when the true concentration is greater than 1.0 f/cc? From a measurement system design perspective, we want both probabilities to be large. We demonstrate below that the probability values associated with a measurement with sensitivity equal to 0.008 f/cc will be larger than the probability values for a measurement with sensitivity equal to 0.08 f/cc.

Based on the usual assumption that the fiber count from an air sample is described by the Poisson probability distribution, we have calculated and plotted probabilities that a measurement will indicate noncompliance with the STEL (i.e., the measurement will be larger than 1 f/cc) versus the true mean airborne concentration, for a range of sensitivity values. Let Y represent the measured concentration for one sample with sensitivity equal to θ. The probabilities of exceeding the STEL and of meeting compliance with the STEL are $P(Y \geq 1 \mid \mu, \theta)$ and $P(Y<1 \mid \mu, \theta) = 1 - P(Y \geq 1 \mid \mu, \theta)$, respectively, where μ is the true average airborne asbestos concentration. The calculations are accomplished in terms of the Poisson distribution applied to fiber counts, $X=Y/\theta$, as follows:

$$P(Y \geq 1 \mid \mu, \theta), = P(X \geq (1/\theta) \mid \lambda = \mu/\theta) \text{ and}$$
$$P(Y<1 \mid \mu, \theta) = P(X < (1/\theta) \mid \lambda = \mu/\theta)$$
$$= 1 - P(X \geq (1/\theta) \mid \lambda = \mu/\theta)$$

where $\lambda=\mu/\theta$ is the mean of the Poisson distribution

Figure 1 displays P ($Y \geq 1 \mid \mu, \theta$) versus μ for $\theta = 0.008, 0.050, 0.080, 0.250, 0.500$, and 1.00. Independent of the sensitivity level, the probability of a noncompliance result rises as μ, the true mean airborne concentration increases. Also, the probability of a correct compliance decision is higher for low values of sensitivity than for high values of sensitivity. (Recall that a measurement system with a low sensitivity value contains more information than a measurement system with a high sensitivity value.) For example, where sensitivity is 0.008 f/cc, the probability of noncompliance is virtually zero when the true mean is less than 0.75 f/cc and rises rapidly to virtual certainty (i.e., probability equal to one) when the true mean is 1.15 f/cc (Figure 1). In contrast, with sensitivity equal to 0.08 f/cc, the probability of a noncompliance result when the true mean is 0.75 f/cc is approximately 0.15, and only about 0.50 when the true mean is 1.15 f/cc (Figure 1). As the sensitivity value takes on larger values, the probabilities of correct compliance decisions deteriorate rapidly.

In summary, although sensitivity principally is associated with deterministic uncertainty, it also has implications for stochastic uncertainty. The analysis described above may be applied to designing asbestos measurement systems for objectives other than determining compliance with OSHA's STEL. The value selected for sensitivity initially limits deterministic uncertainty. However, a complete analysis to specify sensitivity also must address stochastic uncertainty in terms of an appropriately stated objective. The objective may be the likelihood of a correct compliance decision for a PEL, STEL, or cleanup after asbestos abatement, or it may be a test to compare indoor and outdoor levels of asbestos. Greater sensitivity (i.e., a lower value) is always preferable, but, the costs of producing measurements increases as the sensitivity value is reduced (i.e., more time spent collecting an air sample and more time inspecting the filter under the microscope). Ultimately, therefore, the choice of sensitivity is based on a tradeoff between better measurement system characteristics and cost.

Limit of Detection (DL)

DL Defined - The DL is one of the most misunderstood concepts in the field of environmental measurement. A brief review of EPA and NIOSH regulatory literature concerning chemical measurement methods reveals a variety of conceptual and operational definitions of DL. These definitions are variations on the same theme, namely an attempt to inform a data user about levels of stochastic uncertainty high enough that it is considered impossible to claim with a high degree of confidence that a non-zero measurement correctly indicates the presence of the chemical in the sampled medium. Under these circumstances, the analyst is instructed to disregard the quantitative measurement and report the measurement only as "less than the detection limit."

Figure 1: Air Sampling Design Characteristics

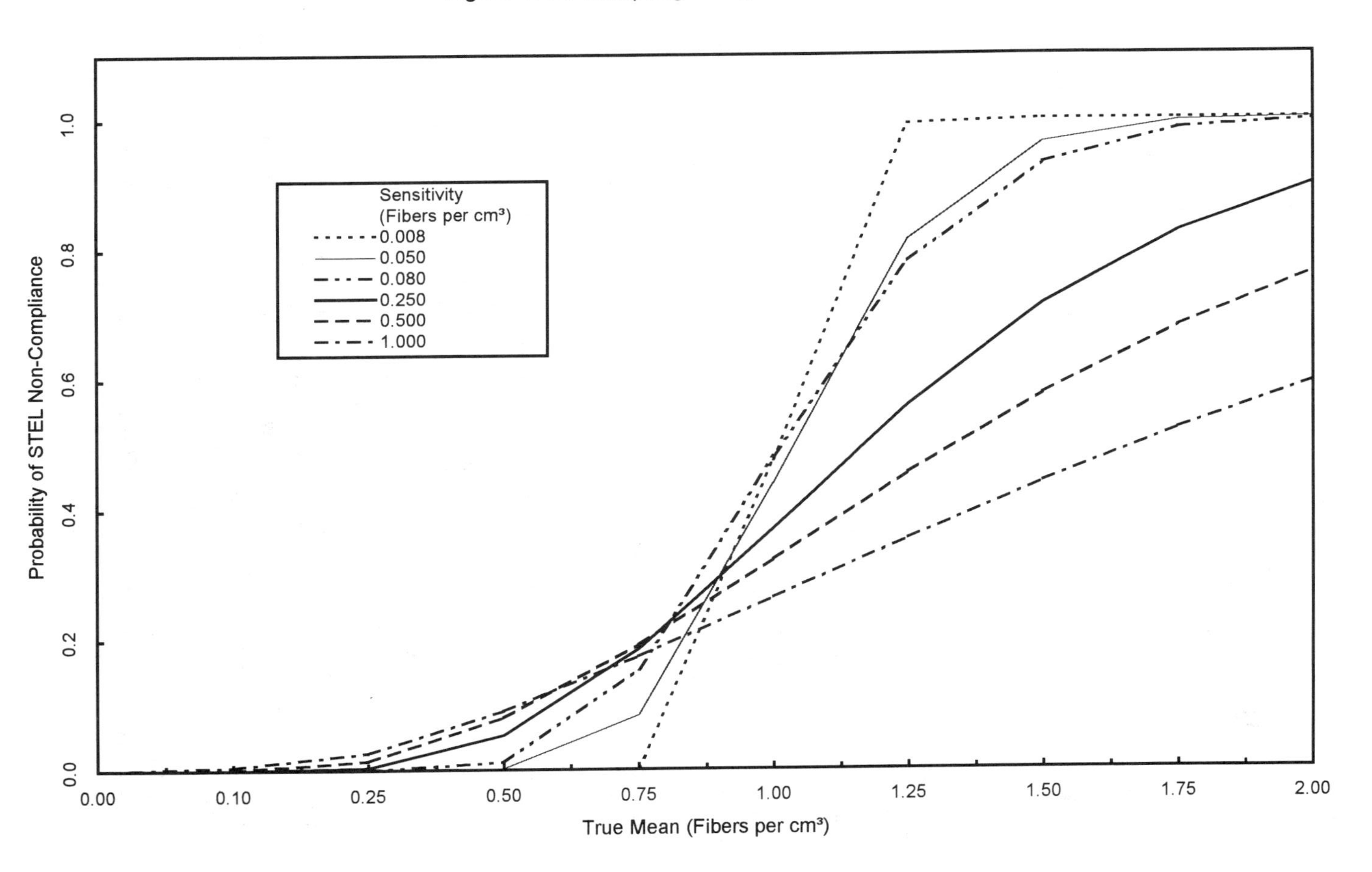

Much of the confusion surrounding the DL is caused by the failure to define the concept with sufficient clarity for translation into operational terms. For example, in EPA's Series 600 analytical methods to analyze for organic chemicals in surface waters, the detection limit is "*the minimum concentration of a substance that can be identified, measured, and reported with 99% confidence that the analyte concentration is greater than zero and is determined from replicate analyses of a sample of a given matrix containing the analyte*." Although this definition contains a glimmer of the correct concept, that concept is well hidden.

The detection limit concept and operational implementation are correctly presented in publications by Brodsky [*1*], Altshuler and Pasternack [*2*], Currie, [*3,4*] and Fowler, [*5*]. The DL is a theoretical value, the true mean concentration of a substance in the sampled medium. The DL must be large enough to lead to a high probability (e.g., 0.95 or larger) for concluding based on one or more measurements from a sample that the true concentration in the medium is, in fact, greater than zero. The DL, therefore, is a parameter in a statistical decision problem concerning whether the concentration of chemical in a sample is zero or greater than zero. This is a legitimate statistical decision problem because replicate measurements of a sample do not yield identical results, and, at least for analyses intended to quantify concentrations of chemicals, a positive measurement may be realized even if the true concentration of chemical in the sample is zero. (That is, the analysis of a "blank" sample may yield a positive value other than zero.) Differences in replicate results are characterized as stochastic variation subject to description by a probability distribution. The decision concerning zero or non-zero concentration, therefore, is a statistical decision that fits the standard hypothesis testing problem in statistics. The statistical testing problem provides the necessary structure for determining a numerical value for the DL as well as guidance concerning the reporting of measurements as "below detection."

As discussed by Currie [3,4], the detection limit is only one of three reporting standards that may be applicable to decisions regarding whether a sample result is considered to be "reportable" or not. The lowest of these is the "critical level" or "decision level." This number is determined by the variability in the concentrations reported in blanks. As discussed in greater detail by Fowler [5], the decision level is a number that is large enough to make a false positive decision unlikely (e.g., a probability less than 1% or 5%.) That is, one avoids a false positive decision, (or Type I Error) where a positive decision is that the sample result truly indicates a non-zero concentration. An example of a decision level in the guise of a DL is given in the NIOSH Standard Operating Procedure 018 ("Limits of Detection and Quantitation") where DL is defined as: "*...the mass of analyte which gives a signal* $3\sigma_b$ *above the mean blank signal, where* σ_b *is the standard deviation of the blank signal.*"

The detection limit, a value larger than the decision level, is the second reporting standard, and is determined as follows. Consider a test of the null hypothesis that the true mean concentration of chemical in a sample, μ, is zero versus the alternative hypothesis that μ is greater than zero. The typical decision rule (such as in the NIOSH procedure cited above) leads to a choice of $\mu>0$ over $\mu=0$ if the measurement is larger than the decision level, a multiple of the standard deviation of the statistical distribution for the measurement of the blanks.[4] The multiple is chosen to limit the False Positive or Type I Error rate for the test (e.g., 5% or 1%).

However, as the basis for specifying the true DL we require a high degree of confidence, or a high probability, that, a decision in favor of $\mu>0$ over $\mu=0$ is correct. This probability is referred to as "power" in statistical test terminology. The power of a statistical test is the probability that a measurement exceeds the decision level (i.e., the probability that the measurement leads to the choice, $\mu>0$) where the true mean concentration is a value larger than zero. The power of the test is an increasing function of the true mean. The DL is the value of μ, the true mean, that makes the power acceptably large. Ordinarily, at least 95% confidence that the measurement correctly indicates a nonzero concentration is required. This statement means that power, the probability of a true positive result, should be 0.95 or greater.

The third reporting standard that is used in some circumstances to define the lower end of the data spectrum is the Limit of Quantification (LOQ). The LOQ is a number such that a sample result at or above it can be viewed as reasonably stable and reliable. [3,4,5]

Given the structure outlined above, reporting data that would be subject to DL considerations would ordinarily be done as follows:

1. Determine the value in the statistical test for determining if a measurement is large enough to conclude that $\mu>0$ is correct and determine the value of μ, say μ_2, for an acceptable level of power.

2. If the measurement passes the test, report it. If the measurement fails the test, report that it is below the DL of μ_2.

DL Applied to Airborne Asbestos Measurements - The DL concept described above embodies a significant feature not shared by asbestos measurements. The concept applies to measurement methods capable of producing positive measurements when the sample contains none of the substance of interest. For chemicals in water or air, false positive signals may be produced by gas chromatography (GC) when none of the target chemical is present. In radiation measurements focused on a particular target, background radiation may be responsible for the nonzero measurement. However, when measuring asbestos by counting discrete fibers under a microscope, a count of one or more fibers is impossible if there are no fibers in the sample.

---

4 The decision rule usually is more complicated than a simple multiple of the standard deviation of the measurement distribution. The statistical distribution of a chemical measurement includes only nonnegative values and typically has a long tail (skewed toward large values). The log-normal distribution is an example that is applicable to many types of measurements. A mathematical function may be chosen to transform the original distribution to the Normal distribution, where the decision rule consisting of a multiple of the standard deviation is appropriate.

"Background" may exist for asbestos measurements, but this is not a DL issue. One form of background may be from collection filters that are contaminated during production, or from contamination introduced during the sampling/analysis process. These are QA/QC issues. Asbestos sampling and measurement protocols should recommend either quantifying the contamination and deducting it from all measurements, or discarding the filters before sampling and proceeding with clean filters. Another form of background may be from the ambient environment when the purpose of the air sampling program is to measure air concentrations inside a building. This type of background may be quantified and deducted from measurements inside the building. Background is not a detection limit issue. DL is generally a meaningless concept for methods used to measure asbestos concentrations and, therefore is generally unnecessary in ASTM methods for measuring asbestos.

Where contamination appears to be unavoidable, and where background/blank correction is deemed impossible, the procedure given in Fowler [5] may be used. In so doing, the analyst should be careful to preserve the distinction between 'decision level' or 'critical level' and detection limit.

Notwithstanding their usual lack of meaning, DLs currently are presented in ASTM methods for asbestos measurement. However, these DLs emerge from an error in translating the DL concept to a procedure for quantifying the DL. In the asbestos measurement literature we find the following statements about the DL:

> **29 CFR 1926.1101 Appendix B.** DL is defined as the number of fibers necessary to be 95% certain that the result is greater than zero. The DL is stated as 4 fibers per 100 grid openings or, equivalently, 5.5 fibers per square millimeter (f/mm^2).
>
> **ASTM Method D5755-95, Section 16.9.** "The limit of detection for this method is defined as, at a minimum, the counting of four asbestos structures during the TEM analysis."[5]

These DL values have been derived from confidence intervals for the mean of the fiber count from a Poisson distribution. Specifically, if the count in the sample is zero, the two-sided 95% confidence interval has a lower limit of zero and an upper limit of 3.68. The upper limit in the one-sided 95% confidence interval is 2.99. The upper limit in the one-sided 99% confidence interval is 4.61.[6] However, confidence intervals alone do not determine DLs. For example, the one-sided upper 99% confidence interval is properly interpreted as follows: "***...if sampling and analysis of the target environment were conducted resulting in a measured structure count of zero, the interval from zero to***

[5] The DL concept is evolving within ASTM Subcommittee D22.07; the next generation of asbestos sampling standards is expected to give the DL as three asbestos structures.

[6] Where DLs for asbestos measurements are discussed in the literature, it is unclear whether they are based on one-sided or two sided confidence intervals.

4.61 would cover the true mean in 99% of the cases. " Confidence intervals provide useful information for interpreting asbestos measurements, but there is no direct logical bridge between the confidence interval statement and a DL.

The stated detection limits in current ASTM asbestos methods will only lead to the obscuration of useful information, since some analysts may take the stated detection limits as representing the lowest point at which valid information can be reported. It is far better to report both sample and blank results rather than censoring certain measurements because they are "below detection."

Asbestos in Dust Measurements

Sampling And Analysis Methods - For the sake of simplicity, we will limit our discussion to those methods in which analysis is by electron microscopy.

Sampling by "Microvacuum" — Examples of this sampling approach are found in ASTM's Standard Test Method for Microvacuum Sampling and Indirect Analysis of Dust by Transmission Electron Microscopy for Asbestos Structure Number Concentrations (ASTM D 5755-95) and Standard Test Method for Microvacuum Sampling and Indirect Analysis of Dust by Transmission Electron Microscopy for Asbestos Mass Concentration (ASTM D 5756-95). The sampling portions of these methods are similar to a method first published in 1985 [*6*] as a method for lead dust sampling. Surface dust is collected by moving a standard air sampling cassette that has been modified by addition of a nozzle on the inlet over a defined area - usually 100 cm^2 - while drawing air through the cassette. The filter with collected dust is then analyzed as an air sample filter after indirect sample preparation. The original sampling filter is dissolved, and a portion (typically 10-100%) of the solute is refiltered. The analysis will be of the secondary filter, with the analytical techniques similar to those used for air samples. The results may be reported as either s/cm^2 or mass units/cm^2.

Wipe Sampling — [An example is ASTM Draft Standard Test Method for Wipe Sampling and Indirect Analysis of Dust by Transmission Electron Microscopy for Asbestos Structure Number Concentrations (ASTM D 22.07.Pxxx.Dxx).] Samples are collected by wiping a surface, and the collected surface dust is transferred to a filter for analysis. Indirect preparation is always required for TEM analysis, and the wiping object must be carefully selected to be compatible with the analysis selected, since it must typically be incorporated in the entire preparation process. Once again, once the collected dust is placed on a filter, it can be treated as an air sample filter. Results are reported as asbestos structures per unit area.

Adhesive lift sampling — [An example is ASTM Draft Standard Test Method for Direct Analysis of Dust by Electron Microscopy for Mineral Fibers (ASTM D 22.07. P009.D05).] In this approach, a suitable adhesive material (specialized commercial devices, or forensic tape) is placed on a surface and the removable surface dust that adheres to the sampling device is taken for analysis. Usually, an attempt is made to perform direct analysis on the collected dust by transferring the sampling surface directly to an SEM stub, and the results are reported as asbestos structures per unit area.

Passive Sampling — [An example is Draft Standard Test Method for Passive Sampling and Indirect Analysis of Dust by Transmission Electron Microscopy for Asbestos Structure Number Concentration (ASTM D 22.07.P010.D07).] A suitable container is placed in an area where the sedimentation of asbestos-containing particles of sufficient size and density to fall by virtue of their own weight is of interest. The container is left in its selected location for an appropriate period of time, and the collected particles are then transferred to a filter, with further preparation by a direct method. The results may be reported as either asbestos mass or asbestos structures per unit area per unit time, by analogy with the dustfall techniques used in the early days of air pollution evaluations.

These methods have been used for several purposes. Among those uses are:

- Determining the presence of asbestos on a surface;
- Defining the extent of spread of asbestos from a suspected source;
- Determining the quality of a cleanup (is "too much" asbestos left?);
- Comparing a surface or surfaces to other surfaces with regard to asbestos contamination.

While the uses above are most likely to rely on qualitative or semi-quantitative comparisons, some users also contend that dust sampling and analysis can be used to predict potential exposures to airborne asbestos, by such devices as calculation of redispersion indices and the like.[*7*] Some of the fundamental problems associated with such attempts to use quantitative approaches to dust sampling results have been discussed elsewhere. [*8,9*] Those problems include disintegration of particles during indirect preparations thus falsely increasing structure counts and precluding any understanding of the nature of the original deposit, and inhomogeneity of surface deposits thus making quantitative direct analyses unlikely to provide reasonably precise estimates of original surface deposition patterns. Nonetheless, it is worthwhile to explore the sensitivity of these techniques in order to assist the potential user.

As can be seen from the brief summaries above, the same concepts regarding sensitivity as were applied to the air sample examples may be applied to some of these dust sample methods, with the substitution of surface area for air volume. That is, the sensitivity (in structures/cm^2) for the microvacuum method (with a direct preparation) will be:

$$\text{Sensitivity} = [(FCA/(\#Fs*FA)]/A*D$$

where:

FCA is filter collection area in square millimeters (mm^2);
#Fs is number of microscopical fields;

FA is average field area in mm^2;
A is surface area sampled in square centimeters (cm^2); and,
D is the dilution factor (~0.01 - 1.0)

As an example, if the surface area sampled is 100 cm^2, the area of a grid opening (field) is 0.009 mm^2, 10 grid openings are counted, the dilution factor is 0.1, and the area of the filter is 385 mm^2:

$$\text{Sensitivity} = [385/(10*0.009)]/100*0.1 = 427.8 \text{ s/cm}^2$$

If the surface area sampled is increased to one (1) square foot (= 929 cm^2), with all other factors held constant:

$$\text{Sensitivity} = [385/(10*0.009)]/929*0.1 = 46 \text{ s/cm}^2$$

It is noted that the sensitivity of ASTM D 5755-95 is given in that method as "1000 s/cm^2" which would imply a dilution factor of ~0.04. It should also be noted that the actual (equivalent) surface area examined at that dilution would be ~1/100,000 of 100 cm^2, or 0.001 cm^2, or 0.1 mm^2 or ~ 0.00016 in^2. It should be noted that a relatively "clean" surface area may yield samples that can be analyzed without dilution (D = 1.0). If a small (25 mm) secondary filter is also used, the sensitivity of the method may be reduced to $<$ 10 s/cm^2.

A dilution factor correction will also always be needed for those other methods (passive and wipe sampling) for which indirect preparation is used. As an example, the sensitivity of the draft passive sampling method ASTM D 22.07.P010.D07 is given as 260 s/cm^2, which implies a dilution factor of about 0.08, if it is assumed that 10 fields are to be examined, and that the original collection area is about 50 cm^2.

For the special case of adhesive lift sampling with direct preparation, the sensitivity can be calculated directly, since the intermediate step of collection onto a filter is not needed. As an example, assuming 10 grid openings analyzed, the sensitivity will be (assuming a single structure counted):

$$\text{Sensitivity} = 1/(10*0.009 \text{ mm}^2*1\text{cm}^2/100 \text{ mm}^2) = 1/0.0009 \sim 1100 \text{ s/cm}^2$$

It is noted that the sensitivities and the actual effective areas examined in the microvacuum method and the direct lift method are not significantly different, if the kinds of dilutions often used for "dirty" surfaces are assumed for the microvacuum method.

Figure 2 shows the same kind of comparison for the dust sampling methods (using 1000 s/cm^2 as the target) as was shown in Figure 1 for the air sampling example. This selected target value of 1000 s/cm^2 has no regulatory or other standard status, but is selected merely as an example. In the past, a surface dust loading of 1,000,000 s/ft^2 has sometimes been used to differentiate between "contaminated" and "uncontaminated"

Figure 2: Dust Sampling Design Characteristics

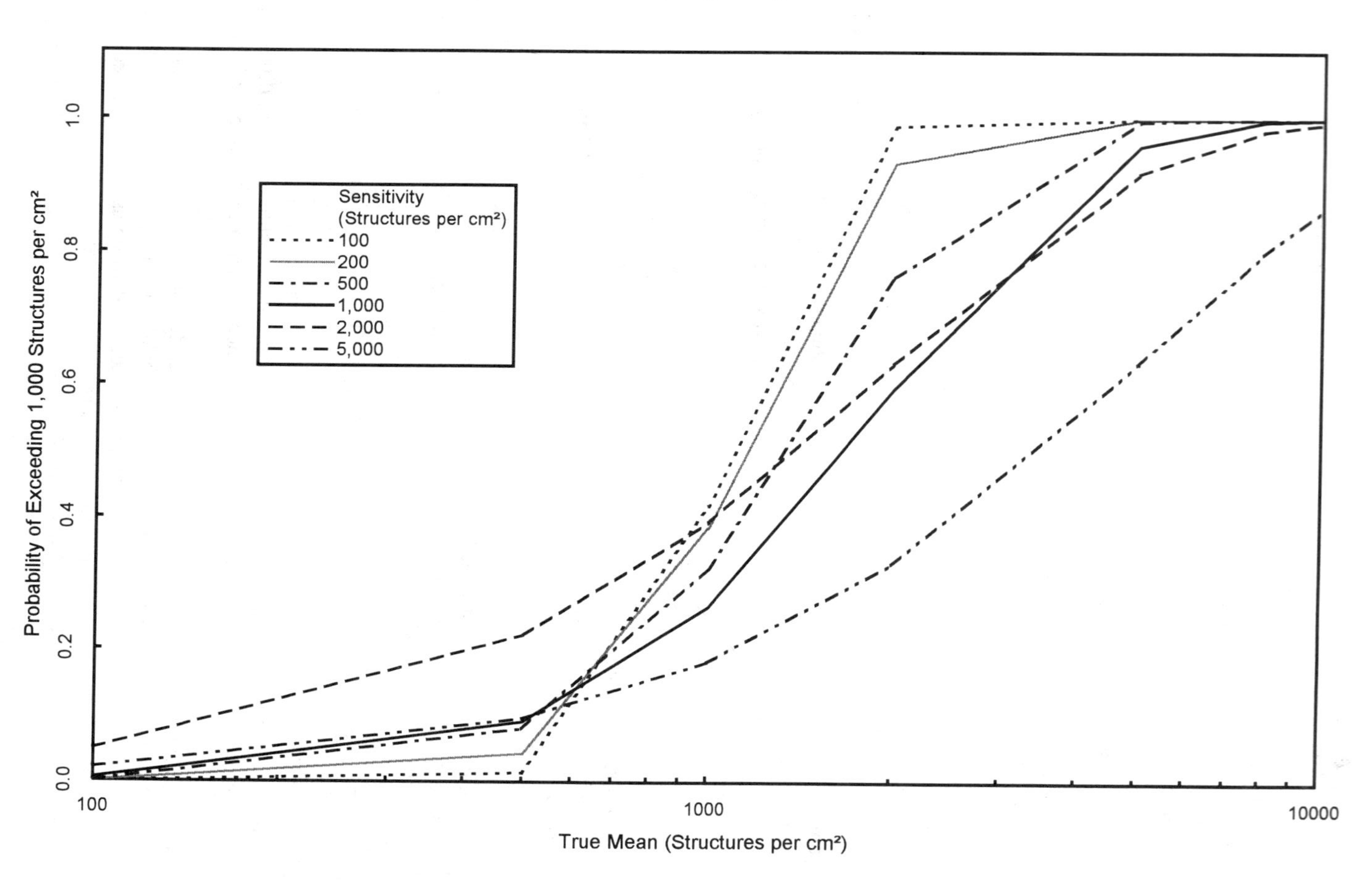

surfaces, and 1000 s/cm^2 = 929,000 s/ft^2.

As with the sensitivity of air sampling methods, the sensitivity of dust sampling methods may be used to determine the reliability and power of conclusions drawn from dust sample data, and to set up the sampling strategy. Assume that a goal of 1000 s/cm^2 has been set, that is, we wish to determine with some reasonable reliability whether or not a dirty surface has more than 1000 asbestos s/cm^2 on it. Assume further that we are willing to rely on one single measurement using ASTM D5755-95 to make that determination. (Note that the method calls for multiple samples, and that use of a single sample for any surface would not be compliant with the method. However, note also that most of the dust-sampling results that the authors have seen rely on single samples for single surfaces.) Figure 2 shows that the probability of a correct decision, using the methods available, is relatively small. In other words, the currently used dust sampling methods cannot reliably determine asbestos in dust concentrations of 1000 s/cm^2 on dirty surfaces.

Quantitatively, even with a sensitivity of 100 s/cm^2, near certainty of a correct noncompliance decision (we say that the surface concentration is $\geq$1000 s/cm^2) is not attained until the true surface concentration is above 2000 s/cm^2. By analogy to the air example for the STEL, a sensitivity of ~ 10 s/cm^2 would be needed to give the same level of probability. That is, given a dirty surface to be evaluated, the analysis of 1000 grid openings would be required, rather than the currently required 10 grid openings, if we were to modify the microvacuum/structure count method (ASTM D5755-95) for the reliable detection of 1000 s/cm^2. However, if the surface to be evaluated is "clean" (e.g. a surface that has been decontaminated, is visually clean, and is being checked for residual asbestos) ASTM D 5755-95 may be perfectly suitable for such use.

The probabilities shown in Figures 1 and 2 and their interpretations are correct subject to the validity of the Poisson assumption. Fibers (structures) may be found in clumps on the filter causing increased measurement variability. The effect on planning a study with the assumption that the Poisson distribution is apt if variability is larger is that the numbers of samples, or the sample sizes, may be inadequate to achieve the study objectives. Such increased variability is often best described by the Negative Binomial distribution. No significant increased effort is required at the electron microscope in order to ensure that appropriate data are obtained to determine the goodness of fit of the Negative Binomial. All that is required is the distribution of counts per microscope field, which is easily obtainable from the count sheet.

However, the data may not have sufficient statistical power to differentiate between the Poisson and Negative Binomial Distributions. In general, the Poisson assumption will almost always be adequate for indirect preparations, but may not be adequate for direct preparations. If the number of fields inspected is large (e.g., 20 or 30 or more), the Normal distribution usually will be an acceptable approximation for counts per field regardless if the true underlying distribution is Poisson or Negative Binomial.

CONCLUSIONS

Sensitivity is an important planning tool for air studies and dust studies, but you have to know what you are attempting to achieve to use the planning tool properly. None of the current ASTM dust sampling methods have adequate sensitivity to reliably detect surface concentrations of 1000 s/cm^2 or below from dirty surfaces, but they may be suitable for such detection on visually clean surfaces, where the only particulate material of interest may be asbestos.

Many statements regarding detection limit in the analytical literature are improper, and lead to improper decisions. With regard to asbestos counting, the concept of DL is typically useless, and leads only to obscuration of valid and useful information. Except in the case of irremediable sample and blank contamination, all visualized structures or fibers in an asbestos analysis are real and convey valid information about the environment from which the sample was taken.

REFERENCES

[1] Brodsky, A. "Exact Calculation of Probabilities of False Positives and False Negatives for Low Background Counting," Health Phys. 63, 198-204, 1992.

[2] Altshuler, B, and B. Pasternack. "Statistical Measures of the Lower Limit of Detection of a Radioactivity Counter," Health Phys. 9, 293-298, 1963.

[3] Currie, L.A., "Limits for Qualitative Detection and Quantitative Determination: Application to Radiochemistry," Analytical Chemistry, 40, 586-593, 1968.

[4] Currie, L.A., Lower Limit of Detection: Definition and Elaboration of a Proposed Position for Radiological Effluent and Environmental Measurements. Publication NUREG/CR4007, National Bureau of Standards. 1984. Available from National Technical Information Service (NTIS), Alexandria, VA.

[5] Fowler, D.P., "Definition of Lower Limits for Airborne Particle Analyses Based on Counts and Recommended Reporting Conventions," Inhaled Particles VIII: Proceedings of an International Symposium Organised by the British Occupational Hygiene Society, pp. 203-209, Editors N. Cherry and T. Ogden, Elsevier Science, Ltd., Oxford, 1997. [Also published as Ann. Occup. Hyg. Vol. 41, Suppl. 1.]

[6] Que Hee, S.S., B. Peace, C.S. Clark, J.R. Broyle, R.L. Bornschein, and P.B. Hammond, "Evolution of Efficient Methods to Sample Lead Sources, Such as House Dust and Hand Dust, in the Homes of Children," Environmental Research, 38, 77-95, 1985.

[7] Millette, J. and S. Hays, Settled Asbestos Dust Sampling and Analysis, Lewis Publishers, CRC Press, Inc., 1994.

[8] Fowler, D.P., and E.J. Chatfield. "Surface Sampling for Asbestos Risk Assessment," Inhaled Particles VIII: Proceedings of an International Symposium Organised by the British Occupational Hygiene Society, pp. 279-286, Editors N. Cherry and T. Ogden, Elsevier Science, Ltd., Oxford, 1997. [Also published as Ann. Occup. Hyg., Vol 41, Suppl. 1.]

[9] Sahle, W. and I. Laszlo, "Airborne Inorganic Fibre Level Monitoring by Transmission Electron Microscope (TEM): Comparison of Direct and Indirect Sample Transfer Methods," Ann. Occup. Hyg., 40, 29-44, 1996.

Owen S. Crankshaw[1], Robert L. Perkins[1], and Michael E. Beard[2]

AN OVERVIEW OF SETTLED DUST ANALYTICAL METHODS AND THEIR RELATIVE EFFECTIVENESS

REFERENCE: Crankshaw, O.S., Perkins, R.L., and Beard, M.E., **"An Overview of Settled Dust Analytical Methods and Their Relative Effectiveness,"** *Advances in Environmental Measurement Methods for Asbestos, ASTM STP 1342,* M.E. Beard, H.L. Rook, Eds., American Society for Testing and Materials, 2000.

ABSTRACT: Methods for sampling and analyzing asbestos in settled dust can be beneficial to document past (and potentially ongoing) episodes of asbestos contamination and to predict potential problems presented by asbestos-containing dust.
Research Triangle Institute conducted an evaluation of several methods for dust collection and analysis, utilizing samples from industrial settings, samples from residential settings, and samples created in a laboratory dust-generation chamber. Sample collection techniques included microvacuuming, wipe sampling, tape sampling, and passive sampling. Analytical methods tested included fiber counting/sizing, fiber mass determination, qualitative analysis, and indirect and direct sample preparation procedures.
The test results help illustrate the advantages and disadvantages of each technique. Each of the methods tested has specific attributes and limitations. Because of the inherent complexity of the methods and the typical variability found in real-world samples, numerous samples of each sample type are recommended, including side-by-side duplicates, representative sampling throughout the target area, and repeat sampling to determine temporal effects.

KEYWORDS: asbestos, settled dust, analysis, transmission electron microscopy, fibers.

The presence of asbestos in settled dust indicates that asbestos has historically been present in the localized atmosphere as airborne particles, and the possibility exists that this asbestos could be re-entrained in the air by future disturbances and activities. Therefore, capable methods for sampling and analyzing settled dust would be beneficial to document past (and potentially ongoing) episodes of asbestos contamination and

---

[1]Research Environmental Scientist and Manager, respectively, Center for Environmental Measurements and Quality Assurance, Research Triangle Institute, Research Triangle Park, NC 27709.

[2]Consultant, 4004 Brewster Drive, Raleigh, NC 27606

to predict potential problems presented by asbestos-containing dust.

Over the past several years, Research Triangle Institute (RTI), under contract to the US Environmental Protection Agency (USEPA) has engaged in research to test the efficacy of various methods for the sampling and analysis of asbestos in settled dust. Some of these methods were originally developed at the grass roots level; some eventually ended up being refined and formalized as ASTM methods. The goals of the research were to determine the strengths and weaknesses of the various methods, both in real-world residential and industrial settings, and in an artificial laboratory environment where reproducibility, precision, and accuracy could be tested.

The four methods tested in this study include ASTM Standard Test Method for Microvacuum Sampling and Indirect Analysis of Dust by Transmission Electron Microscopy for Asbestos Structure Number Concentrations (D5755), ASTM Standard Test Method for Microvacuum Sampling and Indirect Analysis of Dust by Transmission Electron Microscopy for Asbestos Mass Concentration (D5756), Draft ASTM Standard Test Method for Passive Sampling and Indirect Analysis of Dust Fall by Transmission Electron Microscopy (D22.07.P010.D03), and Draft ASTM Standard Test Method for Sampling and Direct Analysis of Dust for Asbestos by Transmission Electron Microscopy (D22.07.P009.D01).

Description of Dust Sampling and Analysis

For all of the studies, sampling was performed using a variety of techniques to assess type(s) of asbestos present in the settled dust and to quantify asbestos structures per unit area on dust-covered surfaces. Settled dust sampling techniques included aggressive wipe sampling, vacuum sampling, passive dustfall sampling, and adhesive tape sampling. Table 1 shows an overview of each analytical method.

Microvacuum Sampling

Microvacuum sampling utilized a low-volume (2 liters per minute) sampling pump with a 37-mm diameter mixed-cellulose ester (MCE) filter (0.8 μm pore size) attached to the inlet with a length of clear flexible tubing. Attached to the inlet of the cassette was a 2-cm length of tubing to serve as the vacuum inlet. The pump was turned on and the tubing inlet/cassette assembly was swept over a 10-cm by 10-cm sampling area for a 2-minute period, collecting all visible dust. The cassette was then capped, and the cassette and inlet tube were enclosed in an individual plastic bag for storage. A separate cassette and inlet tube were used for each sample. Cassettes and inlet tubes were not reused.

Passive Sampling

Passive sampling involves placing a portable substrate in the path of falling dust. Various media can be used for passive sampling; these studies employed petrislides, which are small round dishes on a rectangular substrate, with a tight-fitting lid. The round sampling portion of the petrislide represents a 17-cm^2 sampling area. Petrislides, with lids removed, were placed on the substrate in order to collect all dust falling on that location. Following the sampling period, each sampler was removed from the location, and the sampler lid was replaced.

TABLE 1-- *Overview of sampling, preparation, and analytical procedures for settled dust methods*

	Sample Collection	Sample Preparation	Sample Analysis	Reporting of Results
Microvacuum	37 mm MCE filter in cassette with inlet vacuum tube connected to low-vol pump	Suspension of dust and filtering of aliquots followed by AHERA direct preparation	Standard AHERA analysis	Asbestos structures/cm^2
Microvacuum Gravimetric	37 mm MCE filter in cassette with inlet vacuum tube connected to low-vol pump	Weighing of dust and filtering of aliquots followed by AHERA direct preparation	Measurement of each structure and calculation of mass	Asbestos mass/cm^2, percent asbestos in dust by mass
Passive	Open face passive collector, e.g. petrislide or metal film canister	Suspension of dust and filtering of aliquots followed by AHERA direct preparation	Standard AHERA analysis	Asbestos structures/cm^2
Wipe	Ashless cellulose filter dampened with water wiped over surface	Ashing of filter; suspension of dust and filtering of aliquots followed by AHERA direct preparation	Standard AHERA analysis	Asbestos structures/cm^2
Direct Tape Lift	Post-it, commercial tape lift, or others applied to surface and removed	Carbon coat and mount for SEM observation or dissolve sampler and mount for TEM analysis	Examination of fiber/matrix relationship; type of asbestos and other materials; visual estimation	Type of asbestos, visual estimate of quantity, qualification of matrix materials

Wipe Sampling

Wipe sampling used ashless cellulose filters dampened with distilled water to collect the dust on the substrate. Previous studies (Crankshaw et al. 1994) have shown that the use of other wipes or organic filter media results in poor sample preparations because of the presence of residue from the ashing process, which obscures the fibers during analysis. After dampening the cellulose filter with water, the filter was wiped over a 10-cm by 10-cm area of the substrate, collecting all visible dust. The filter was then placed in a glass vial for storage.

Direct Tape Sampling

Direct tape sampling involves removing the dust from the substrate using a sticky sheet of material; the intention is to preserve the dust particle spatial relationships and individual particle structural integrity. Two materials were employed for tape sampling: a commercially available tape sampler and Post-it® notes. The commercial tape lift consists of a dried adhesive applied to a plastic sheet. The adhesive was activated by application of an aqueous solution. The sticky side of the sheet was applied to the dust surface and then peeled off. The sample was then placed in a plastic box (dust side up) for storage. Post-it® note samples were collected in the same manner, except that they required no wetting.

Sample Preparation

Microvacuum structure count samples, microvacuum gravimetric samples, passive samples, and proprietary tape lift samples were all prepared for TEM examination by the appropriate methodology. Tape lift samples and Post-it® note samples were also prepared for scanning electron microscope (SEM) examination.

Microvacuum Samples

The microvacuum samples were prepared for analysis and analyzed by the current ASTM draft settled dust protocol. This involves rinsing the particulate material from the cassette, redepositing an aliquot of the suspension on a secondary MCE filter, and preparing the second filter by the standard Asbestos Hazard Emergency Response Act (AHERA) direct preparation technique.

Microvacuum Gravimetric Samples

Sample preparation began with the desiccation and taring of a blank MCE filter. The sample cassette was then rinsed and the suspension was filtered through the tared MCE filter. This filter was dried, weighed, and then ashed in a muffle furnace at 480 °C for 12 hours. The ash was then rinsed into a laboratory bottle, and the suspension was ultrasonicated for 15 minutes. Aliquots were then filtered onto MCE filters and prepared for TEM examination as described above.

Passive Samples

The passive samples collected on petrislides were prepared by sequential rinsing of the material in the petrislide to create an aqueous suspension of the total material. Aliquots of each suspension were then deposited onto 0.45 µm pore size MCE filters for subsequent direct preparation.

Wipe Samples

The wipe samples were ashed at 480°C for approximately 12 hours to remove all organic material. The residue from each sample was then dispersed by ultrasonication in distilled water. Aliquots of the suspensions were redeposited onto 0.45 µm pore size MCE filters, which were then prepared by the standard AHERA direct preparation technique.

Direct Tape Lift Samples

The commercial tape lift samples were prepared for TEM and SEM examination, while the Post-it® note samples were prepared only for SEM examination. TEM preparation of the tape lift samples involved coating the dust side of the sample with a carbon layer in a vacuum evaporator and then excising portions of the sample for placement on a copper TEM grid. SEM preparation of both the commercial tape lift samples and the Post-it® note samples involved excising a portion of the sample and mounting it, dust side up, on an aluminum SEM stub using conductive paint. The entire sample was then carbon-coated in a vacuum evaporator for sample conductivity.

Sample Analysis

The samples were analyzed by TEM examination; the direct tape lift samples were also analyzed by SEM examination. For the microvacuum structure count samples, the passive samples, and the wipe samples, the goal was to determine the number of structures per unit area. For the gravimetric samples, the goal was to determine the mass of asbestos per unit area and the mass percent of asbestos in the dust. For the direct tape lift samples, the goal was to determine the number of structures per unit area, the relationship of the asbestos to other materials in the sample, and the structural appearance of the asbestos in the sample.

Microvacuum Structure Count, Passive, and Wipe Samples

The microvacuum structure count samples, the passive samples, and the wipe samples were all analyzed by TEM examination using standard AHERA TEM counting rules. This process typically involves noting each asbestos fiber, bundle, matrix, or cluster observed, including classification of structure type and identification. The number of asbestos structures per TEM grid opening were totaled, and the results were calculated incorporating dilution factor, effective filter area, area sampled, aliquot filtered, number of grid openings analyzed, and grid opening area.

Microvacuum Gravimetric Samples

Microvacuum gravimetric samples were analyzed by measuring the length and diameter of each asbestos fiber or bundle. In order to optimize the representativeness of the analysis, the analysis continued until the mass of the largest bundle in 50 grid openings comprised no more than 10% of the total mass of all structures observed or until the total number of grid openings analyzed was either 100 or [200 000/selected magnification], whichever was smaller.

The size measurements were recorded, and a mass for each fiber or bundle was determined, incorporating the density of chrysotile. A typical analysis consisted of recording the dimensions of 300 to 500 asbestos fibers and bundles and then determining the total asbestos mass for the sample. The total mass for each sample and the mass per unit area were calculated incorporating dilution factor, effective filter area, area sampled, aliquot filtered, number of grid openings analyzed, and grid opening area.

Direct Tape Lift Samples

Only the commercial tape lift samples were analyzed by TEM. The intention of the analysis was to determine the number of structures per unit area on the TEM grid. However, the loading of the samples was so high that it prevented any quantitative analysis due to a method requirement that the grid openings be less than 25% covered with particulate material. Qualitative analysis was done, however, to determine the presence of asbestos, to identify other materials present, and to characterize the structural condition of the asbestos. The commercial tape lift samples and the Post-it® note samples were both analyzed by SEM to determine the structural relationships of the asbestos and the other materials present.

Industrial Study Site

In order to evaluate dust sampling and analytical techniques in an industrial environment, a tire-brake repair shop was chosen as a study site. This location would be expected to generate airborne asbestos because of the long-time use of asbestos in brake linings and the common practice of using jets of compressed air to remove accumulated dust from brake assemblies. No asbestos-containing materials (ACMs) were found in the building, as determined by polarized light microscopy (PLM) analysis of potential asbestos-containing materials. The business had been in operation for about five years at the time of the study. Fortunately for this study, their "housekeeping", or lack thereof, assured an ample supply of dust-covered surfaces. To the best of the employees' memories, the dust had been accumulating since the opening day of business.

Analytical Results

The brake shop represented an industrial environment with the typical combination of airborne dust, oil mist, etc., along with a known source of asbestos dust. All results are summarized in Table 2.

TABLE 2-- *Results of analysis for asbestos in settled dust in an industrial setting*

	Microvacuum Structures/cm²	Wipe Structures/cm²	Passive Structures/cm²	Passive Structures/cm² per day
Location 1	120	1 300 000	41 000	1 500
Location 1 duplicate	*	1 400 000	*	*
Location 2	230	1 600 000	89 000	3 200
Location 2 duplicate	*	1 700 000	*	*
Location 3	810	7 400 000	75 000	2 700
Location 3 duplicate	*	6 000 000	*	*
Location 4	*	*	55 000	2 000
Location 5	*	*	62 000	2 100

*samples not collected in these locations

Microvacuum samples--The microvacuum samples all showed low levels of asbestos concentration (below 1,000 structures/cm^2).

Wipe samples--The wipe samples had concentrations ranging from 1.3 to 7.4 million structures/cm^2. Of additional note is the high level of agreement between the three paired samples at each sampling location.

Passive samples--Following one month of sampling, all passive samples showed measurable amounts of asbestos, ranging from 41 000 to 89 000 structures/cm^2. The concentration of asbestos on the passive samples multiplied by the 60 months the shop had been in operation is roughly equivalent to the results from the wipe samples.

Air Samples--Both the inside and outside air samples had no asbestos fibers detected (sampled and analyzed using AHERA procedures for area samples).

Discussion

The low level of asbestos in the commercial site microvacuum samples was notable since asbestos was found in much higher concentrations in the passive samples. This apparent discrepancy can be attributed to two potential causes: (1) Asbestos settling in the sampling areas may not remain in place; it may later relocate to lower level surfaces or be carried outside the building, and (2) Other substances are also being deposited on the surface, such as oil mist, dust, etc. (not unexpected for an automotive garage) which may cause the asbestos dust to bind to the surface to a degree that it cannot be removed by the light suction of the microvacuum sampling device.

The high level of asbestos in the wipe samples may be attributable to a more complete sampling of the various layers of accumulated material on the substrates in the brake shop.

The presence of asbestos in the passive samples indicates that asbestos is continuously and measurably accumulating on the surfaces at the brake shop. In addition, because the intersample concentration variability was low, the accumulation appears to be rather even throughout the shop. These passive sample results appear to be more relevant to the evaluation of this environment than microvac samples; they demonstrate the dynamic aspect of dust accumulation and serve as an indicator of potential hazard from airborne asbestos dust.

Residential Study Sites

In order to evaluate dust sampling and analytical techniques in a residential environment, two residential study sites were chosen. The first residence, denoted Residence One, contains a known source of asbestos fibers, and the second residence, denoted Residence Two, contains no known asbestos fiber sources.

Residence One was built in 1927; the basement contains a boiler and pipes which are wrapped with asbestos-containing plaster and paper. The asbestos wrap was encapsulated in 1993 with canvas/adhesive, though no overall cleanup of accumulated basement dust was done. Because asbestos was previously identified in this dust, the home has a known source of asbestos fibers for redistribution and accumulation in settled dust.

Residence Two was built in 1970; no ACMs were found in the residence, as determined by PLM analysis of building materials. Theoretically, any asbestos found in the settled dust in this home would have come from outside sources or from items brought into the home subsequent to construction.

Residence One Analytical Results

Residence One represented the location with a friable asbestos source. Of particular concern was the degree to which the asbestos had dispersed throughout the residence, including the upstairs living areas. All analytical results are summarized in Table 3.

TABLE 3-- *Results of analysis for asbestos in settled dust in residential settings*

	Microvacuum Structures/cm²	Wipe Structures/cm²	Passive Structures/cm²	Passive Structures/cm² per day
Residence One				
Basement 1	62 000	210 000	77 000	1 700
Basement 1 duplicate	33 000	140 000	*	*
Basement 2	6 800 000	7 000 000	150 000	3 300
Basement 3	350 000	600 000	42 000	910
Basement 4	120 000	310 000	14 000	300
Basement 5	*	*	280 000	6 100
Upstairs 1	7 100	5 300	11 000	240
Upstairs 2	11 000	4 900	7 000	150
Upstairs 3	240	1 800	13 000	280
Residence Two				
Location 1	16 000	23 000	650	6.9
Location 1 duplicate	14 000	9 800	*	*
Location 2	6 700	3 600	330	3.5
Location 3	440	890	330	3.5
Location 4	*	*	<330	<3.5
Location 5	*	*	<330	<3.5
Location 6	*	*	650	6.9

*samples not collected at these locations

Microvacuum samples--The five microvac samples collected in the basement showed levels of asbestos ranging from 33 000 to 6 800 000 structures/cm². Two microvac samples taken side-by-side on the duct surface gave results of 62 000 and 33 000 s/cm², respectively. The three microvac samples collected in the upstairs occupied portion of the house showed levels of asbestos ranging from 240 (at the detection limit) to

11 000 s/cm^2.

Wipe samples--The five wipe samples collected in the basement showed levels of asbestos ranging from 140 000 to 7 000 000 s/cm^2. All basement wipe samples were taken at side-by-side locations with respect to the five microvac samples, and had, on average, 2.6 times more asbestos than their microvac counterparts. The two wipe samples taken side-by-side on the duct surface gave results of 210 000 and 140 000 s/cm^2, respectively (compared to 62 000 and 33 000 s/cm^2 for the microvac samples in the same location). The three wipe samples collected upstairs showed levels of asbestos ranging from 1 800 to 5 300 s/cm^2. The wipe samples upstairs were comparable to their microvac counterparts from the side-by-side sampling.

Passive samples--The five passive samples collected in the basement showed levels of asbestos ranging from 7 000 to 280 000 s/cm^2. There appeared to be no pattern regarding level of asbestos and proximity to the boiler (the sample taken on top of the boiler had 77 000 s/cm^2). The three passive samples collected upstairs showed levels of asbestos ranging from 7 000 to 13 000 s/cm^2.

Air samples--The two air samples collected downstairs showed asbestos concentrations of 0.077 and 0.098 structures/cm^3 (s/cm^3), respectively. Both upstairs air samples had concentrations below the detection limit (samples were collected and analyzed using the AHERA methodology for area samples).

Residence Two Analytical Results

Residence Two represented the residence for which there is no known asbestos source inside the home. Any asbestos present would presumably have entered the home after the completion of construction.

Microvacuum samples--The four microvac samples showed asbestos levels ranging from 440 to 16 000 s/cm^2 (Table 3). The two samples taken side-by-side had concentrations of 16 000 and 14 000 s/cm^2, respectively.

Wipe samples--The four wipe samples showed asbestos levels ranging from 890 to 23 000 s/cm^2. The two samples taken side-by-side had concentrations of 23 000 and 9 800 s/cm^2, respectively (compared to 16 000 and 14 000 s/cm^2 for the microvac samples in the same location).

Passive samples--The six passive samples taken throughout the residence showed asbestos levels ranging from below the detection limit of 330 to 650 s/cm^2.

Air samples-- No air samples were collected at this site.

Discussion

The presence of substantial asbestos in the basement of Residence One indicates that the release of asbestos from the boiler and pipes resulted in widespread dissemination of fibers throughout the basement areas. This most likely occurred primarily before the boiler and pipes were encapsulated or during encapsulation activities; however, as the passive samples demonstrate, the dispersion of asbestos fibers is continuing to result in measurable accumulations or redistributions. The effectiveness of the wipe sampling technique was higher than the microvac sampling technique in the basement where concentrations were high. The level of asbestos in the upstairs was substantially lower than in the basement. The lack of accumulation of asbestos upstairs over the 46-day period indicates that asbestos dispersion from the basement to the occupied areas of the residence is minimal to absent.

The levels of asbestos in Residence Two (mean of 9 300 s/cm^2 from the microvac and wipe samples) were comparable to those in the upstairs areas of Residence One. The source of this asbestos is not known; possibilities include items brought into the home subsequent to construction, ambient airborne asbestos, and introduction of asbestos from the clothing of residential occupants. The passive samples had levels 100 times below those from the occupied locations of Residence One (results corrected for sampling time differences), indicating that current accumulation rates are very low for Residence Two.

Where side-by-side sampling was conducted for microvac and wipe samples, the methods show good precision. Wipe sampling appears to be a more effective collection technique than microvac sampling on surfaces containing high asbestos concentrations, but on cleaner surfaces, the microvac and wipe sampling techniques appear to be relatively equivalent.

The highest residential asbestos levels were those from samples taken in close proximity to a known source of asbestos (Residence One). Those concentrations dropped rapidly, however, on samples taken on other floors. While this is partly a function of the ventilation system and air movement in the house, it does demonstrate differential asbestos concentrations in the residence as a probable function of distance.

The concentrations of asbestos in Residence Two indicate that, regardless of construction materials used, asbestos can be found in measureable concentrations even in presumably "clean" locations.

Lab-Created Samples

In order to perform an objective evaluation of dust sampling/analytical methods, a set of standardized, characterized samples was needed. It was determined that real-world settled dust samples were unlikely to provide the necessary uniformity of concentration or the minimization of other variables, including asbestos concentration, matrix type, homogeneity, asbestos fibrous structure size, emission source, emission rate, and dust thickness. For this reason, the dust used for this project was custom-mixed and dispersed in an environmental chamber prior to sampling.

The goal in the creation of the dust for this study was to create a uniform small fiber dimension which would mimic the typical small fibril sizes found with settled dust particles which have traveled in an airstream for a period of time and then settled out onto a horizontal surface. Chrysotile asbestos was combined with the matrix materials by gravimetric addition to yield the following mixtures:

- 0.1% chrysotile in calcium carbonate,
- 1.0% chrysotile in calcium carbonate,
- 10% chrysotile in calcium carbonate, and
- 1.0% chrysotile in vermiculite.

The dust was dispersed in an environmental chamber using a dry powder insufflator attached to a pump operating at 10-liters-per-minute pressure. The dust was allowed to settle for 48 hours in the environmental chamber, at which time the samples were collected. Prior to using the chamber/dust application system for evaluation of the settled dust methods, the homogeneity of the created dust layer was tested gravimetrically and determined to be acceptable.

Analytical Results and Discussion

Eighty-six samples were analyzed, yielding a total of 33 605 chrysotile asbestos structures counted. The median size for the asbestos fibers/bundles was 1.8 μm in length by 0.067 μm in diameter. The mean size was 2.5 μm in length by 0.10 μm in diameter.

Sampling efficiency--Sampling efficiency was determined by subjective means only. The microvacuum sampling method utilized a low-volume pump capable of picking up most loose particles. However, following sampling, there was noticeable material remaining on the sampling surface, indicating incomplete dust collection. The wipe sampling appeared to be more effective, collecting all dust from the sampling surface. The passive samples in effect became the sampling surface, so their efficiency relied on the sample preparation technique. For this experiment and the specific thickness of dust used throughout this study, the direct tape lift sample collection technique was quite inefficient; the samplers only collected the top monolayer of dust, leaving the majority of the material behind.

Microvacuum structure count results--Results from the microvacuum structure count samples indicate that the method adequately tracks the concentration of asbestos in the dust (see Table 4) and that variability is quite low. When the concentration of asbestos was increased ten-fold, as from 0.1% to 1% or from 1% to 10%, the number of structures per area increased proportionately. Also, the variability from sample to sample (intersample variation) within one deposition run should be small, indicating low introduction of variables from sampling and preparation. Low variability from grid opening to grid opening (intrasample variation) indicates low sample preparation variability. Intersample variability was typically less than ±15%. The replicate deposition of 1% chrysotile also produced results very close to the original 1% chrysotile results.

TABLE 4-- *Microvacuum structure count results for lab-created samples*

	Asbestos Structures/cm^2 x 10^6		
Formulation	**Median Result**	**Intersample Range**	**Inter-Grid Opening Range**
0.1% in $CaCO_3$	1.2	0.80 - 1.4	0.80 - 1.6
1% in $CaCO_3$	7.5	6.2 - 8.4	5.3 - 9.3
1% in $CaCO_3$ (replicate)	7.8	6.3 - 8.8	5.7 - 9.7
3x1% in $CaCO_3$	9.8	9.0 - 11	7.9 - 11
10% in $CaCO_3$	150	140 - 180	120 - 180
1% in Vermiculite	5.6	4.8 - 5.9	4.2 - 7.1

Microvacuum gravimetric results--The gravimetric samples did not perform as well at tracking asbestos concentration (Table 5) and had considerably higher intersample variability (typically ±80%). The asbestos mass percent concentration found in the gravimetric samples did not correspond well with the known concentration of the asbestos dust. Though some results, especially at the higher concentration levels were reasonably close, the intersample variability caused the average result for each concentration to be significantly biased.

TABLE 5-- *Microvacuum gravimetric sample results for lab-created samples*

	ng Asbestos/cm^2		
Formulation	**Median Result**	**Intersample Range**	**Percent Asbestos**
0.1% in $CaCO_3$	69	49 - 86	0.0018 - 0.0028
1% in $CaCO_3$	3 300	730 - 7 600	0.032 - 0.26
3x1% in $CaCO_3$	22 000	1 100 - 55 000	0.044 - 2.4
10% in $CaCO_3$	140 000	45 000 - 290 000	1.5 - 13

Most of the variability in the gravimetric results was due solely to the effect of the single large initial structure. Only 3 of the 12 analyses continued until the relative mass of the first large structure diminished to 10%; the others were stopped automatically after 100 grid openings, as specified by the method. An average of 2 000 grid openings would have been required to reduce the first structure to the 10% level, and in one case 10 000 openings would have been required. This has the effect of proportionately increasing the bias caused by the first large structure, thereby increasing method variability.

Passive results--The passive samples also tracked the asbestos concentration well and showed low variability (Table 6). Intersample variability was typically less than ±15%. The quantitative results of the passive samples were reasonably equivalent to the microvacuum structure count samples. Though the passive sampling was more efficient, these samples represented a smaller collection area, and any loss of fibers due to sample preparation would have been proportionately larger than with the microvacuum structure count samples, which would have balanced out the final concentration to some degree.

TABLE 6-- *Passive sample results for lab-created samples*

	Asbestos Structures/cm^2 x 10^6		
Formulation	**Median Result (normalized to 7 g dust/application)**	**Intersample Range**	**Inter-Grid Opening Range**
0.1% in $CaCO_3$	0.87	0.79 - 0.98	0.56 - 1.2
1% in $CaCO_3$	7.0	6.0 - 7.8	5.6 - 8.3
3x1% in $CaCO_3$	10	9.7 - 11	9.2 - 11
10% in $CaCO_3$	130	120 - 140	110 - 150
1% in Vermiculite	6.6	5.3 - 7.9	5.0 - 8.4

Wipe results--The wipe samples had the best performance in tracking asbestos concentration (Table 7), with 1.9 million structures/cm^2 (Ms/cm^2) at 0.1%, 17 Ms/cm^2 at 1%, and 200 Ms/cm^2 at 10%. Intersample variability was also less than ±15%. In terms of absolute concentration, the wipe samples were the most effective at fiber collection and retention during preparation, as evidenced by their higher concentration of fibers per unit area compared to the microvacuum structure count samples and the passive samples.

Direct tape lift results--As indicated above, the commercial tape lift samples were unable to be analyzed by TEM as specified by the method, which requires appropriate loadings of less than 25%. SEM examination of the two types of tape lift samples (commercial tape lift samples and Post-it® samples) was carried out, and appeared to be a more appropriate method of examination for these sample types. With SEM analysis, identification of the asbestos fibers and other particles by x-ray spectroscopy and fiber morphology was feasible, and structural relationships between the asbestos and non-asbestos particles (as they appeared on the tape lift sampler surface) was readily observable.

TABLE 7-- *Wipe sample results for lab-created samples*

	Asbestos Structures/cm^2 x 10^6		
Formulation	**Median Result**	**Intersample Range**	**Inter-Grid Opening Range**
0.1% in $CaCO_3$	1.9	1.6 - 2.1	1.6 - 2.3
1% in $CaCO_3$	17	16 - 18	15 - 19
3x1% in $CaCO_3$	17	16 - 18	16 - 19
10% in $CaCO_3$	200	190 - 200	170 - 230
1% in Vermiculite	7.4	7.3 - 7.5	6.5 - 8.4

Summary and Conclusions

The RTI studies have demonstrated that the settled dust methods have low internal variability, and with the exception of the microvacuum gravimetric method, are quite precise. However, each of the methods tested has specific attributes and limitations. Depending upon the environment, some sampling and analysis techniques for settled dust may have greater utility than others, and assumptions made for one environment may not hold true for another. For instance, where settled dust is very loose and is not currently accumulating, the microvac sampling technique should be more appropriate than the passive sample approach. Where dust is still accumulating, and adhering tightly to the surface as it collects (as appeared to be the case at the industrial site), a combination of passive sampling and wipe sampling would be more appropriate than microvacuum sampling.

Passive collection techniques and active collection techniques (microvac and wipe) are both necessary to determine current asbestos accumulation and historical asbestos accumulation. Microvac samples tend to more accurately reflect potential re-entrainable asbestos, wipe samples tend to more accurately reflect all accumulated asbestos, and passive samples provide a measure of current accumulation rates. Air sampling provides a snapshot in time of airborne fiber levels.

It should be pointed out that real-world samples will be likely to have substantial variability, though under the best circumstances (little or no bias or variability introduced during sample collection or preparation and analysis) the variability found should be attributable to the samples themselves and not to the method or to the personnel collecting or preparing the samples.

A comprehensive, effective approach to settled dust analysis would utilize more than one method in order to determine historical accumulation, loose versus bound dust, source location, and current accumulation. Because of the inherent complexity of the methods and the typical variability found in real-world samples, numerous samples of each sample type are recommended, including side-by-side duplicates, representative sampling throughout the target area, and repeat sampling to determine temporal effects.

References

[1] Crankshaw, O.S., Perkins, R.L., and Beard, M.E. "Quantitative Evaluation of the Relative Effectiveness of Various Methods for the Analysis of Asbestos in Settled Dust," *Environmental Information Association Technical Journal,* Vol. 4, No. 1, Summer 1996, pp.6-12.

[2] Crankshaw, O. S., Perkins, R. L., and Beard, M. E. "An Evaluation of Sampling, Sample Preparation, and Analysis Techniques for Asbestos in Settled Dust in Commercial and Residential Environments." *Environmental Information Association Technical Journal*, Vol. 2, No. 3, 1994.

James R. Millette[1] and Michael D. Mount[2]

APPLICATIONS OF THE ASTM ASBESTOS IN DUST METHOD D5755

REFERENCE: Millette, J. R. and Mount, M. D., **"Applications of the ASTM Asbestos in Dust Method D5755,"** *Advances in Environmental Measurement Methods for Asbestos, ASTM STP 1342,* M. E. Beard and H. L. Rook, American Society for Testing and Materials, 2000.

ABSTRACT: The American Society for Testing and Materials Standard Method D5755 (Asbestos Count Dust Method) is used by a variety of industrial hygienists, architects, engineers, laboratories and environmental consultants in their assessments of asbestos fibers in a building. Dust sampling provides information that cannot be obtained with bulk sampling or air sampling. Information obtained using Method D5755 has been used to assess the extent of contamination resulting from an accidental or unauthorized disturbance of asbestos containing material and to assist in making decisions about cleaning a potentially contaminated area. In some cases, the asbestos values in the dust have been compared to the values determined in the dust from other areas deemed to be non-contaminated. In other situations, the levels of 1,000 str/cm^2 or 10,000 str/cm^2 have been used in decision making.

KEYWORDS: Asbestos, Dust, ASTM D5755, Contamination, Loading Index

---

Asbestos may cause adverse health effects when it is inhaled. Asbestos in surface dust, therefore, presents a potential health risk if it is possible for the dust to be disturbed and the fibers become airborne. In industrial processing, the term "surface contamination" is used to describe the fouling of a surface with an undesired or unexpected material. In building assessments, "surface contamination" generally refers to the part of the settled or accumulated surface dust that contains materials that are considered to be toxic (potential health risks). According to the Occupational Safety and Health Administration (OSHA) Field Operations Manual, removable surface contamination may need to be assessed because accumulated toxic materials (i.e., asbestos, lead or beryllium) may become resuspended in air and contribute to airborne exposures [*1*]. In this article, the term "asbestos dust contamination" refers to removable

[1] Executive Director, MVA, Inc., 5500 Oakbrook Parkway, Suite 200, Norcross, GA 30093

[2] CIH, Cape Environmental Management, Inc., 2302 Parklake Dr., Suite 200, Atlanta, GA 30345

surface dust that contains asbestos fibers that might become suspended in the air and contribute to airborne exposures.

In 1988, a precursor to the American Society for Testing and Materials (ASTM) analysis method for asbestos in settled dust entitled: "Standard Test Method for Microvacuum Sampling and Indirect Analysis of Dust by Transmission Electron Microscopy for Asbestos Structure Number Concentrations" (D5755-95) was applied to the problem of asbestos-contaminated furnishings in a multistory office building.[*2*] The office areas occupied by the tenant comprised several floors in the upper part of a downtown high-rise building. Asbestos-containing dust (ACD) from asbestos in the building and an asbestos abatement project on a lower floor had contaminated the tenant's space. In addition, a fire occurring during the asbestos abatement project had distributed smoke containing ACD and soot throughout the tenant's space. In order to move furniture out of the contaminated space into offices in a new building, the prospective landlord required that the tenant ensure that asbestos-contaminated furnishings would not be moved directly into the new space without cleaning. In order to meet this requirement, an extensive decontamination program was designed and monitored. The extent of the project was large; the offices contained over 4,000 desks, chairs, file cabinets, credenzas, wall partitions, computers, telephones, typewriters and calculators. There were also objects of art and personal items that could not be discarded. The area air samples taken during the initial post-fire inspection generally were low, less than 0.02 structures per cubic centimeter (str/cc) by transmission electron microscopy (TEM); but personal samples taken during this initial walkthrough (0.07 str/cc and 0.08 str/cc by TEM) were elevated over a predetermined baseline value of 0.01 str/cc. The microvacuuming technique of dust sampling was used to determine whether surface contamination by asbestos fibers existed in the tenant spaces and to what level. Dust collection and analyses showed that some of the surfaces contained over 60,000 str/ cm^2. In comparison, microvacuum samples collected from the new space that contained no ACM showed levels below 150 str/cm^2. Within fully contained decontamination units, cleaning procedures for smooth surfaces, fabrics, and electronics were found that could achieve the surface contamination criterion of less than 150 str/cm^2 as tested by the microvacuum asbestos dust procedure.

In 1997, a telephone survey of TEM laboratories performed by one of the authors (MM) showed that the microvacuum asbestos dust procedure in the form of ASTM D5755 was routinely applied across the country. Of forty-eight TEM laboratories (not including multiple sites belonging to one company) surveyed from the National Institute of Standards and Technology (NIST) TEM asbestos laboratory list, thirty-five reported using ASTM D5755. Most of the laboratories (30 of 35) indicated that their clients used it for assessment of asbestos in buildings; some (9 of 30) indicated that their clients used the results for clearance activities. Discussions with the laboratories and other consultants provided some additional anecdotal information on how D5755 was being applied. Some consultants are using it for quarterly building operations and maintenance (O&M) assessment. Before purchasing a building, some buyers are contacting laboratories to use the method D5755 to assess the asbestos in dust in the building. In one situation, government office employees are given regular reports of the results of periodic dust sampling to allay fears of the build-up of asbestos contamination. In another situation, regular dust sampling is used to monitor the build-up of asbestos levels in a facility that is no longer in use. The facility area has been cleaned and is awaiting dismantling. The monitoring for elevated levels of asbestos in the dust gives an indication of either deterioration of the material or possible vandalism. One consultant used the results of D5755 analyses which included detailed information about non-asbestos fibers for fiberglass contamination assessment in the dust samples.

The Vacuum (Microvac) Method

The American Society for Testing and Materials (ASTM) analysis method for asbestos in settled dust entitled: "Standard Test Method for Microvacuum Sampling and Indirect Analysis of Dust by Transmission Electron Microscopy for Asbestos Structure Number Concentrations" is designated D5755-95. As described in the "Summary of Test Method" for D5755-95, a sample is collected by vacuuming a known surface area with a standard 25 or 37 mm air sampling cassette using a plastic tube that acts as a nozzle. The plastic tube is attached to the inlet orifice. The sample is transferred from inside the cassette to an aqueous suspension of known volume. Aliquots of the suspension are then filtered through a membrane. A section of the membrane is prepared and transferred to a TEM grid using the direct transfer method. The asbestiform structures are identified, sized, and counted by transmission electron microscopy (TEM), using selected area electron diffraction (SAED) and energy dispersive x-ray spectroscopy analysis (EDXA) at a magnification of 15,000 to 20,000 x.

An interest in assessing asbestos in surface dust can be traced back at least to 1935 where, to evaluate hygiene conditions in asbestos plants, Hurlbut and Williams [*3*] analyzed dust that had settled on rafters for asbestos fibers. Over the years, various procedures such as scooping thick dust into canisters, wiping up dust with wet cloths and vacuuming it into cassettes have been used to collect dust for the analysis of asbestos. At first, much of the analysis of samples was qualitative; was asbestos present or not? However, the need for building owners and consultants to know how much asbestos was present led to procedures in which transmission electron microscopy was used to quantify the number of fibers present [*4,5*]. In the spring of 1989, a group of scientists engaged in dust monitoring presented a session at a meeting of the National Asbestos Council. The members of this group proposed their protocol to ASTM for development as an ASTM standard method. ASTM met the following month and the draft was introduced as a work item under Committee D22. In July of that year, asbestos experts from all over the U.S. and Canada met at the U.S. EPA Research Center in Cincinnati and discussed various methods for the analysis of asbestos in settled dust. An EPA interim draft method based on modifications of the first draft of the ASTM microvacuum method was prepared during that meeting. This draft method specified collecting a sample of settled dust from a known area on a surface and then rinsing out the collection cassette to put all the collected dust particles into suspension. The suspension was dispersed in an ultrasonic bath in a manner developed for the preparation of water samples. A portion of this suspension was filtered through a membrane filter. The filter was prepared in the same way an air sample was prepared for clearance of an asbestos project. The resulting grids were analyzed by transmission electron microscope (TEM). A value for the asbestos on the original surface was calculated from the fibers identified by the TEM method and a result reported in terms of asbestos structures per unit area. The microvacuum procedure was approved by the ASTM in 1995. A second microvacuum procedure that results in a value of asbestos in terms of mass per unit area was also approved at the same time. It is listed as ASTM D5756-95 Method: "Standard Test Method for Microvacuum Sampling and Indirect Analysis of Dust by Transmission Electron Microscopy for Asbestos Mass Concentration". A guidance document for the two methods and other methods for assessing asbestos in dust is being developed by ASTM.

The microvacuum method has also been developed independently in other areas of environmental concern such as lead. The vacuum-filter sampling apparatus described by Que Hee in 1985 consisted of a 37 mm, 0.8 micrometer MCE filter in a plastic cassette attached to a standard industrial hygiene vacuum pump [*6*]. A 5-cm length of plastic tubing was attached to the inlet of the filter cassette, and a 45 degree angle was cut into the sampling end. Samples were collected over a 225-cm^2 sample area at a flow rate of 2 L/min. For lead-containing dust collection, the microvacuum method was selected by some researchers over other more powerful vacuum methods and over wipe methods for several reasons: It is applicable to a variety of surfaces (wood, linoleum, carpets, etc.); it tends to remove only the surface dust most readily available to disturbance; and is light-weight, portable, easy to use, and not dependent on household electricity [*7*]. A recent comparison of vacuum and wipe samplers for lead-containing dusts showed that the vacuum method had a significantly higher recovery from carpet but that the recovery of lead was consistently highest for the HUD wipe method from most other surfaces [*8*]. The HUD wipe method [*9*] uses commercial wipes (respirator wipe pads-alcohol free) wetted with distilled water moved over the sampling area (one square foot) in an "S" pattern using maximum pressure. The wipe is then folded in half, exposed side inward, and the procedure repeated at a 90 degree angle to the first "S".

A wet wipe procedure may also be more efficient in the collection of asbestos fibers from smooth surfaces especially when the dust may be associated with oil or grease, but the vacuuming procedure would tend to collect that dust that is most readily disturbed from all surfaces. A wet wipe asbestos dust collection procedure is being developed by ASTM D22.07.

Because dust layers over a single layer thick cannot be analyzed directly, the microvacuum procedure includes an indirect procedure to disperse the dust particles so they can be examined. Some procedures that have been developed for direct collection and analysis of dust are very useful for determining the source of the asbestos fibers in the dust but are not generally used for quantitative determinations because of overloading problems.

Loading vs. Concentration Indices

When the consideration is surface contamination, the results of tests for many environmental pollutants are given in terms of an amount of the pollutant per unit area of surface. Radioactive contamination measurements are given in terms of dpm/100 cm^2, PCBs in terms of ug/100 cm^2 and lead in terms of ug/ ft^2. These are generally considered to be in terms of a loading index, that is, how much material is present per area of surface. For asbestos, the results for D5755 are generally given in terms of asbestos structures per cm^2. Values in terms of structures per sq. ft. or per sq. meter are used on occasion. Multiplying the value in str/cm^2 by 929 gives the value in terms of str./ft^2 and multiplying the value in str/cm^2 by 10,000 gives the value in terms of str./M^2. Mass or weight percentage values can be calculated with ASTM Method D5756. This value is considered a concentration index. There are important differences between the two.

The number or mass of structures per area sampled (Loading) is very different from the mass percentage of asbestos to total dust (Concentration). Let us suppose that a sample of dust is taken from a building and found to contain 1% asbestos in the dust and in a control building nearby the dust is found to contain 0.001% asbestos. The building owner given this information may assume that his building is not clean compared to the control building. He will then instruct the cleaning crew to clean until his dust is the same as that in the control building. The cleaning crew cleans 90% of the dust out of his building. The 10% of the dust that is left still contains 1% asbestos unless the crew's procedure preferentially takes out asbestos (which is not likely). The building owner then sends them back to clean further. After taking out more dust, they still have the same percentage of asbestos in the dust that is left: 1%. In theory they will never be finished. If the same building owner had been told that his building dust contained 100,000 str/cm^2 and the control building contained 100 str/cm^2, he could have called his cleaning crew and seen some progress with each cleaning. After the removal of 90% of the dust, the asbestos level in his building would be 10,000 str/ cm^2. An additional cleaning with the same efficiency would result in a level of 1000 str/ cm^2. Clearly one more cleaning would reach the desired level. The story illustrates that using a loading (number or mass of structures per area sampled) measure of asbestos in the dust is much more useful in monitoring cleaning activities than a concentration measure.

When Not to Apply D5755

It is not necessary to collect and analyze surface dust samples if pieces of building material are present that can be identified as to a known asbestos-containing material source. In the Asbestos Hazard Emergency Response Act (AHERA) documents, "asbestos debris" is defined as pieces of asbestos containing building material (ACBM) that can be identified by color, texture, or composition [*10*]. If chunks of material are present that clearly appear to be the broken pieces of an in-place ACBM, it may be assumed that an area is contaminated and surface sampling is not needed. It should be noted that both D5755-95 and D5756-95 exclude debris by using a 1 mm screen to filter the washings from the microvac cassette.

Interpretation of Results

If the presence of asbestos is detected in dust on a given surface, the relevant question becomes whether activities associated with the area in which the dust resides will create an elevated airborne concentration of asbestos. A published standard method for dust sampling and analysis is important for comparing sets of dust data.

Based on years of observations and experience, Millette and Hays [*11*] considered levels of asbestos in settled dust as determined by the microvac technique to be low if less than 1,000 str/cm^2 (1 million str/ft^2). Levels above 10,000 str/cm^2 were considered above general background. Levels above 100,000 str/cm^2 were considered high and in the range of what had been observed on surfaces outside of leaky containment areas around abatement sites. A value of 5,100 str/cm^2 was reported by Ewing, Dawson and Alber [*12*] for microvac dust samples collected outside 5 buildings located in a large city.

As resuspension of asbestos is a concern, several studies have attempted to determine a level of airborne asbestos resulting from the disturbance of a measured level of asbestos in the settled dust. The ratio between the air level (in structures/cm^3) and the surface loading level (in structures/cm^2), called the K-factor, was apparently first determined for asbestos by Carter in 1970 [*13*]. Using phase contrast microscopy (PCM) to measure chrysotile levels in both the dust on contaminated materials and in the air during handling of these materials, he determined a ratio of the air level to the surface dust level for handling chrysotile asbestos contaminated material. More recent studies [*11,14*] in which the asbestos air and surface dust levels were measured by TEM have been used to calculate additional K-factors (Table 1). These data suggest that the amount of asbestos fiber suspended into the air during a specific activity may be primarily dependent on the amount of asbestos on the surface initially and the energy of the activity. The activity in which a propane powered forklift was used caused the highest relative air levels. These data also suggest that if the settled dust levels are under 1,000 str/cm^2 that air levels over 0.01 str/cm^3 will not occur for dust disrupting procedures such as general cleaning. The data determined from the apartment cleaning test of Kelman et al. [*15*] tend to support that finding. In that study, an average loading of asbestos in settled dust of 4,000 str/ cm^2 did not produce levels of asbestos in the air above 0.02 str/cc when a vigorous, dry spring cleaning effort was pursued in an unfurnished apartment.

Three laboratory studies performed by the authors also generally support this relationship. These studies were performed in an abatement type enclosure measuring eight feet by eight feet by seven feet high. A person in a protective suit dry swept the floor for 15 minutes in three different test periods. The researcher wore a respirator and a personal air sampling unit for each test. Air samples were collected for the 15 minute activity period. There was no exhaust or air exchange during the testing. Chrysotile asbestos-containing dust was distributed onto the floor of the test chamber enclosure for each of the three tests. The chamber was cleaned between tests. The asbestos dust levels were determined using D5755 and the air levels were determined by TEM using the Asbestos Hazard Emergency Response Act (AHERA) procedure (*16*). The geometric mean asbestos dust loading for the first test was 560 str/cm^2. The air level was below 0.0024 str/cc on the individual performing the sweeping. For a mean surface dust level of 113,000 str/cm^2 in the second test, similar dry sweeping produced 1.2 str/cc. When a dust level of 78,000 str/cm^2 was swept vigorously with a broom and blown with a leaf blower in the third test, the air level was found to be 124 str/cc.

Table 1. Asbestos Levels for Dust and Air in Controlled Studies

Activity	Surface Dust Level (str/cm^2)	Air Level (str/cm^3) Indirect	K-factor (cm^{-1})	Data Source [Ref]
Gym-Athletic Activities	9,700	0.23	2x 10^{-5}	[*11*]
Cleaning Storage Area	870,000	2.7	3 x 10^{-6}	[*11*]
Warehouse (with forklift)	8,200	30	4 x 10^{-3}	[*11*]
Cable Pull	2,000,000	29	1 x 10^{-5}	[*11*]
Broom Sweeping	760,000	54	7 x 10^{-5}	[*11*]
		Air Level (str/cm^3) Direct		
Conventional Carpet Cleaning	23,000	0.09	4 x 10^{-6}	[*14*]
Apartment Cleaning	4,000	<0.02	<5 x 10^{-6}	[*15*]

Application of D5755 to assessment of possible asbestos contamination

The concern over asbestos contamination in public buildings and residences has prompted a variety of actions by a number of different asbestos consultants. The "tools" available to the consultant to aid in making decisions include: visual inspection, bulk sampling, air monitoring and surface dust analysis. Individual buildings may differ considerably and therefore each situation requires the development of a specific plan with possibly a different combination of tools to measure, evaluate and control the contamination, if present.

Those items necessary for a proper building assessment will vary somewhat by the particular situation, as well as what criteria to use in making decisions about the contamination. Determining where the contamination came from may also be important.

In addition to the fact that an assessment requires knowledge of asbestos inspection techniques and should be done in accordance with applicable local, state and federal regulations, the following steps could be included in the assessment plan in making decisions about possible cleanup options:

1. A visual assessment of the building should be performed by someone on the assessment team. Use a floorplan or sketch aids to make certain that all

rooms are examined. Look for suspect asbestos-containing material. Document obvious damage to the material.

2. Review information about previous bulk sampling and if necessary collect bulk samples of each suspect asbestos-containing material. Generally, there should be at least 3 samples collected because to be considered negative for asbestos under OSHA, three samples of a material must show a negative finding. AHERA regulations require at least 3 samples collected from a homogeneous area.

3. Collect samples of debris from each room in which it is present. Note possible source material by comparing the debris with in-place building material on the basis of color, texture and appearance.

4. If greater than 50% of rooms have debris, then it might be cost effective to assume that the entire area should be cleaned if the debris and source material are confirmed to be asbestos-containing.

5. If less than 50% of the rooms have debris, then assume rooms with debris should be cleaned and collect dust samples from the remaining rooms. These microvac dust samples could be collected as three samples in different rooms in 3 different types of situations. Three samples could be collected off high contact horizontal surfaces (tables, desks, etc.) in different rooms. Three samples could be collected from low contact horizontal areas in different rooms where there is accumulated dust. Three samples could be collected from carpeted areas or tile floor, also in different rooms.

6. Analyze the samples from high contact (low accumulation) areas first. If these are over 1000 asbestos str/ cm^2, it is probable that the other samples will also be elevated. If these high contact samples show low concentrations, then analyze the low contact (high accumulation) samples. If the low contact samples show low asbestos concentrations (less than 1000 str/ cm^2), it is probable that the concentration of asbestos in the overall building dust is low. If the high contact samples are low and the low contact samples are high, additional samples could be analyzed to provide a more complete database.

7. Review the data collected in the assessment and determine if additional information is necessary. For instance, if all the samples are high, it may not be necessary to collect samples from the HVAC system. It can be assumed that the HVAC system needs to be cleaned.

There are several ways to use the assessment information to aid in decisions on what to do about asbestos contamination in buildings. One of the best ways is to attempt to compare assessment values with regulatory values. Unfortunately, currently there are no numerical Environmental Protection Agency (EPA) regulations that set the maximum permissible level of asbestos in the air or settled dust of a building. The EPA National Emission Standards for Hazardous Air Pollutants (NESHAP) regulates visible dust in an area where asbestos-containing material (ACM) has been identified. Any visible dust must be cleaned as ACM. However, EPA does not require sampling and analysis by any technique such as D5755 for making this decision. Visible dust in an area with ACM is treated as ACM. The Occupational Safety and Health Administration (OSHA) regulations apply to a worker if demolition or renovation involving ACM is taking place. However, the OSHA regulations may not be protective for young persons or those with health problems.

If the asbestos structure levels in the settled dust in an area are significantly higher than the levels outside the area, the area could be considered contaminated. If there are no differences between the levels of asbestos present in the dust from the area and what is found outside, the area can be concluded to be no more contaminated than other outside areas. In addition, it would not seem possible to clean the area to any better level than is found outside. This type of comparison is used in the Asbestos Hazard Emergency Response Act (AHERA) regulations to determine clearance of school classrooms in the absence of an EPA permissible air level (if the inside air samples are above the 70 str/mm^2 level). In AHERA the airborne concentration of asbestos in the work area after abatement is compared to the airborne concentration of asbestos in an area outside the abatement work area. The abatement work area 'passes' if it is no worse than the area outside the work area. A similar procedure could be used to compare asbestos in settled dust levels.

Under the AHERA clearance procedure, after a school room is abated and cleaned of asbestos debris and dust, aggressive air sampling is done by using a leaf blower to disturb any settled dust that might remain. This is considered a worse case disturbance of the dust. The testing is done in a contained work area that prevents any asbestos from being released into other parts of the building. It is not reasonable to use this aggressive testing in a furnished residence or occupied building as a way to assess whether or not an area is contaminated. If there is any asbestos contamination present, it may be spread over a larger area or contaminate previously uncontaminated items if vigorous dust disturbance with leaf blowers is performed. Therefore, aggressive air testing using a leaf blower is only recommended for clearance testing after cleaning and only within a proper enclosure.

With settled dust it is difficult to define where outside samples can be collected. Settled dust does not stay in one place outdoors when subjected to the elements of weather. Accumulations in protected areas may have resulted from contamination inside an area having been carried outside. The greatest difficulty in using an indoor/outdoor comparison for each area is that different areas could have very different levels of asbestos outside. Therefore one area might be considered contaminated while another with the same level would not because the comparisons are done on different levels.

A collection of data from a number of non-asbestos areas could be a useful approach in determining a baseline value. The two major difficulties with this approach are the difficulties in making certain that each non-asbestos area has never contained any asbestos-containing building materials and in gaining access to the non-asbestos buildings.

To make judgments about what contamination occurred because of a natural disaster, information about the conditions before the disaster event is most helpful. Some information about the physical condition of an asbestos-containing ceiling may be found in old photographs but information about debris or dust levels is generally not available. However, it may be possible to perform some tests to help draw conclusions about whether or not asbestos contamination in a building resulted from the natural disaster.

In some situations it may be possible to find horizontal surface areas that were covered on the day of the disaster event. For instance, a newspaper placed on a desk on the day before the disaster can be sampled both on top and underneath for a comparison of asbestos in dust levels before and after the event. Areas that are known to have been cleaned by wet methods on the day before the event may also be useful.

Microscopical analysis of debris samples can usually establish if the debris particles are of the same combination of constituents as the apparent source material. The usual analysis of a microvac dust sample does not include information about the possible source of the asbestos in the dust. Additional microscopical analysis either with a small portion of the microvac dust preparation or another sample collected by a direct sampling procedure often can provide information about the particles associated with the asbestos fibers (matrices) and thereby provide information about the source of the fibers in the dust. This information does not establish that the asbestos levels resulted from one certain event, but rather may establish that the source of many of the fibers is inside or outside of the residence.

Discussion and Conclusions

Asbestos analysis laboratories are using D5755 to help clients in the assessment of where asbestos is present in the building environment and, in some cases, for clearance activities. Information provided by ASTM D5755-95 is a useful adjunct to data from bulk surveys and air monitoring in making decisions about how to handle asbestos in buildings. The method can be used to assess the extent of contamination resulting from an accidental or unauthorized disturbance of asbestos-containing material and to assist in making decisions about cleaning a potentially contaminated area. The method D5755 results in a loading index of asbestos in the dust described in terms of asbestos structures per square centimeter. Dust loading index values determined using D5755 can be compared to the values determined from the dust in other areas deemed to be non-contaminated.

Current research efforts, although limited, suggest that low levels of asbestos in the settled dust (in the range of 1,000 str/cm^2) do not give rise to significantly elevated levels of airborne asbestos when the dust is disturbed during normal disturbance activities such as cleaning.

References

[1] Occupational Safety and Health Administration (OSHA), Sampling for Surface Contamination, *OSHA Instruction CPL 2-2.20*, Washington, D.C., March 30, 1984, Chapter VIII.

[2] Hays, S.M. and Millette, J.R. DECON: a case study in technology, Asbestos Issues, Feb. 1990. Reprinted as Appendix 2, *In*: Millette, J.R. & Hays, S.M. Settled Asbestos Dust Sampling and Analysis. Lewis Publishers, Boca Raton, 236 pages, 1994

[3] Hurlbut, C.S. and Williams, C.R. The Mineralogy of Asbestos Dust, *Journal Industrial Hyg*iene, 17, 289, 1935

[4] Millette, J. R., Kremer, T., and Wheeles, R. K., Settled Dust Analysis Used in Assessment of Buildings Containing Asbestos, *Microscope*, Vol. 38, pp. 215-219, 1990.

[5] Wilmoth, R. C., Powers, T. J., and Millette, J.R., "Observations on Studies Useful to Asbestos O&M Activities", *Microscope*, Vol. 39, pp. 299-312, 1991.

[6] Que Hee, S.S., B. Peace, C.S. Clark, J.R. Boyle, et al.: Evolution of efficient methods to sample lead sources, such as house dust and hand dust, in the homes of children. *Environmental Research*, 38:77-95 (1985)

[7] Clark, C.S., Bornschein, R.L., Pan, W., Menrath, W., et al. An examination of the relationships between the U.S. Department of Housing and Urban Development floor lead loading clearance level for lead-based paint abatement, surface dust lead by a vacuum collection method, and pediatric blood lead. *Applied Occupational and Environmental Hygiene*, 10:107-110 1995

[8] Reynolds, S.J., Etre, L., Thorne, P.S., Whitten, P., Selim, M. and Popendorf, W.J., Laboratory comparison of vacuum, OSHA, and HUD sampling methods for lead in household dust. *American Industrial Hygiene Association Journal*, 58:439-446 (1997)

[9] Department of Housing and Urban Development (HUD): Lead-based paint: Interim Guidelines for Hazard Identification and Abatement in Public and Indian Housing, Washington, DC:HUD May 1991.

[10] USEPA. Asbestos-Containing Materials in Schools; Final Rule and Notice. 40 CFR Part 763, October 30, 1987.

[11] Millette, J.R. & Hays, S.M. Settled Asbestos Dust Sampling and Analysis. Lewis Publishers, Boca Raton, 236 pages, 1994

[12] Ewing, W.M., Dawson, T.A., and Alber, G.P.Observations of Settled Asbestos Dust in Buildings, *EIA Technical Journal*, Summer, 1996

[13] Carter, R.F., The measurement of asbestos dust levels in a workshop environment. United Kingdom Energy Authority, A.W.R.E. Report N0.028-70, Aldermaston, UK, 1970

[14] Kominsky, J.R., Freyberg, R.W., Chesson, J., Cain, W.C., Powers, T.J. and Wilmoth, R.C., Evaluation of Two Cleaning Methods for the Removal of Asbestos Fibers from Carpet, *American Industrial Hygiene Association Journal*, 51, 9, 500, 1990

[15] Kelman, B. J., Millette, J.R., and. Bell, J.U. Resuspension of Asbestos in Settled Dust in an Apartment Cleaning Situation, *Applied Occupational and Environmental Hygiene*, 9(11):878-878, 1994

[16] Appendix A to Subpart E - "Interim Transmission Electron Microscopy Analytical Methods", U.S. EPA, 40 CFR Part 763. Asbestos-Containing Materials in Schools, Final Rule and Notice. *Federal Register* 52(210): 41857-41894, 1987

Eric J. Chatfield[1]

Correlated Measurements of Airborne Asbestos-Containing Particles and Surface Dust

---

REFERENCE: Chatfield, E. J., "**Correlated Measurements of Airborne Asbestos-Containing Particles and Surface Dust**", *Advances in Environmental Measurement Methods for Asbestos, ASTM STP 1342*, M. E. Beard and H. L. Rook, Eds., American Society for Testing and Materials, West Conshohocken, PA, 2000.

ABSTRACT: Airborne dust from each of several common types of chrysotile-containing materials was generated by abrasion of the material at the base of a vertical elutriator. Air samples were collected on membrane filters situated at the top of the elutriator. The upward linear velocity within the elutriator was selected such that respirable particles and fibers were collected on the membrane filters, while larger, non-respirable particles and fibers fell to the horizontal surface below the base of the elutriator. Transmission electron microscope (TEM) specimens were prepared from the membrane filters by both ISO 10312 and ASTM D5755-95. TEM specimens were also prepared from the settled dust using the technique specified in ASTM D5755-95. The results show that most of the chrysotile structures detected on the TEM specimens prepared from the settled dust samples were of such sizes that their falling speeds would be very much lower than the vertical air velocity in the elutriator. Individual chrysotile structures of these dimensions, therefore, could not have been present in the original dust and debris which settled. The numbers, sizes and characteristics of the chrysotile structures reported by the ASTM D5755-95 method were therefore not representative of the particles as they existed on the original surface, nor of the airborne respirable particles generated by abrasion of the original material. It is also concluded that simple resuspension factor calculations using surface dust measurements, made by either ASTM D5755-95 or D5756-95, do not provide a valid scientific basis for prediction of airborne chrysotile concentrations.

KEYWORDS: Asbestos, chrysotile, air, dust, fiber, transmission electron miscroscopy, resuspension, k-factor, falling speed

Introduction

Measurement of radioactivity in surface dust is accepted as a means of monitoring the presence of radioactive contamination in facilities where radioactive materials are

---

[1]President, Chatfield Technical Consulting Limited, 2071 Dickson Road, Mississauga, Ontario, Canada L5B 1Y8

handled [*1,2*]. In such facilities, definition of an acceptable level of surface radioactivity is a necessary component of routine monitoring. For loose contamination which can be removed from the surface by wipe testing, acceptable levels of surface radioactivity are derived by calculating the value which, under typical conditions of disturbance, could lead to an airborne concentration equal to the maximum permitted level. The calculation is made using the formula:

$$C_a = K.C_s$$

where C_a = air concentration in radioactivity units/unit volume
K = a resuspension factor
C_s = surface concentration in radioactivity units/unit area

It should be noted that the resuspension factor K has the dimension length^{-1}, and the numerical value of K will be different when different sets of units are selected.

Values of K for various scenarios in the nuclear industry have been determined [*3*], both indoors and outside, and a range of 10^{-5} to 10^{-6} m^{-1} is commonly assumed. All of these determinations of resuspension factors were based on measurement of mass, and there was also an implicit assumption that the material measured on the surfaces had settled from airborne suspension and was therefore in a size range that could be resuspended. Another study included measurements of particle size [*4*].

It is clear that no inhalation hazard is presented by material on a surface if it exists only in the form of large fragments, the dimensions of which exceed those capable of remaining suspended in air for substantial periods of time. In such a situation, although the total amount of material per unit area of surface may be very large, disturbance of the material would not give rise to any airborne material, and therefore the resuspension factor would be zero. For decreasing sizes of particles on the surface, the same degree of disturbance would lead to increasing air concentrations, and therefore higher values of resuspension factor. At some point, however, particle aggregation and the scavenging effects of other types of particle must be considered as factors which impose a ceiling value on the resuspension factor. The interaction of other particles with the particle species under consideration is an important effect, as evidenced by the efficiency of commercially-available dust suppression sweeping compounds. The resuspension factor, therefore, is a function of various parameters, including the particle size and state of aggregation of the species being measured, the type and amount of other particles present, the type of surface from which the material is being resuspended, and the nature of the disturbance.

There are two ASTM standard methods for measurement of asbestos on surfaces using a microvacuum device to collect the dust from a known area, followed by analysis of the collected dust by transmission electron microscopy (TEM). These two standards are: Test Method for Microvacuum Sampling and Indirect Analysis of Dust by Transmission Electron Microscopy for Asbestos Structure Number Concentrations (D5755-95); and, Test Method for Microvacuum Sampling and Indirect Analysis of Dust by Transmission Electron Microscopy for Asbestos Mass Concentration (D5756-95). In measurements of asbestos in surface dust, D5755-95 provides results in terms of the number of asbestos structures/unit area of surface. D5756-95 provides results either in terms of the mass of asbestos/unit area of surface, or as the weight percent of asbestos in the collected dust. D5756-95 includes ashing of the collected dust to remove organic materials, and the procedure is specifically designed to disperse complex aggregates of asbestos fibers and particles into their constituent fibers and fiber bundles, so that fiber mass can be determined more reliably. This mass measurement clearly cannot provide information concerning the morphological nature, sizes, and aerodynamic characteristics of the original asbestos-containing particles on the surface, and therefore the method was not included in this study. D5755-95 is also an indirect procedure for preparation of TEM specimens, but the method does not include ashing.

The earliest use of a microvacuum sampling device to collect asbestos from surfaces appears to be that reported by Carter in 1970 [*5*]. In those measurements, the dust originated in a workshop environment by machining of two materials, one containing amosite and the other containing chrysotile. Fibers in the air samples and dust samples were both measured by optical microscopy directly on the collection filters. The resuspension factors calculated from these measurements were generally consistent with those obtained for other types of particles.

Measurements of resuspension factors for various types of particles, including asbestos, have been reviewed by Millette and Hays [*6*]. The possibility that air concentrations of asbestos can be predicted, using surface dust measurements made by D5755-95 in combination with a resuspension factor, has been discussed [*6*]. However, some of the measurements available for asbestos in buildings yield resuspension factors several orders of magnitude lower than the ranges published in the literature for other types of particles [*7*]. D5755-95 is an indirect-transfer TEM specimen preparation procedure, and the question arises as to how any alteration of the particle size distribution which might be introduced by the analytical procedure could affect the calculation of resuspension factors.

The experiments reported here were performed to examine the nature of the respirable particles generated by abrasion of several different chrysotile-containing materials, and to compare these particles to the chrysotile structures which would be counted in the TEM specimens prepared from the dust samples by D5755-95.

Experimental Procedure

The experiments were conducted using a vertical elutriator, as illustrated in Figure 1. The vertical part of the elutriator had an internal diameter of approximately 15 cm and a height of 46 cm. Two 37 mm air sampling cassettes were situated at the top of the elutriator; one cassette contained a pre-cleaned 0.4 μm pore size polycarbonate (PC) filter backed by a 5 μm porosity mixed cellulose ester (MCE) diffusing filter, and the other contained a 0.45 μm porosity MCE filter backed by a 5 μm porosity MCE filter. Each cassette was connected by plastic hose to an air sampling pump. The volumetric flow rate through each filter cassette was accurately controlled at a calibrated value of 0.89 liter/minute by a fixed critical orifice. The base of the elutriator consisted of an adaptor which reduced the internal diameter from 15 cm to 5 cm. The volumetric flow rate through the filters was selected to provide a vertical air velocity of 0.163 cm/s in the

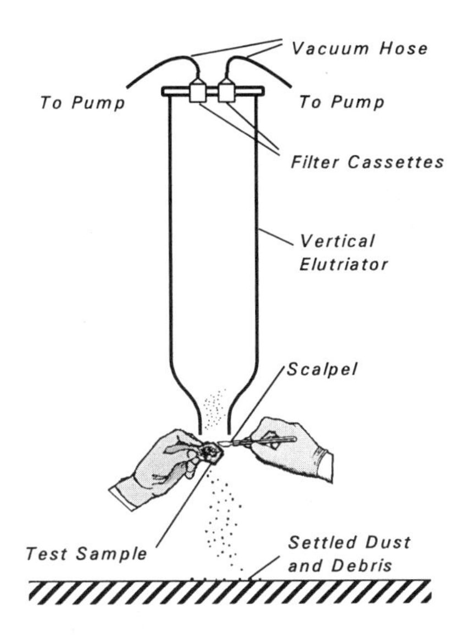

Figure 1--*Vertical elutriator system used for abrasion experiments*

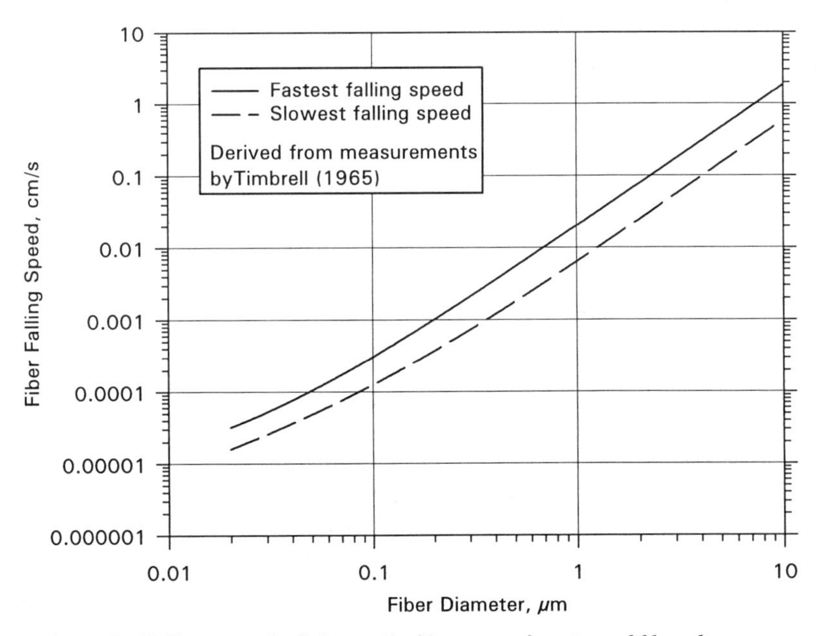

Figure 2--*Falling speed of chrysotile fibers as a function of fiber diameter*

large diameter part of the elutriator, which corresponds to the fastest falling speed of a chrysotile fiber with a diameter of 3 μm, equal to the upper limit of respirability. The vertical air velocity in the small diameter base of the elutriator was 1.5 cm/s.

Figure 2 shows the falling speeds of chrysotile fibers as a function of their diameters, and was derived from experimental measurements by Timbrell [*8*], with the addition of a Cunningham slip factor correction for fiber diameters comparable with the mean free path of air. These measurements of falling speeds are in close agreement with falling speeds derived theoretically, by Sawyer and Spooner [*9*], which formed the basis of guidance issued in 1978 by the U.S. Environmental Protection Agency (EPA) concerning sprayed asbestos-containing materials in buildings.

Prior to each experiment, the test sample was weighed. After starting the pumps, the test sample was held within 1 - 2 cm of the base of the elutriator where it was abraded using the edge of a No. 22 scalpel blade. Abrasion by the scalpel blade was always in an upwards direction, so that any detached particles were driven into the entrance of the elutriator. The air velocity at the entrance to the elutriator was such that smaller particles were drawn up into the elutriator and particles which were too large to be drawn up fell to a clean sheet of aluminum foil on the horizontal surface approximately 40 cm below the base of the elutriator. After the abrasion activity was concluded, the pumps were operated for an additional 15 minutes, in order to ensure that any particulate material which would rise and still remained within the elutriator was collected on the filters.

After completion of the experiment, the remaining portion of the test material was weighed, so that, for quantitative evaluation, the amount of material abraded from the sample could be calculated. The settled dust and debris were transferred to a petri dish.

Direct-transfer TEM specimens were prepared from the air filters using the direct-transfer procedure ISO 10312 [*10*]. Indirect-transfer TEM specimens were also prepared from the air filters by the procedures of D5755-95 to determine the nature of any changes in the size distribution of the airborne particles and fibers when this indirect-transfer analytical procedure was conducted on the airborne size fraction of the material generated by abrasion of the test samples. TEM specimens were prepared from the settled dust and debris samples using the D5755-95 analytical procedure. TEM micrographs were recorded to illustrate the nature of typical particles and fibers found on the direct-transfer TEM specimens, as compared with the nature of typical particles and fibers found on the TEM specimens prepared from the dust and debris by D5755-95. Quantitative chrysotile structure counts, in which structure sizes were recorded, were made on the TEM specimens prepared from the air filters by ISO 10312 and by D5755-95, in order to determine the size distributions of the chrysotile structures measured using the two different TEM specimen preparation methods. For each structure count, attempts were made to accumulate more than 100 structures in the count, but for the direct-transfer preparations this was not feasible within a reasonable amount of effort.

Experiments were performed on six different test samples, representative of some of the types of chrysotile-containing materials often found in buildings. The samples were: ceiling tile containing approximately 1% chrysotile with mineral wool, kaolinite and starch; vinyl-asbestos floor tile containing approximately 15% chrysotile; acoustic surfacing material containing approximately 5% chrysotile with perlite and gypsum; fireproofing containing approximately 15% chrysotile with vermiculite and gypsum;

fireproofing containing approximately 20% chrysotile with mineral wool, gypsum and Portland cement; and, pipe elbow cement containing approximately 60% chrysotile.

Results

For each test sample, an electron micrograph showing one or more typical particles found on a TEM specimen prepared by direct-transfer from the PC air filter is compared with an electron micrograph showing the appearance of the TEM specimens prepared by indirect-transfer from the settled dust samples. The particulate loadings on the air filters were very low; therefore the TEM specimens obtained from the direct-transfer (ISO 10312) preparation of the air filters exhibited only a few chrysotile structures on each grid opening, and generally only one chrysotile structure was visible in a field of view when the magnification was sufficient to illustrate the nature of the structure. The preparation of TEM specimens from the dust and debris samples using the indirect-transfer preparation (D5755-95) includes removal of particles larger than 1 mm and allows adjustment of particulate loadings; therefore micrographs at an appropriate magnification could be obtained showing a number of chrysotile structures in a field of view. However, these loadings meant that many of the chrysotile fibers and small bundles overlapped on these TEM specimens, creating the impression of large clusters. For each test sample, the size distributions measured for chrysotile structures on the TEM specimens prepared from the air sample PC filter by ISO 10312 and by D5755-95 were calculated in terms of the airborne concentrations (in chrysotile structures/mL) for structures within a series of size categories, ranging from 0.5 μm up to the largest chrysotile structure measured, and these data are compared in a table.

Ceiling tile containing approximately 1% chrysotile with mineral wool, kaolinite and starch

The results for the experiment on a sample of ceiling tile are shown in Figures 3 and 4 and Table 1. Figure 3 shows a typical airborne particle observed on the TEM specimens prepared from the PC air filter by the direct-transfer method ISO 10312. Many of the particles on the filter were similar, each consisting of a matrix containing, so far as can be determined by TEM, some or all of the components of the original ceiling tile. Figure 4 shows the appearance of the structures on a TEM specimen prepared by D5755-95 from the settled dust. Comparison of Figures 3 and 4 shows that the morphological characteristics of the airborne particulate released as a result of abrasion of this ceiling tile, and collected on the air filters, bear little resemblance to the chrysotile structures found on the TEM specimens prepared by an indirect-transfer procedure from the settled dust.

Table 1 shows a comparison of the airborne chrysotile structure concentrations calculated for the air sample when TEM specimens were prepared from the PC filter by the direct-transfer procedure ISO 10312 and by the indirect-transfer procedure D5755-95. It can be seen that the preparation of this air filter using D5755-95 caused a large increase to occur in the reported concentrations of chrysotile structures in the shorter size categories. The ratio between the chrysotile structure concentrations derived from the

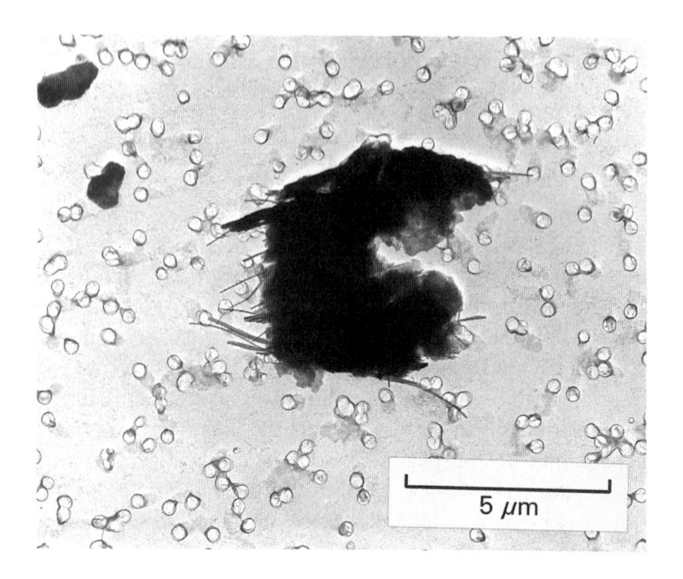

Figure 3--*Airborne chrysotile-containing particle produced by abrasion of ceiling tile*

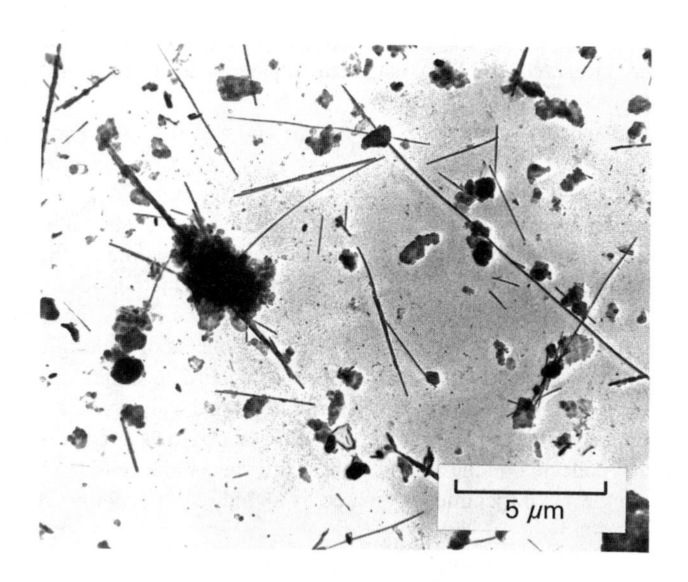

Figure 4--*TEM specimen prepared from settled dust produced by abrasion of ceiling tile*

Table 1--*Ceiling tile: analysis of air sample by ISO 10312 and by D5755-95*

Structure Size Range, µm	Direct-Transfer (ISO 10312)		Indirect-Transfer (D5755-95)	
	Number Counted	Concentration Structures/mL	Number Counted	Concentration Structures/mL
0.50 - 0.73	2	0.55	41	60.15
0.73 - 1.08	7	1.91	43	63.09
1.08 - 1.58	6	1.64	42	61.62
1.58 - 2.32	8	2.18	30	44.01
2.32 - 3.41	9	2.46	13	19.07
3.41 - 5.00	5	1.37	11	16.14
5.00 - 7.34	17	4.64	12	17.61
7.34 - 10.77	15	4.10	5	7.34
10.77 - 15.81	1	0.27	0	0.00
15.81 - 23.21	0	0.00	0	0.00
23.21 - 34.06	0	0.00	0	0.00
Total Structures	70	19.11	197	289.03
Fibers and Bundles	20	5.46	180	264.10
% Fibers and Bundles	28.6		91.4	

indirect-transfer preparation and the direct-transfer preparation decreased for larger structure sizes. Approximately 71% of the chrysotile structures on the TEM specimens prepared using ISO 10312 were complex matrices and clusters, the other 29% being short chrysotile fibers and bundles, whereas 91% of the chrysotile structures in the preparation of the air sample PC filter by D5755-95 were either single fibrils or small fiber bundles.

Vinyl-asbestos floor tile containing approximately 15% chrysotile

The results for the experiment on a sample of vinyl-asbestos floor tile are shown in Figures 5 and 6 and Table 2. The particle illustrated in Figure 5 is typical of the majority of the particles observed on the TEM specimens prepared from the PC air filter by the direct-transfer method ISO 10312. This type of particle exhibited an energy dispersive x-ray (EDXA) spectrum consistent with the inorganic components of floor tile (calcite, dolomite, chrysotile, and titanium dioxide), and some short ends of chrysotile fibrils can be seen emerging from the edges of the particle. No isolated chrysotile fibrils were detected during the examination of the TEM specimens prepared by ISO 10312. Figure 6 shows the appearance of the TEM specimens prepared by D5755-95 from the settled dust. Most of the chrysotile structures were single fibrils of chrysotile.

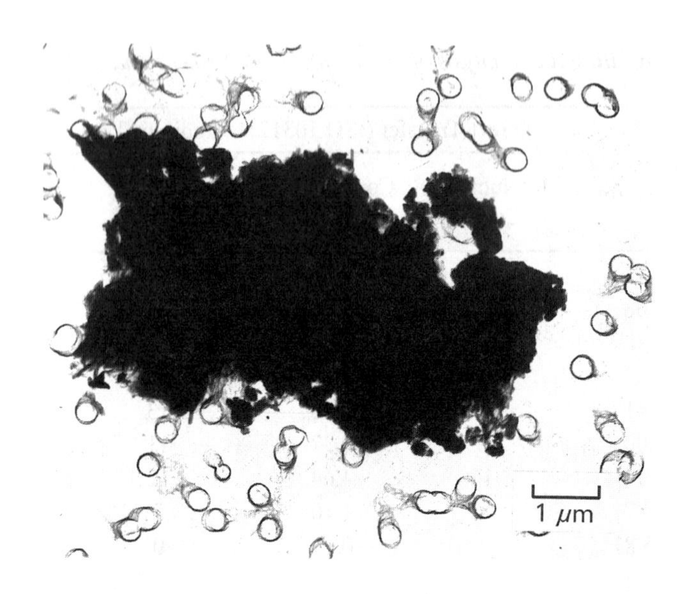

Figure 5--*Airborne chrysotile-containing particle produced by abrasion of floor tile*

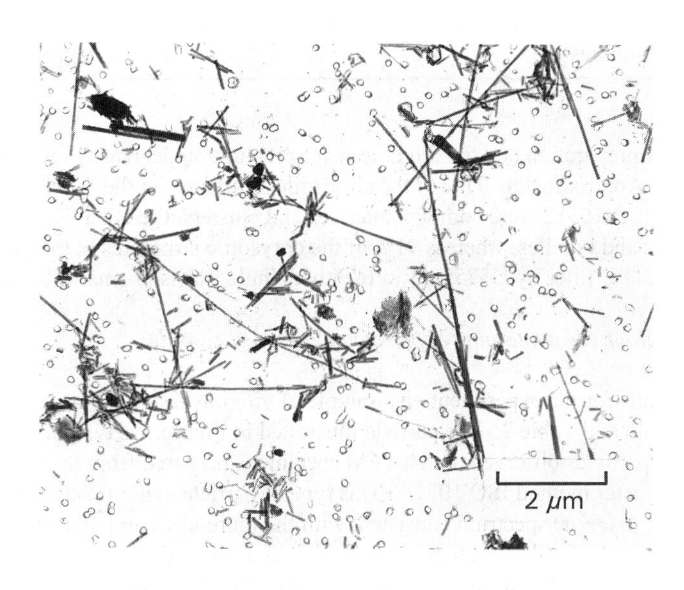

Figure 6--*TEM specimen prepared from settled dust produced by abrasion of floor tile*

Table 2 shows the comparison of the airborne chrysotile structure concentrations calculated for the air sample when TEM specimens were prepared from the PC filter by the direct-transfer procedure ISO 10312 and by the indirect-transfer procedure D5755-95. Only 4 chrysotile structures were detected during examination of 45 grid openings of the TEM specimens prepared by direct-transfer, and all of these particles were matrices. Substantial concentrations of chrysotile structures, particularly in the size categories less than 5 μm, resulted from the indirect-transfer preparation of the PC air filter by D5755-95. Approximately 84% of the chrysotile structures in the preparation of the air sample PC filter by D5755-95 were single fibrils or small bundles.

Table 2--*Floor tile: analysis of air sample by ISO 10312 and by D5755-95*

Structure Size Range, μm	Direct-Transfer (ISO 10312)		Indirect-Transfer (D5755-95)	
	Number Counted	Concentration Structures/mL	Number Counted	Concentration Structures/mL
0.50 - 0.73	0	0.00	48	4.90
0.73 - 1.08	1	0.09	28	2.86
1.08 - 1.58	0	0.00	16	1.63
1.58 - 2.32	0	0.00	10	1.02
2.32 - 3.41	0	0.00	5	0.51
3.41 - 5.00	0	0.00	7	0.71
5.00 - 7.34	2	0.18	1	0.10
7.34 - 10.77	0	0.00	0	0.00
10.77 - 15.81	1	0.09	0	0.00
15.81 - 23.21	0	0.00	0	0.00
23.21 - 34.06	0	0.00	0	0.00
Total Structures	4	0.36	115	11.73
Fibers and Bundles	0	<0.09	96	9.79
% Fibers and Bundles	0		83.5	

Acoustic surfacing material containing approximately 5% chrysotile with perlite and gypsum

The results of the experiment on a sample of acoustic surfacing material are shown in Figures 7 and 8 and Table 3. The particle shown in Figure 7 is typical of the chrysotile-containing matrices found on the TEM specimens prepared by the direct-transfer method ISO 10312 from the air sample collected on the PC filter. These

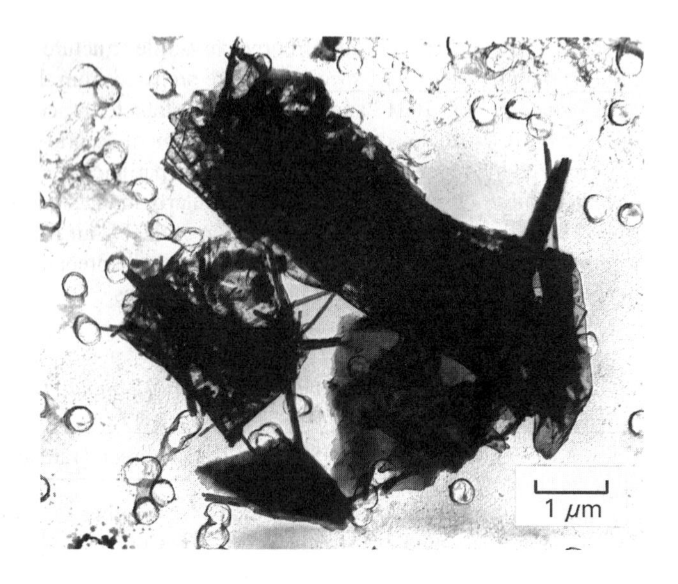

Figure 7--*Airborne chrysotile-containing particle produced by abrasion of acoustic surfacing material*

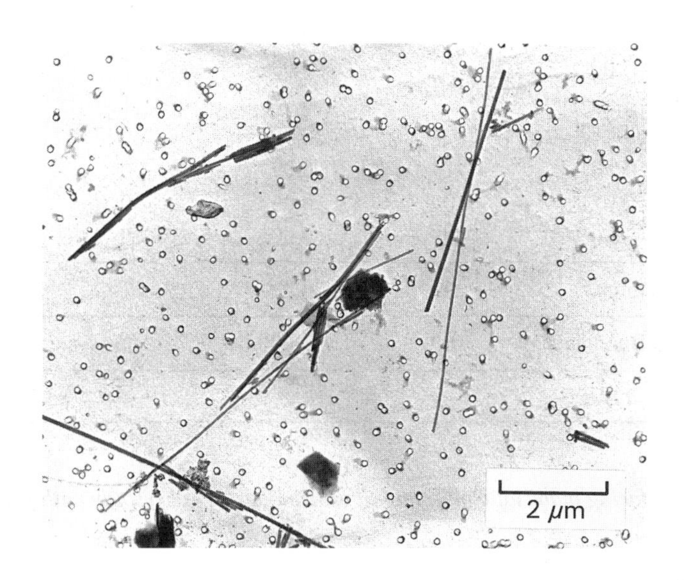

Figure 8--*TEM specimen prepared from settled dust produced by abrasion of acoustic surfacing material*

particles consisted mostly of gypsum with embedded fibers of chrysotile. The ISO 10312 procedure results in some dissolution of gypsum, as can be seen at the edges of the particle where there are areas of empty replica of the original particle. Figure 8 shows the appearance of a TEM specimen prepared by D5755-95 from the settled dust. The preparation of TEM specimens by D5755-95 includes dispersal of the dust in water, resulting in dissolution of the gypsum contained in the original material, and therefore gypsum does not appear on these TEM specimens. The majority of the chrysotile found on the TEM specimens prepared from the settled dust was in the form of single fibrils or small bundles.

Table 3 shows a comparison of the airborne chrysotile structure concentrations calculated for the air sample when TEM specimens were prepared from the PC filter by the direct-transfer procedure ISO 10312 and by the indirect-transfer procedure D5755-95. It can be seen that the preparation of this air filter using D5755-95 caused a large increase to occur in the reported concentrations of chrysotile structures in the shorter size

Table 3--*Acoustic surfacing material: analysis of air sample by ISO 10312 and by D5755-95*

Structure Size Range, μm	Direct-Transfer (ISO 10312)		Indirect-Transfer (D5755-95)	
	Number Counted	Concentration Structures/mL	Number Counted	Concentration Structures/mL
0.50 - 0.73	2	0.37	50	13.73
0.73 - 1.08	8	1.47	42	11.54
1.08 - 1.58	6	1.10	30	8.24
1.58 - 2.32	7	1.28	20	5.49
2.32 - 3.41	7	1.28	8	2.20
3.41 - 5.00	8	1.47	4	1.10
5.00 - 7.34	5	0.92	5	1.37
7.34 - 10.77	6	1.10	1	0.27
10.77 - 15.81	1	0.18	0	0.00
15.81 - 23.21	1	0.18	0	0.00
23.21 - 34.06	1	0.18	0	0.00
Total Structures	52	9.52	160	43.95
Fibers and Bundles	23	4.21	150	41.20
% Fibers and Bundles	44.2		93.8	

categories and a decrease in the reported concentrations of structures in the longer size categories. On the TEM specimens prepared from the PC air filter by ISO 10312, approximately 56% of the chrysotile structures were matrices similar to that illustrated in Figure 7, and approximately 44% of the chrysotile structures were single fibrils or bundles. Approximately 94% of the chrysotile structures observed on the TEM specimens prepared from the PC air filter by D5755-95 were single fibrils or small bundles.

Fireproofing containing approximately 15% chrysotile with vermiculite and gypsum

The results for the experiment on a sample of fireproofing containing vermiculite and gypsum are shown in Figures 9 and 10 and Table 4. Figure 9 shows a typical chrysotile-containing matrix found on the TEM specimens prepared by the direct-transfer method ISO 10312 from the air sample collected on the PC filter. As for the experiment using a sample of acoustic surfacing material, there were regions of empty replica where some of the gypsum had been dissolved away during the TEM specimen preparation. In

Table 4--*Fireproofing (containing vermiculite and gypsum): analysis of air sample by ISO 10312 and by D5755-95*

Structure Size Range, μm	Direct-Transfer (ISO 10312)		Indirect-Transfer (D5755-95)	
	Number Counted	Concentration Structures/mL	Number Counted	Concentration Structures/mL
0.50 - 0.73	0	0.00	33	3.63
0.73 - 1.08	6	0.86	50	5.49
1.08 - 1.58	7	1.01	33	3.63
1.58 - 2.32	10	1.44	21	2.31
2.32 - 3.41	2	0.29	14	1.54
3.41 - 5.00	3	0.43	2	0.22
5.00 - 7.34	2	0.29	2	0.22
7.34 - 10.77	0	0.00	0	0.00
10.77 - 15.81	0	0.00	0	0.00
15.81 - 23.21	0	0.00	0	0.00
23.21 - 34.06	0	0.00	0	0.00
Total Structures	30	4.31	155	17.03
Fibers and Bundles	22	3.16	147	16.15
% Fibers and Bundles	73.3		94.8	

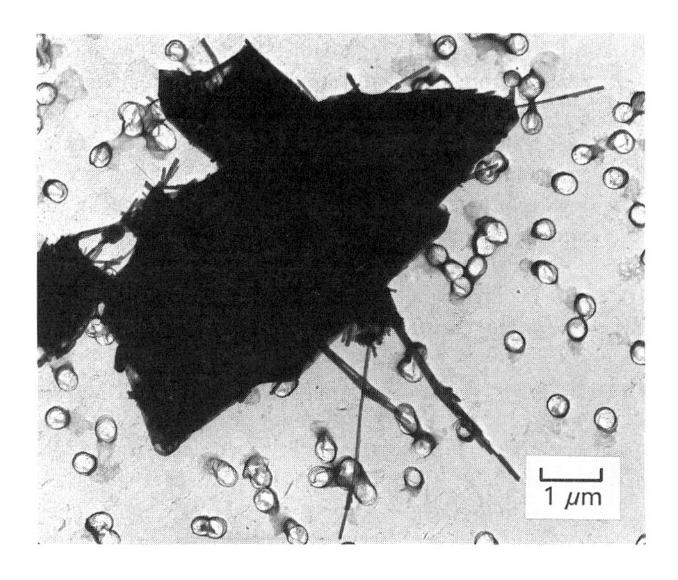

Figure 9--*Airborne chrysotile-containing particle produced by abrasion of fireproofing (containing vermiculite and gypsum)*

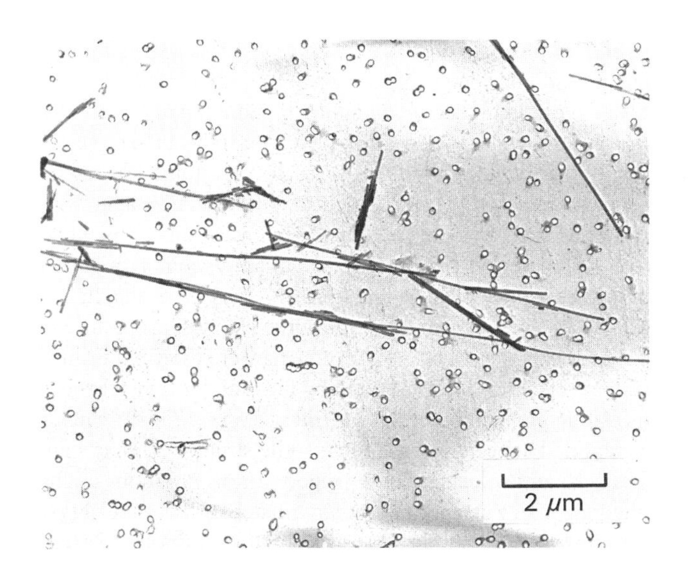

Figure 10--*TEM specimen prepared from settled dust produced by abrasion of fireproofing (containing vermiculite and gypsum)*

addition to the matrices, single fibrils and bundles of chrysotile were detected on the TEM specimens prepared from the PC air filter by ISO 10312. Figure 10 shows the appearance of a TEM specimen prepared by the indirect-transfer method D5755-95 from the settled dust. The preparation of TEM specimens by D5755-95 includes dispersal of the dust in water, resulting in dissolution of the gypsum contained in the original material, and therefore gypsum does not appear on these TEM specimens. The majority of the chrysotile structures found on the TEM specimens prepared from the settled dust were in the form of single fibrils or small bundles.

Table 4 shows a comparison of the airborne chrysotile structure concentrations calculated for the air sample when TEM specimens were prepared from the PC filter by the direct-transfer procedure ISO 10312 and by the indirect-transfer procedure D5755-95. It can be seen that the preparation of this air filter using D5755-95 caused a large increase to occur in the reported concentrations of chrysotile structures in the shorter size categories. Approximately 27% of the chrysotile structures on the TEM specimens prepared from the air sample PC filter by ISO 10312 were complex matrices and clusters similar to that illustrated in Figure 9, and approximately 73% were single fibrils or bundles of chrysotile. Approximately 95% of the chrysotile structures observed on the TEM specimens prepared from the air sample PC filter by D5755-95 were single fibrils or small fiber bundles.

Fireproofing containing approximately 20% chrysotile with mineral wool, gypsum and Portland cement

The results for the experiment on a sample of fireproofing containing mineral wool, gypsum and Portland cement are shown in Figures 11 and 12 and Table 5. Figure 11 shows a typical chrysotile-containing particle found on the TEM specimens prepared by the direct-transfer method ISO 10312 from the air sample collected on the PC filter. Many of the particles were similar, each consisting of a matrix containing chrysotile fibers with attached gypsum, calcite and Portland cement. Figure 12 shows the appearance of a TEM specimen prepared by the indirect-transfer method D5755-95 from the settled dust. The majority of the chrysotile structures found on the TEM specimens prepared from the settled dust were single fibrils or small fiber bundles.

Table 5 shows a comparison of the airborne chrysotile structure concentrations calculated for the air sample when TEM specimens were prepared from the PC filter by the direct-transfer procedure ISO 10312 and by the indirect-transfer procedure D5755-95. It can be seen that the preparation of this air filter using D5755-95 caused a large increase to occur in the reported concentrations of chrysotile structures in the shorter size categories and a substantial decrease in the reported concentrations of chrysotile structures in the longer size categories. Approximately 67% of the chrysotile structures on the TEM specimens prepared from the air sample PC filter by ISO 10312 were complex matrices and clusters similar to that illustrated in Figure 11, and approximately 33% were single fibrils or bundles of chrysotile. Approximately 89% of the chrysotile structures observed on the TEM specimens prepared from the air sample PC filter by D5755-95 were either single fibrils or small fiber bundles.

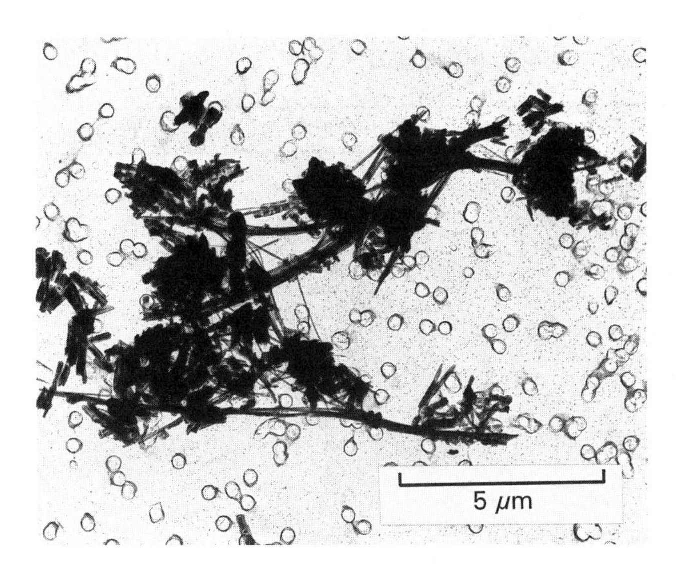

Figure 11--*Airborne chrysotile-containing particle produced by abrasion of fireproofing (containing mineral wool, gypsum and Portland cement)*

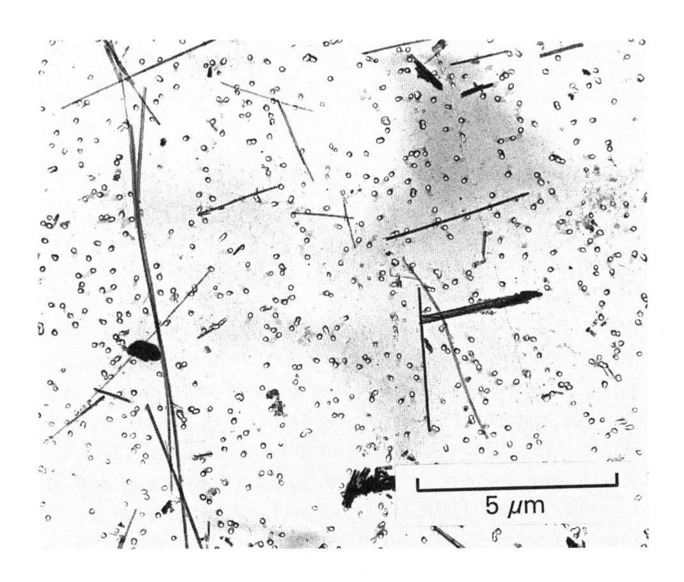

Figure 12--*TEM specimen prepared from settled dust produced by abrasion of fireproofing (containing mineral wool, gypsum and Portland cement)*

Table 5--*Fireproofing (containing mineral wool, gypsum and Portland cement): analysis of air sample by ISO 10312 and by D5755-95*

Structure Size Range, µm	Direct-Transfer (ISO 10312)		Indirect-Transfer (D5755-95)	
	Number Counted	Concentration Structures/mL	Number Counted	Concentration Structures/mL
0.50 - 0.73	0	0.00	15	12.03
0.73 - 1.08	4	0.96	40	32.08
1.08 - 1.58	4	0.96	21	16.84
1.58 - 2.32	11	2.64	21	16.84
2.32 - 3.41	8	1.92	15	12.03
3.41 - 5.00	7	1.68	10	8.02
5.00 - 7.34	12	2.88	1	0.80
7.34 - 10.77	5	1.20	0	0.00
10.77 - 15.81	1	0.24	0	0.00
15.81 - 23.21	2	0.48	0	0.00
23.21 - 34.06	1	0.24	0	0.00
Total Structures	55	13.18	123	98.66
Fibers and Bundles	18	4.32	110	88.20
% Fibers and Bundles	32.7		89.4	

Pipe elbow cement containing approximately 60% chrysotile

The results for the experiment on a sample of pipe elbow cement are shown in Figures 13 and 14 and Table 6. Figure 13 shows a typical chrysotile-containing particle found on the TEM specimens prepared by the direct-transfer procedure ISO 10312 from the air sample collected on the PC filter. Many of the chrysotile structures were matrices in which the chrysotile was associated with components of the cement. Figure 14 shows the appearance of a TEM specimen prepared by the indirect-transfer procedure D5755-95 from the settled dust. The majority of the chrysotile structures found on the TEM specimens prepared from the settled dust were single fibrils or small bundles.

Table 6 shows a comparison of the airborne chrysotile structure concentrations calculated for the air sample when TEM specimens were prepared from the PC filter by the direct-transfer procedure ISO 10312 and by the indirect-transfer procedure D5755-95. It can be seen that the preparation of this air filter using D5755-95 caused a large increase to occur in the reported concentrations of chrysotile structures in the shorter size categories. Approximately 70% of the chrysotile structures on the TEM specimens prepared from the PC air filter by ISO 10312 were matrices similar to that illustrated in

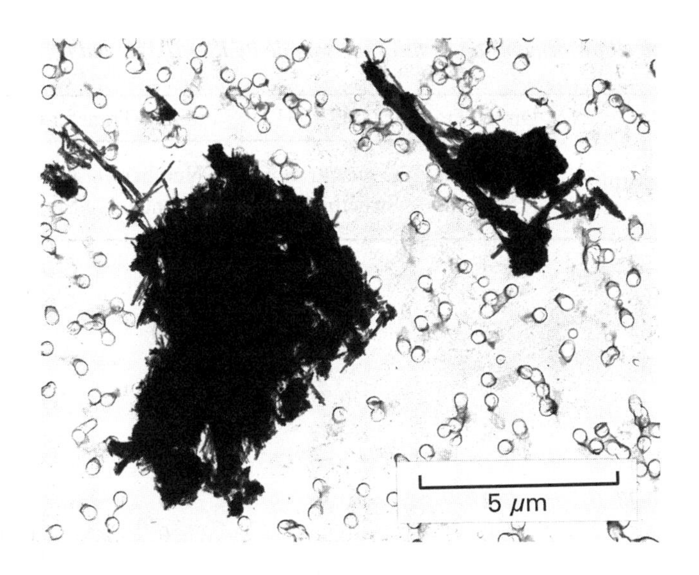

Figure 13--*Airborne chrysotile-containing particles produced by abrasion of pipe elbow cement*

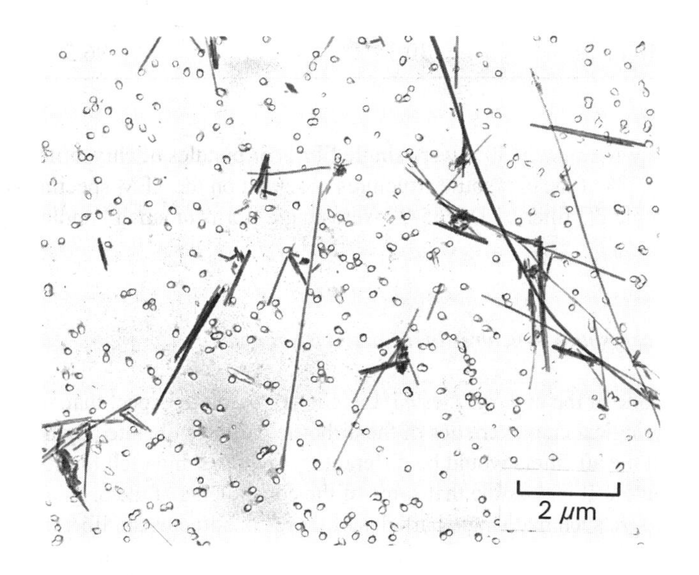

Figure 14--*TEM specimen prepared from settled dust produced by abrasion of pipe elbow cement*

Table 6--*Pipe elbow cement: analysis of air sample by ISO 10312 and by D5755-95*

Structure Size Range, μm	Direct-Transfer (ISO 10312)		Indirect-Transfer (D5755-95)	
	Number Counted	Concentration Structures/mL	Number Counted	Concentration Structures/mL
0.50 - 0.73	4	0.72	34	21.25
0.73 - 1.08	2	0.36	34	21.25
1.08 - 1.58	7	1.26	20	12.50
1.58 - 2.32	6	1.08	20	12.50
2.32 - 3.41	4	0.72	3	1.88
3.41 - 5.00	4	0.72	2	1.25
5.00 - 7.34	4	0.72	1	0.63
7.34 - 10.77	2	0.36	0	0.00
10.77 - 15.81	0	0.00	0	0.00
15.81 - 23.21	0	0.00	0	0.00
23.21 - 34.06	0	0.00	0	0.00
Total Structures	33	5.93	114	71.26
Fibers and Bundles	10	1.80	100	62.50
% Fibers and Bundles	30.3		96.5	

Figure 13, and approximately 30% were single fibrils or bundles of chrysotile. Approximately 97% of the chrysotile structures observed on the TEM specimens prepared from the air sample PC filter by D5755-95 were single fibrils or small bundles.

Discussion

Morphology of chrysotile structures in air and dust samples

For any one of the materials tested, there is no reason to expect that, other than size, the morphological characteristics of the airborne particles liberated by the abrasion and collected on the air filters would be different from those which fell to the surface below the elutriator. It is possible that some of the constituents of the original material are of a particle size such that single particles of these constituents are liberated, and this is more likely for some of the materials tested than for others. For example, in the abrasion of the floor tile which contains large (up to approximately 500 μm) calcite crystals embedded in a well-homogenized matrix consisting of short chrysotile fibers and micrometer-sized titanium dioxide in polyvinyl chloride, it would be expected that some individual particles of calcite would be released by the abrasion, but it would be most unlikely that separation of the homogenized matrix material into its individual constituents would be a significant effect. This was confirmed by the observation that the chrysotile structures observed on the TEM specimens prepared by the direct-transfer method ISO 10312 from the air filter after abrasion of the floor tile consisted totally of fragments, each of which contained all of the constituents of the floor tile matrix, and some of which also contained calcite. Other than the floor tile, all of the other test samples were friable materials, and although some short chrysotile fibers were observed on the TEM specimens prepared by ISO 10312 from the air filters, most of the chrysotile on the air filters was present as larger, but respirable, matrices or clusters containing some or all of the other constituents of the original material. It would be expected that the fragments and dust which fell to the surface under the elutriator were larger matrices or clusters, similar in all respects other than size to those collected on the air filters.

In buildings, the accumulation of surface dust from asbestos-containing materials occurs as a result of various mechanisms, physical damage by abrasion being one of them. It is likely that the morphological characteristics of surface dust originating from asbestos-containing materials by this mechanism would be similar to those of the particles generated in these experiments. In the absence of a vertical air velocity, some of the particles of respirable size would also settle, and some would remain suspended for a sufficient period to be drawn out of the building by air exchange. Interaction and mixing of the asbestos-containing surface dust with other extraneous dust which normally accumulates on surfaces [*11*] would further modify the morphology of asbestos structures as they exist on surfaces.

Modification of the morphology of chrysotile structures by the analytical methods

For all six materials tested, many of the chrysotile structures observed on the TEM specimens prepared by the direct-transfer method ISO 10312 from the PC air filters were complex matrices or clusters. Analysis of the same PC air filters by the indirect-transfer method D5755-95 invariably produced TEM specimens on which most of the chrysotile structures observed were single fibrils or small bundles. The numerical chrysotile structure concentrations for the air samples when TEM specimens were prepared from the PC filters by D5755-95 were much higher than those obtained when TEM specimens were prepared from the same filters by ISO 10312. Moreover, analysis of the PC filters by D5755-95 invariably increased the reported concentration of chrysotile fibrils and bundles, by a factor ranging from approximately 5 to more than 100, over the reported concentrations when the same filters were analyzed by ISO 10312, depending on the nature of the test sample. For example, in the experiment on the ceiling tile, the concentration of chrysotile fibers and bundles derived from the TEM specimens prepared by ISO 10312 from the air sample collected on the PC filter was 5.46 structures/mL, whereas that derived from the TEM specimens prepared by D5755-95 from the same air filter was 264.1 structures/mL. It is therefore obvious that the difference of 258.64 structure/mL between these concentrations must have been generated as a result of disintegration of matrices and clusters by the D5755-95 preparation procedure. The same reasoning holds for each of the other materials tested, the most extreme being the experiments on the vinyl-asbestos floor tile. For this material, no fibers or bundles were detected on the TEM specimens prepared by ISO 10312 from the air sample collected on the PC filter, corresponding to an airborne structure concentration of less than 0.09 structure/mL, whereas fibers and bundles dominated the TEM specimens prepared by D5755-95 from the same air filter, yielding a structure concentration of 11.72 structures/mL of which 9.79 structures/mL were fibers and bundles.

Most of the chrysotile structures on the TEM specimens prepared by D5755-95 from the settled dust, from each of these experiments, consisted of single fibrils and small bundles of chrysotile; very few matrices or clusters were present on the TEM specimens. In view of the results obtained from the analyses of the air filters, it is a logical conclusion that many of these fibrils and small bundles of chrysotile were generated as a result of disintegration of matrices and clusters during preparation by the D5755-95 procedure. Dispersal of chrysotile-containing matrices and clusters in water is generally sufficient by itself to cause large numbers of single fibrils and small bundles of chrysotile to be detached from more complex structures, even if the structures do not completely disintegrate. The limited amount of ultrasonic treatment specified by D5755-95 further accelerates the disintegration of matrices and clusters.

Some of the isolated single fibrils and small bundles of chrysotile observed on the TEM specimens prepared by ISO 10312 were in very close proximity to complex chrysotile-containing aggregates on the TEM specimens. The aggregates themselves were usually widely separated from each other. This suggests that at least some of these fibers had become detached from the complex aggregates, either at the time of sample collection when the aggregate impacted the surface of the air filter, or as a result of release from the chrysotile structures during specimen preparation. All of the routine "direct-transfer" TEM specimen preparation methods are potentially capable of modifying particles in the sample to some extent, depending on the interaction between the particles and the reagents used.

Aerodynamic considerations

In Figure 2, it can be seen that the fastest falling speed for a single fibril of chrysotile with a diameter of 0.05 μm is approximately 0.0001 cm/s. The time required to perform each experiment was between 20 and 30 minutes, and in still air, a single fibril could fall a distance of less than 2 mm in this period of time. The fastest time, in still air, for a single fibril to fall the 40 cm from the test sample to the collection surface is approximately 110 hours. The single fibrils, observed on the TEM specimens prepared by D5755-95 from the settled dust samples, could therefore not have fallen to the collection surface within the time period of the experiment. Any single fibril released from the test sample by abrasion would have been captured by the upward air velocity of 1.5 cm/s at the entrance of the elutriator, and should have been collected on the air sample filters. With the exception of the experiment on floor tile, some single fibrils and small bundles of chrysotile were, in fact, detected on the TEM specimens prepared from the air sample filters by the direct-transfer method ISO 10312. The single fibrils which constituted the majority of the chrysotile structures observed on the TEM specimens prepared by the indirect-transfer method D5755-95 from the settled dust samples must, therefore, have been deposited on the collection surface as constituents of much larger aggregates with sufficiently high falling speeds that they would have fallen to the surface within the time period of the experiment. These large aggregates of particles and fibers in the settled dust and debris are the large particle size fraction of a distribution of such aggregates, the smaller ones of which had falling speeds lower than the vertical air velocity in the elutratior and were collected on the air filters at the top of the elutriator, and the larger ones which fell to the surface below.

In the TEM micrographs of chrysotile structures on the air filters, some of the chrysotile structures illustrated appear be larger than respirable sizes. The vertical air velocity in the elutriator was selected such that respirable chrysotile fibers, defined as fibers with a maximum diameter of 3 μm, would be collected on the air filters at the top of the elutriator. Under these conditions, compact spherical particles of density 2500 kg/m^3 and diameters up to 4.72 μm would also be collected on the air filters. It has been demonstrated experimentally by Hamilton [*12*] that particles with a particular falling speed exhibit a wide range of sizes when measured by microscopy, and many have dimensions which significantly exceed the projected area of the equivalent spherical particle. The chrysotile structures illustrated are of random shapes and some are connected groups of fibers and particles, all of which would have falling speeds significantly lower than those calculated for compact spheres of equivalent mass. The linear dimensions of the structures illustrated in the TEM micrographs can therefore significantly exceed 4.72 μm and still have falling speeds less than that of a 3 μm diameter chrysotile fiber.

Although mineral wool is a major component of two of the test materials studied, it was rarely observed on the TEM specimens prepared from the air filters by either procedure. This was probably because the diameters of most of the mineral wool fibers were large, and consequently had high falling speeds. Mineral wool was also rarely observed on the TEM specimens prepared from the dust samples, probably because D5755-95 specifies that the larger particles in the aqueous suspension of dust be allowed to settle for 2 minutes before aliquots of the suspension are filtered. Large diameter mineral wool fibers would therefore settle out, and only the small diameter fibers would be present on the filters used for preparation of TEM specimens.

Effect of analytical methods on resuspension calculations

It might be thought that, if the analytical procedures used for determination of the surface and airborne chrysotile concentration remain constant, but not necessarily the same, the individual biases introduced by each of the analytical methods would be of no consequence, because the calculated resuspension factor would incorporate these biases. Although this would be the case if all parameters were the same, it is unlikely that situations would be sufficiently similar to permit scientifically-valid use of such a calculated resuspension factor. Because both D5755-95 and D5756-95 include indirect-transfer preparation of the collected dust and debris, the same structure concentration can be reported from a measurement of surface dust when there are significant differences. Moreover, it would be very difficult to demonstrate that situations were sufficiently similar to warrant use of this approach when there is no determination of the particle size distribution of the asbestos-containing dust and debris on the surface, the amount and nature of any associated extraneous dust, or the nature of disturbance of the dust. All of these parameters affect the nature and amount of airborne particulate material which can be generated.

The size range of particles and fibers as collected on air sample filters represent only a very small fraction of the particle size range found in the surface dust, which, by definition, may extend up to 1 mm in dimension. Therefore, an indirect-transfer analytical method used to determine the concentration of chrysotile on an air filter can only include contributions from particles at the small end of the size distribution, whereas the same analytical method, used to analyze a dust sample, includes contributions from the entire size range of the surface dust up to 1 mm in dimension.

Conclusions

These experiments show that when several different chrysotile-containing materials were abraded, particles of the materials were dispersed into the air. For each of the materials tested, some of the particles released were in the respirable size range. When the airborne respirable particles were collected on a filter and TEM specimens were prepared from the filter using the direct-transfer procedure ISO 10312, it was found that much of the chrysotile on the filter was present as matrices or clusters. When TEM specimens were prepared from the same filter using the indirect-transfer procedure D5755-95, it was found that the calculated concentration of chrysotile structures was much higher than the concentration reported from analysis of specimens prepared by ISO 10312, and that over 80% of the chrysotile structures observed on the specimens were single fibrils and small bundles. Considering chrysotile fibers and bundles only, the numerical concentrations calculated from the TEM specimens prepared by D5755-95 exceeded those calculated from the TEM specimens prepared by ISO 10312 by a factor between approximately 5 to 100, depending on the material under test. The single fibrils and small bundles in the D5755-95 preparation dominated the measurement, and could only have originated as a result of disintegration of the matrices and clusters during specimen preparation. Therefore, a quantitative chrysotile structure count made on TEM specimens prepared from an air filter using the procedures of D5755-95 provides little information concerning the morphological nature, sizes, and aerodynamic characteristics of the original chrysotile-containing particles as they were collected on the filter.

There is no reason to expect that the effects of the D5755-95 analytical procedure on the morphological characteristics of the settled dust, representing the large size fraction of the abraded material, would be substantially different from those observed in the analysis of the small size fraction of the abraded material which had been collected on the air filters. Since the majority of the chrysotile structures observed in the TEM specimens prepared from the settled dust by D5755-95 were single fibrils and small bundles, it can reasonably be concluded that a quantitative chrysotile structure count on a TEM specimen prepared from settled dust by D5755-95 provides little information concerning the morphological nature, sizes, and aerodynamic characteristics of the original chrysotile-containing particles on the surface.

These experiments have shown that, for settled dust particles which were originally demonstrably non-respirable by virtue of their high falling speeds, a measurement made by D5755-95 would indicate incorrectly that most of the chrysotile structures in the settled dust were in the respirable size range and were therefore capable of being resuspended. Therefore, there is no scientific basis for prediction of the potential airborne concentration of chrysotile-containing matrices and clusters in air, as could be generated by resuspension of surface dust, from a numerical structure count made on TEM specimens prepared by D5755-95, because the majority of the chrysotile structures observed on these TEM specimens consist of single fibrils and small fiber bundles which were generated by the analytical procedure, and which have entirely different aerodynamic characteristics from the particles as they existed on the surface.

More research is required to determine the relevance to human health of specific levels of asbestos-containing dust and debris on surfaces. Experiments in which well-characterized deposits of dust and debris prepared from materials such as those tested are resuspended in an enclosure by application of standardized methods of disturbance, and collection of the resulting airborne material for analysis by a direct-transfer analytical method, could form the basis of a meaningful measurement of the inhalation exposure potential of asbestos-containing surface dust.

References

[*1*] Dunster, H. J., "The Concept of Derived Working Limits for Surface Contamination", *Surface Contamination, Proceedings of a Symposium Held at Gatlinburg, Tennessee.* B. R. Fish, Ed., Pergamon Press, 1967, pp. 139-147.

[*2*] Brunskill, R. T., "The Relationship Between Surface and Airborne Contamination", *Surface Contamination, Proceedings of a Symposium Held at Gatlinburg, Tennessee.* B. R. Fish, Ed., Pergamon Press, 1967, pp. 93-105.

[*3*] Stewart, K., "The Resuspension of Particulate Material from Surfaces", *Surface Contamination, Proceedings of a Symposium Held at Gatlinburg, Tennessee.* B. R. Fish, Ed., Pergamon Press, 1967, pp. 63-74.

[*4*] Mitchell, R. N., and Eutsler, B. C., "A Study of Beryllium Surface Contamination and Resuspension", *Surface Contamination, Proceedings of a Symposium Held at Gatlinburg, Tennessee.* B. R. Fish, Ed., Pergamon Press, 1967, pp. 349-352.

[*5*] Carter, R. F., "The Measurement of Asbestos Dust Levels in a Workshop Environment", United Kingdom Atomic Energy Authority, A.W.R.E. Report No. O28/70, Aldermaston, Berkshire, U.K., 1970.

[*6*] Millette, J. R., and Hays, S. M., "*Settled Asbestos Dust Sampling and Analysis*", Lewis Publishers, CRC Press, Inc., Boca Raton, Florida 33431, 1994.

[*7*] Fowler, D. P., and Chatfield, E. J., "Surface Sampling for Asbestos Risk Assessment", Annals of Occupational Hygiene, Vol. 41, Supplement 1, 1997, pp. 279-286.

[*8*] Timbrell, V., "The Inhalation of Fibrous Dusts", Annals of the New York Academy of Sciences, Vol. 132, 1965, pp. 255-273.

[*9*] Sawyer, R. N., and Spooner, C. M., "Sprayed Asbestos-Containing Materials in Buildings: A Guidance Document", Report EPA-450/2-78-014 (OAQPS No. 1.2-094), United States Environmental Protection Agency, Office of Air and Waste Management, Office of Air Quality Planning and Standards, Research Triangle Park, North Carolina 27711, March 1978.

[*10*] International Organization for Standardization, "Ambient air - Determination of asbestos fibres - Direct-transfer transmission electron microscopy method", ISO 10312, 1995.

[*11*] Schneider, T., "Cleaning and the Indoor Environment", *Cleaning in Tomorrow's World, The First International Congress on Professional Cleaning*, Finnish Association of Cleaning Technology, Kolmas linja 7 B 37, 00530 Helsinki, 1995.

[*12*] Hamilton, R. J., "The Relation Between Free Falling Speed and Particle Size of Airborne Dusts", *The Physics of Particle Size Analysis*, British Journal of Applied Physics, Supplement No. 3, The Institute of Physics, London, 1954, pp. S90-S95.

Steve M. Hays[1]

INCORPORATING DUST SAMPLING INTO THE ASBESTOS MANAGEMENT PROGRAM

REFERENCE: Hays, S.M., **"Incorporating Dust Sampling into the Asbestos Management Program,"** *Advances in Environmental Measurement Methods for Asbestos, ASTM STP 1342,* M.E. Beard, H.L. Rook, eds., American Society for Testing and Materials, West Conshohocken, PA, 2000.

ABSTRACT: Measurement of asbestos in dust has been done extensively in the last ten years. Consequently, a large data base is now available to assist in interpreting results. A widely accepted ASTM measurement protocol exists for counting asbestos structures in dust. This author believes that this protocol is one of the most important quantitative measurements available for predicting risks associated with managing certain asbestos-containing (AC) materials in place. Research has clearly documented the exposures associated with disturbance of AC dust. Federal guidance exists for implementing the administrative portions of a proper asbestos management program, and a private sector consensus document exists to guide users in developing asbestos specific work practices necessary to operate and maintain buildings.

This paper will describe the use of dust measurements in evaluating buildings as candidates for in-place management of asbestos. It will also describe how to use these data in designing management strategies and in developing proper O&M work practices.

KEY WORDS: dust, operations & maintenance, asbestos management program, asbestos-containing dust, work practices, dust measurements

---

Introduction

The measurement of asbestos (structure count) in settled dust is one of the most important sampling and analytical protocols available for determining the best way to manage asbestos in the built environment. "Best" in this context incorporates the concepts of proper protection of health and the environment and of economic feasibility for those who must bear the cost. The author's perspective on the use of dust measurements to evaluate facilities which have certain types of asbestos-containing materials (ACMs) is

---

[1]Chairman, Gobbell Hays Partners, Inc., 217 Fifth Avenue North, Nashville, TN 37219

presented. Additionally, the use of these data for designing a management strategy and for developing operations and maintenance (O&M) work practices (WPs) are discussed.

This study cites published literature and the author's experiences, both published and unpublished, to demonstrate that the disturbance of asbestos-containing (AC) dust may result in extremely high exposures to airborne asbestos for workers and others in the vicinity of the disturbance. The objectives of an asbestos management program, dust measurements, the feasibility of in-place management, and specific ways to incorporate dust measurement data into a management program are outlined. These observations are based on 15 years' experience in asbestos management programs in over 10,000 facilities nationwide and experience in dust measurements as an assessment tool since 1987.

Objective of the Asbestos Management Program

The United States Environmental Protection Agency (EPA) states that asbestos is hazardous and that risk of asbestos-related disease is a function of airborne exposure [*1*]. The Agency recommends an in-place management program whenever ACMs are discovered in any facility [*2*] and requires such a program in schools [*3*]. These requirements and recommendations are to protect human health and the environment. According to the Agency's publication, Managing Asbestos In Place, the objective of a management program is to "...prevent asbestos fiber releases or resuspension of already-released fibers, . . ." [*4*]. The Agency also recognized the importance of "already-released" fibers in some of its earlier guidance documents; for example, Guidance for Controlling Asbestos-Containing Materials in Buildings recommends initial and periodic cleaning of surfaces in the vicinity of spray-applied ACMs [*5*]. It is clear that asbestos which is no longer bound in its original construction material matrix is of paramount concern and is a primary focus of an asbestos management program.

The National Emission Standards for Hazardous Air Pollutants (NESHAP) regulations for asbestos require removal of certain ACMs before renovation or demolition. In this context, the objective of a management program is to provide interim controls for these ACMs until they are eventually removed. These interim controls must address AC dust to effectively protect health and the environment. The Agency's position is that disturbance of regulated ACMs or disturbance of AC dust and debris above the NESHAP quantities triggers the regulation, as discussed in detail in a letter dated December 29, 1992, from W.G. Rosenberg of the EPA to H.J. Singer of the General Services Administration [*6*].

In short, the fundamental objective of an asbestos management program is to prevent exposure to airborne asbestos. This includes controlling disturbance of AC dust and debris, as well as disturbances of in-place ACMs.

Dust Measurements

Significant experience measuring asbestos in dust has been gained by the asbestos consulting community over the past ten or twelve years. This has occurred because surface contamination is typically assessed by scientists as part of investigating risk from toxic materials [*7*]. Good science required that investigation proceed beyond attention only to the in-place ACMs and, consequently, protocols to measure asbestos in dust were developed, tried, refined, reviewed, debated, and codified. A protocol now exists for measuring the number of asbestos structures in dust, "ASTM Standard Test Method for Microvacuum Sampling and Indirect Analysis of Dust by Transmission Electron Microscopy for Asbestos Structure Number Concentrations" (D 5755-95). A protocol is under development by ASTM's D22.07 Subcommittee for measuring asbestos accumulation, over time, in dust by passive collection, "ASTM Draft Standard Test Method for Passive Sampling and Indirect Analysis of Dust by Transmission Electron Microscopy for Asbestos Structure Number Concentration."

The NESHAP regulation states that no dust or debris should be visible where regulated asbestos-containing material (RACM) is found [*8*]. Critics of asbestos-in-dust measurements maintain that because no regulatory quantification standard exists, use of results in assessing risk is not possible. Lack of a regulated limit for asbestos in dust notwithstanding, a wealth of useful data exists in the literature [*9*], [*10*]. Ewing has reported results from 66 buildings in the United States [*9*]. These data are useful for placing measurements in context. They also document the existence of significant amounts of asbestos in dust where certain ACMs are present. Millette and Hays have reported dust measurements and airborne asbestos levels associated with various disturbances of the contaminated surfaces [*10*]. They have also reported their experience regarding the significance of measured levels of asbestos in dust [*11*]. Additional data exist in the open literature [*12*], [*13*], [*14*], [*15*]. By drawing from the collective knowledge of experienced investigators, interpretation of data as part of a building assessment is straightforward.

The importance of asbestos-in-dust measurements to building assessment is direct and logical to establish. Health risks are primarily associated with inhalation of airborne asbestos. Exposures are measured in terms of airborne concentration, which is the number of asbestos fibers (or structures) per unit volume of air. The number of asbestos structures which exists on a surface and which can be resuspended in air has a direct relationship to exposure potential. The relationship is a function of many things, including the characteristics of the dust, the characteristics of the surface, and the activity which disturbs the dust. While the exact relationship between dust and airborne asbestos concentrations for a given situation may not be known, enough applied research has been published to document that asbestos in dust is a very serious exposure risk and to suggest at what concentrations asbestos in dust becomes a major factor in a management program.

In the author's experience, asbestos is virtually always present in dust associated with spray-applied ACMs, such as fireproofing and acoustical ceiling treatments. It has not, to the author's knowledge, been established when in the life of a material the release of asbestos from the matrix begins. The facilities investigated have not contained "new" installations of these products. However, in the significant number of investigations which have been done, asbestos is typically present in dust where these products are installed. The surfaces which usually have the highest concentrations of asbestos in dust are those, such as the top sides of ceiling tiles, which are installed below spray-applied fireproofing [*12*]. Studies have been done which establish that very high airborne exposures (in some cases, over 100 s/cm^3) result from disturbances, by ordinary maintenance activities, of surfaces contaminated with AC dust [*13*], [*14*], [*15*]. In these studies, the airborne asbestos resulted primarily from resuspension of AC dust. (The OSHA 8-hour time weighted average permissible exposure limit is 0.1 f/cm^3 [*16*]. The EPA "Silver Book" reports ambient airborne concentrations, as measured by transmission electron microscopy, to be less than 0.05 f/cm^3 [*17*].)

The National Institute of Building Sciences (NIBS) has recently released the second addition of the Guidance Manual for Asbestos Operations and Maintenance Work Practices. Dust is discussed in this consensus guidance document, which recommends that "Asbestos-containing dust and debris should be a controlled system under the O&M program [*18*]." NIBS believes that no management program is complete without proper consideration of AC dust.

In-Place Management Feasibility

The information necessary to design a management program is discussed in EPA guidance documents [*19*], [*20*] and regulations [*21*]. Most of this information and data are collected by bulk sampling of suspect ACMs and visual assessment [*22*]. Work procedures are regulated by the Occupational Safety and Health Administration (OSHA), [*23*] and guidance for specific operations and maintenance work practices is provided by NIBS [*24*].

An important consideration in designing a management program is the ability of the O&M staff to understand the program and to be committed to it [*25*]. In the author's experience, making the subjective judgment regarding the ability and commitment of an O&M staff is difficult. When AC dust is present at a facility, the O&M procedures are significantly complicated beyond the procedures necessary to control disturbances to in-place ACMs. The presence of dust must be considered in evaluating potential exposure risk, and the work practices implied by significant amounts of asbestos in dust must be a factor in judging whether in-house staff can adequately and consistently perform under a rigorous management program. Conclusions regarding a staff's ability are crucial in deciding if in-place management is viable for a given facility.

Use of Dust Measurements

Building evaluations are done to gather sufficient information to design a management plan. AC dust may be addressed by measurement, or surfaces under and near ACMs may be assumed to have significant contamination. The assumption of contamination is reasonable for surfaces associated with spray-applied ACMs, especially fireproofing. If measurements are made, concentrations above 1,000 structures per square centimeter (s/cm^2) may warrant further consideration. Levels above 10,000 s/cm^2 are probably above background, and work practices should be designed to carefully control disturbances of these surfaces [*26*].

Asbestos management strategies can generally be grouped into three major categories: 1) managing ACMs in place with an O&M plan until regulations require removal, 2) managing ACMs in place with an O&M plan until a convenient time presents to remove those materials which are most difficult to manage, or 3) removing immediately those ACMs which would have the most negative impact on facility operations and maintenance. Each of these options has many variations, discussions of which are beyond the scope of this paper. Each option has an attendant set of cost and operational considerations unique to the facility and its individual circumstances. The cost of managing certain ACMs in place may be more expensive and disruptive in the long term for some situations than immediate removal. The presence of AC dust adds to the management expense and may add sufficient burden to warrant immediate removal of some ACMs. Measurement of asbestos in dust is the only way to be sure if special considerations for surface dust are necessary or if general decontamination of some building areas is needed. Millette and Hays discuss the particulars of sample collection and data interpretation [*27*]. NIBS provides guidance for design of work practices which are sufficient to deal with contaminated surfaces [*28*]. The use of dust measurements provides the additional necessary data to manage asbestos properly in the most cost effective manner by indicating where resuspension of AC dust is a valid concern and by giving quantification to contamination. This quantification is useful in the overall program design and in the selection of appropriate work practices.

Summary

Measurements of asbestos in dust are appropriate in assessing facilities. A protocol has been established by ASTM, and that procedure has been used often in many facilities. A significant body of data exists, and much knowledge has been gained and published about dust measurements. This collective experience allows useful interpretation of data in the context of facility assessment. Asbestos-in-dust measurements are very important in properly designing an asbestos management program.

References

[*1*] *Managing Asbestos In Place: A Building Owner's Guide to Operations and Maintenance Programs for Asbestos-Containing Materials* (Green Book), U.S. EPA, Office of Pesticides and Toxic Substances, Washington, DC, 20T-2003, July, 1990, p. vii.

[*2*] Ibid., p. viii.

[*3*] USEPA, 40 CFR Part 763, Asbestos Hazard Emergency Response Act (AHERA).

[*4*] *Managing Asbestos In Place: A Building Owner's Guide to Operations and Maintenance Programs for Asbestos-Containing Materials* (Green Book), U.S. EPA, Office of Pesticides and Toxic Substances, Washington, DC, 20T-2003, July, 1990, p. 1.

[*5*] *Guidance for Controlling Asbestos-Containing Materials in Buildings* (Purple Book), U.S. EPA, Office of Pesticides and Toxic Substances, Washington, DC EPA 560/5-85-024, June, 1985, pp. 3-2, 3-3.

[*6*] Letter from W.G. Rosenberg of the EPA to H.J. Singer of the General Services Administration, December 29, 1992.

[*7*] Millette, J.R., Hays, S.M., *Settled Asbestos Dust Sampling and Analysis*, CRC Press, Inc., 1994, p. 1.

[*8*] Federal Register, Vol. 55, No. 224, November 20, 1990, *National Emission Standards for Hazardous Air Pollutants.*

[*9*] Ewing, W.E., "Observations of Settled Asbestos Dust in Buildings," *EIA Technical Journal*, Vol. 4, No. 1, Summer 1996.

[*10*] Millette, J.R., Hays, S.M., *Settled Asbestos Dust Sampling and Analysis*, CRC Press, Inc., 1994, pp. 59-63.

[*11*] Ibid., pp. 49-50.

[*12*] Ewing, W.E., "Observations of Settled Asbestos Dust in Buildings," *EIA Technical Journal*, Vol. 4, No. 1, Summer 1996.

[*13*] Keyes, D.L., Chesson, J., Ewing, W.M., Faas, J.C., Hatfield, R.L., Hays, S.M., Longo, W.E., and Millette, J.R., "Exposure to Airborne Asbestos Associated with Simulated Cable Installation Above a Suspended Ceiling," *American Industrial Hygiene Journal*, November, 1991.

[*14*] Keyes, D.L., Ewing, W.M., Hays, S.M., Longo, W.E., and Millette, J.R., "Baseline Studies of Asbestos Exposure During Operations and Maintenance Activities," *Applied Occupational Environmental Hygiene Journal*, November, 1994.

[*15*] Ewing, W.M., Chesson, J., Dawson, T.A., Ewing, E.M., Hatfield, R.L., Hays, S.M., Keyes, D.L., Longo, W.E., Millette, J.R., and Spain, W.H., "Asbestos Exposure During and Following Cable Installation in the Vicinity of Fireproofing," *Environmental Choices Technical Supplement*, March/April 1993.

[*16*] OSHA, 29 CFR Part 1910, 1915, and 1926, Occupational Exposure to Asbestos.

[*17*] *Measuring Airborne Asbestos Following An Abatement Action* (Silver Book), U.S. EPA, Offices of Research and Development and Toxic Substances, Washington, DC, November 1985, p. 5-5.

[*18*] *Guidance Manual for Asbestos Operations & Maintenance Work Practices*, National Institute of Building Sciences, December, 1996, p. 4-10.

[*19*] *Guidance for Controlling Asbestos-Containing Materials in Buildings* (Purple Book), U.S. EPA, Office of Pesticides and Toxic Substances, Washington, DC EPA 560/5-85-024, June, 1985.

[*20*] *Managing Asbestos In Place: A Building Owner's Guide to Operations and Maintenance Programs for Asbestos-Containing Materials* (Green Book), U.S. EPA, Office of Pesticides and Toxic Substances, Washington, DC, 20T-2003, July, 1990.

[*21*] USEPA, 40 CFR Part 763 Asbestos Hazard Emergency Response Act (AHERA).

[*22*] Millette, J.R., Hays, S.M., *Settled Asbestos Dust Sampling and Analysis*, CRC Press, Inc., 1994, p. 69.

[*23*] OSHA, 29 CFR Part 1910, 1915, and 1926, Occupational Exposure to Asbestos.

[*24*] *Guidance Manual for Asbestos Operations & Maintenance Work Practices*, National Institute of Building Sciences, December, 1996.

[*25*] *Managing Asbestos In Place: A Building Owner's Guide to Operations and Maintenance Programs for Asbestos-Containing Materials* (Green Book), U.S. EPA, Office of Pesticides and Toxic Substances, Washington, DC, 20T-2003, July, 1990, p. 8.

[*26*] Millette, J.R., Hays, S.M., *Settled Asbestos Dust Sampling and Analysis*, CRC Press, Inc., 1994, p. 50.

[*27*] Ibid.

[*28*] *Guidance Manual for Asbestos Operations & Maintenance Work Practices*, National Institute of Building Sciences, December, 1996.

Author Index

Subject Index

E

F

G

H

I

L

M

N